Materials Degradation Caused by Acid Rain

ACS SYMPOSIUM SERIES 318

Materials Degradation Caused by Acid Rain

Robert Baboian, EDITOR
Texas Instruments, Inc.

Developed from a symposium sponsored by
the Division of
Industrial and Engineering Chemistry, Inc.
at the 20th State-of-the-Art Symposium
of the American Chemical Society,
Arlington, Virginia,
June 17–19, 1985

American Chemical Society, Washington, DC 1986

Library of Congress Cataloging-in-Publication Data

Materials degradation caused by acid rain.
(ACS symposium series, ISSN 0097-6156; 318)

"Developed from a symposium sponsored by the Division of Industrial and Engineering Chemistry, Inc., at the 20th State-of-the-Art Symposium of the American Chemical Society, Arlington, Virginia, June 17-19, 1985."

Bibliography: p.
Includes index.

1. Materials—Deterioration—Congresses. 2. Acid rain—Congresses. 3. Weathering—Congresses. 4. Acid Rain Deposition—Congresses.

I. Baboian, Robert. II. American Chemical Society. Division of Industrial and Engineering Chemistry. III. Series.

TA418.74.M38 1986 620.1′1223 86-20560
ISBN 0-8412-0988-X

PRINTED IN THE UNITED STATES OF AMERICA

ACS Symposium Series

M. Joan Comstock, *Series Editor*

FOREWORD

The ACS SYMPOSIUM SERIES was founded in 1974 to provide a medium for publishing symposia quickly in book form. The format of the Series parallels that of the continuing ADVANCES IN CHEMISTRY SERIES except that, in order to save time, the papers are not typeset but are reproduced as they are submitted by the authors in camera-ready form. Papers are reviewed under the supervision of the Editors with the assistance of the Series Advisory Board and are selected to maintain the integrity of the symposia; however, verbatim reproductions of previously published papers are not accepted. Both reviews and reports of research are acceptable, because symposia may embrace both types of presentation.

CONTENTS

Preface ... xi

MEASUREMENT AND MONITORING OF ATMOSPHERIC DEPOSITION

1. **Acidification of Precipitation** ... 2
   B. Ottar

2. **Wet Deposition Chemistry** ... 23
   Richard G. Semonin

3. **Measurement of Dry Deposition onto Surrogate Surfaces: A Review** ... 42
   James S. Gamble and Cliff I. Davidson

4. **Fog and Cloud Water Deposition** ... 64
   Michael R. Hoffmann

5. **Urban Dew: Composition and Influence on Dry Deposition Rates** ... 92
   Patricia A. Mulawa, Steven H. Cadle, Frank Lipari, Carolina C. Ang, and René T. Vandervennet

METALLIC CORROSION

6. **Influence of Acid Deposition on Atmospheric Corrosion of Metals: A Review** ... 104
   Vladimir Kucera

7. **Environmental Effects on Metallic Corrosion Products Formed in Short-Term Atmospheric Exposures** ... 119
   D. R. Flinn, S. D. Cramer, J. P. Carter, D. M. Hurwitz, and P. J. Linstrom

8. **Bronze, Zinc, Aluminum, and Galvanized Steel: Corrosion Rates as a Function of Space and Time over the United States** ... 152
   D. E. Patterson, R. B. Husar, and E. Escalante

9. **Environmental Factors Affecting Corrosion of Weathering Steel** ... 163
   Fred H. Haynie

10. **A Laboratory Study to Evaluate the Impact of NO_x, SO_x, and Oxidants on Atmospheric Corrosion of Galvanized Steel** ... 172
    Edward O. Edney, David C. Stiles, John W. Spence, Fred H. Haynie, and William E. Wilson

11. **A Field Experiment to Partition the Effects of Dry and Wet Deposition on Metallic Materials** ... 194
    John W. Spence, Fred H. Haynie, Edward O. Edney, and David C. Stiles

12. **Effects of Acid Deposition on Poultice-Induced Automotive Corrosion** ... 200
    R. C. Turcotte, T. C. Comeau, and Robert Baboian

13. **Effect of Acid Rain on Exterior Anodized Aluminum Automotive Trim** ... 213
    Gardner Haynes and Robert Baboian

14. **Effects of Acid Rain on Indoor Zinc and Aluminum Surfaces** ... 216
    J. D. Sinclair and C. J. Weschler

MASONRY DETERIORATION

15. **Limestone and Marble Dissolution by Acid Rain: An Onsite Weathering Experiment** 226
Michael M. Reddy, Susan I. Sherwood, and B.R. Doe

16. **Effects of Acid Deposition on Portland Cement Concrete** 239
R. P. Webster and L. E. Kukacka

17. **Deterioration of Brick Masonry Caused by Acid Rain** 250
A. E. Charola and L. Lazzarini

18. **Cultural Resource Monitoring: Concurrent Aerometric and Materials Deterioration Studies at Mesa Verde National Park** 259
D. A. Dolske and W. T. Petuskey

19. **Effects of Atmospheric Exposure on Roughening, Recession, and Chemical Alteration of Marble and Limestone Sample Surfaces in the Eastern United States** 266
C. A. Youngdahl and B. R. Doe

20. **Elemental Analysis of Simulated Acid Rain Stripping of Indiana Limestone, Marble, and Bronze** 285
K. M. Neal, S. H. Newnam, L. M. Pokorney, and J. P. Rybarczyk

21. **Effects of Acid Rain on Deterioration of Coquina at Castillo de San Marcos National Monument** 301
D. G. Rands, J. A. Rosenow, and J. S. Laughlin

DEGRADATION OF ORGANICS

22. **Effects of Acid Rain on Painted Wood Surfaces: Importance of the Substrate** 310
R. Sam Williams

23. **Effect of Acid Rain on Woody Plants and Their Products** 332
Ellen T. Paparozzi

24. **Acid Rain Degradation of Nylon** 343
Karen E. Kyllo and Christine M. Ladisch

ECONOMIC EFFECTS

25. **Materials Damage and the Law** 360
Roger C. Dower and Sarah E. Ball

26. **Economic Features of Materials Degradation** 369
Thomas D. Crocker

27. **Economic Effects of Materials Degradation** 384
E. Passaglia

28. **Model for Economic Assessment of Acid Damage to Building Materials** 397
Thomas J. Lareau, Robert L. Horst, Jr., Ernest H. Manuel, Jr., and Frederick W. Lipfert

29. **Application of a Theory for Economic Assessment of Corrosion Damage** 411
Frederick W. Lipfert and Ronald E. Wyzga

INDEXES

Author Index 433

Subject Index 433

Left side–right side comparison of the Statue of Liberty torch shows darkening of the left side facing Manhattan and the northeast. Darkening of the left side is due to erosion of the green patina by acid deposition and the severe weather from the northeast. (Photo by Robert Baboian.)

Corrosion of bronze, commonly termed "bronze plague," in the Torrey monument in Mount Auburn Cemetery, Cambridge, MA, due to the effects of acid deposition. (Photo by Robert Baboian.)

Corrosion of reinforcing steel in concrete (elevated highway, Providence, RI), where road de-icing salts combine with acid precipitation to produce a severe environment. (Photo by Robert Baboian.)

PREFACE

ACID RAIN IS AN IMPORTANT AND GROWING TOPIC. This book addresses the important materials problems resulting from acid deposition. It is divided into five sections: Measurement and Monitoring of Atmospheric Deposition, Metallic Corrosion, Masonry Deterioration, Degradation of Organics, and Economic Effects.

The section on measurement and monitoring concentrates on the scope of the acid deposition problem. This includes wet deposition chemistry, dry deposition, fog and cloud water, and the composition of dew.

The section on metallic corrosion clearly indicates that the conventional method of classification of environments into marine, industrial, and rural no longer is adequate. More specific information is needed about the actual chemical components in the atmosphere as well as humidity and other factors. Specific environments also are addressed in the metallic corrosion section. For example, the automotive environment in the northeastern United States is particularly severe because of the combination of acid deposition and the use of road de-icing salts. These factors exert a synergistic effect on the corrosion behavior of auto-body steel and on exterior anodized aluminum automobile trim.

The section on masonry deterioration focuses on limestone, coquina, sandstone, marble, concrete, brick, and mortar as related to acid deposition effects on structures such as buildings and on cultural resources such as monuments.

Auto body corrosion adjacent to stainless steel trim occurs in environments where de-icing salts are used. The effects of acid deposition combined with road salts produces a synergystic effect on the degradation of automobiles. (Photo by Robert Baboian.)

The section on degradation of organics deals with paints, plastics, nylon, wood, and architectural organics. The effects of acid deposition on wood and other cellulosic materials are described. Strength losses in wood may be caused by hydrolytic degradation of the hemicelluloses and a sulfonation reaction of the lignin. Thus, the fibrils and matrix structure is affected. Cotton materials can be affected similarly, and soiling will result. The effect of acid deposition of nylon is indicative of a potentially shorter serviceable lifetime for outdoor fabrics.

The section on economic effects presents the methodology used in assessing costs of degradation of materials due to acid deposition. The difficulty in accurately assessing the cost of materials degradation by acid deposition is described in this section. Thus, the various techniques used have a high degree of uncertainty.

In summary, this book serves to provide information on the wide range of materials affected by acid deposition. Although a large amount of information is presented on this subject, it is evident that much remains to be done. A better understanding of the nature and mechanisms of materials damage by acid deposition could lead to a reduction or avoidance of this kind of damage. Thus, expenditures for work in this field could lead to huge annual dollar savings.

ROBERT BABOIAN
Electrochemical and Corrosion Laboratory
Texas Instruments Inc.
Attleboro, MA 02703

Organizing Committee for the Symposium

Robert Baboian, Chairman
Texas Instruments Inc.

Edward Escalante, Session Chairman
National Bureau of Standards

David R. Flinn
U.S. Bureau of Mines

James H. Gibson, Session Chairman
Colorado State University

Ray Hermann
National Park Service

Frederick W. Lipfert, Session Chairman
Brookhaven National Laboratory

Richard A. Livingston
Environmental Protection Agency

Hugh C. Miller
National Park Service

Robert S. Shane, Session Chairman
Shane Associates

Susan Sherwood, Session Chairman
National Park Service

Thaddeus Whyte
The PQ Corporation

William E. Wilson
Environmental Protection Agency

Degradation of paint occurs by reaction at the surface and at the paint interface. Acid deposition can cause paint peeling on wood. (Photo by R. S. Williams.)

Bricks and mortar (New Haven, CT) are susceptible to deterioration through the action of acid deposition. (Photo by Robert Baboian.)

St. Thomas Church in New York City. Areas on the facade accumulate a layer of gypsum (calcium sulfate) produced by acid deposition attack on the limestone and then darken by dirt, soot, and other combustion products. (Photo by Elena Charola.)

Spalling of marble column on Department of Justice building, Washington, DC. (Photo by Bruce Doe.)

MEASUREMENT AND MONITORING OF ATMOSPHERIC DEPOSITION

1

Acidification of Precipitation

B. Ottar

Norwegian Institute for Air Research, P.O. Box 130, N-2001 Lillestrøm, Norway

The acid rain studies which started with the OECD project 1972-77 have since become a major issue in both North America and Europe. The main cause is sulphuric and nitric acid from the use of fossil fuels. The resulting fish kills in acidified rivers and lakes depend on the soil composition and the release of toxic aluminium ions. This may also be a contributing factor to the forest damage in central Europe., but photochemical oxidants are now believed to be more important. Generally the acidification and the increasing oxidation potential of the atmosphere is slowly changing our chemical environment, and it is not known to what extent the resulting ecological changes will be reversed if the emissions are reduced.

In the summer of 1969, OECD called a meeting to discuss evidence on the acidification of the precipitation in Europe. The year before the Swedish scientist S. Odén (1) by analysing precipitation data from a European network of atmospheric chemistry stations, which had been established in 1954-55, had found that a central area with highly acid precipitation had expanded to include the southern parts of Scandinavia. The main acidifying agent was sulphuric acid, and the source was assumed to be the increasing use of fossil fuels with a high content of sulphur. His findings were related to the acidification of rivers and lakes in Scandinavia and the disappearance of fish in these waters.

At this meeting, the OECD countries agreed that this development deserved further attention, and the Nordic countries should produce a coordinated plan to examine the situation. A planning committee and a preliminary research programme were established through NORDFORSK (The Nordic Council for Pure and Applied Research), and after considerable preparations and negotiations the OECD project "Long Range Transport of Air Pollution" was started in July 1972. After a first introductory phase, the OECD countries in 1973 agreed that this was a serious situation and increased the research effort.

0097-6156/86/0318-0002$06.00/0

The OECD project was completed in 1977. The final report (2) showed that the air over the central parts of Europe was substantially polluted. The countries on the continent received as much sulphur pollutants from neighbouring countries as from their own sources, and most of the increased acidity of the precipitation in the surrounding countries was due to the emissions of sulphur dioxide and nitrogen oxides in the central parts of the area (3,4,5).

In order to bring this development under some control, close co-operation would be needed between the countries of both Eastern and Western Europe. The problem was taken up by the U.N. Economic Commission for Europe (ECE) in Geneva, and further progress was made at the first Conference on Security and Co-operation in Europa in Helsinki in 1975, which recommended the development of an extended programme for monitoring and evaluation of the long range transport of air pollution in Europe.

In 1978 the first 3-year phase of the "Co-operative Programme for Monitoring and Evaluation of the Long Range Transmission of Air Pollutants in Europe" (EMEP) was started. The programme was financed by the U.N. Environmental Program (UNEP), and close co-operation was established with the World Meteorological Organization (WMO). EMEP is now part of the ECE convention on long range transboundary air pollution (6). Today 26 countries, with their national air sampling networks, participate in the programme, including Canada and USA as observers. The scientific work is co-ordinated through a Chemical Coordinating Center (CCC) at the Norwegian Institute for Air Research (NILU) and two meteorological synthesizing centers, one at the Norwegian Meteorological Institute in Oslo (MSC-W) and one at the Hydro-Meteorological Institute in Moscow (MSC-E).

So far EMEP has concentrated on measuring sulphur pollutants (7). The exchange of sulphur pollutants between the European countries is now calculated on a routine basis. An example is shown in Table 1. From a scientific point of view, there has long been a need to include measurements of nitrogen oxides and other nitrogen compounds, in order to obtain a more complete data base for evaluation of the acidification of precipitation. At present this is only done on a voluntary basis in some of the countries. For various reasons, it has so far not been possible to include these measurements in the regular programme. More recently, the rapidly increasing forest damage in Europe has also pointed to the importance of measurements of ozone and other photochemical oxidants. There is also an interest in heavy metals. According to plans, measurements of 24-hourly mean values of NOx will begin in 1986, and hourly measurements of ozone are considered for the 4th phase of EMEP, which starts in 1987.

Considering the rapid expansion of the regional air pollution problems in Europe during the later years, as examplified by the forest damage, the appearance of "red tide" in the North Sea, and the increasing mercury content of the freshwater fish, the development of EMEP has been slow. It is interesting to note that when the OECD project started in 1972, Canada participated as an observer, while the US EPA did not believe that acid precipitation would become a problem in North America as well. Similarly, the European countries did not think that photochemical oxidants would become a problem in Europe. There was not enough sunshine, they reasoned.

Today we have the same problems on both sides of the Atlantic, and it has gradually become evident that the two phenomena are closely interrelated with respect to sources and effects. The oxidation of sulphur dioxide to sulphuric acid is largely governed by the photochemical activity in the atmosphere, which also produces the nitric acid found in the precipitation. The damage to vegetation and materials in many cases is the result of acidification, combined with a higher oxidation potential. In southern Sweden, the elevated copper content of drinking water, due to increased corrosion of pipelines, can no longer be tolerated by small children.

As a result of these developments, the term acid precipitation has acquired a wider popular meaning, including both acid precipitation and other regional effects of air pollution.

Sources Of Acid Precipitation

The OECD project showed that the acidification of precipitation was due to an increased content of sulphuric and nitric acid. The main cause was identified as the increased use of fossil fuels, including motor vehicle traffic. In the atmosphere, sulphur dioxide and nitrogen oxides from combustion processes are oxidized to sulphuric acid and nitric acid, which are taken up and deposited by the precipitation.

The growing use of fossil fuels in Europe during this century is illustrated in Figure 1 (5). The early sulphur dioxide emissions in Europe were mainly due to the combustion of sulphur containing coals and in some areas the processing of sulphidic ores. The increased demand for energy after 1950 was met by a wide-spread introduction of petroleum products, and as a result the sulphur dioxide emissions in Europe were doubled in the period 1950-75. Lately they have remained largely unchanged. Many countries have reduced their sulphur dioxide emissions considerably, by energy saving and by changing to other fuels. In France 60% of the electricity demand is now met by nuclear power production, while in other countries increased electricity production based on traditional fuels has led to higher sulphur emissions.

In North America the sulphur dioxide emissions mainly originate from the use of coals with a relatively high sulphur content, and from various special industrial processes. The relative contribution of sulphur from oil combustion is much smaller than in Europe. Figure 2 shows the summer and winter coal consumption in USA since 1940 (8). It shows a peak in 1943, and since 1960 the summer consumption has grown at a rate of 5.8% p.a. as compared to 2.8% p.a. for winter time. In addition to this, the consumption of heavy fuel oil has increased by 50% since 1959.

On a global scale, industrial processes (mainly roasting of sulphidic copper, nickel, lead and zinc ores, manufacturing of sulphur acid, and the paper and pulp industry) account for about 10% of the total sulphur pollution (9).

The emissions of nitrogen oxides are mainly due to oxidation of nitrogen from the air during combustion processes. The main sources are motorized traffic, power stations, and space heating (10). In this large scale picture, process emissions are of lesser significance. High combustion temperature and an excess of air favour the formation of nitrogen oxides. Thus, modern diesel engines have about

Table 1. Example of exchange of sulphur pollutants between European countries (2).

	BELGIUM	DENMARK	FRG	FINLAND	FRANCE	NETHERLANDS	NORWAY	SWEDEN	UNITED KINGDOM	CZECHOSLOVAKIA	GDR	POLAND	OTHER AREAS	UNDECIDED	SUM	EMISSION
Sweden 1974	7	30	30	10	10	6	6	100	40	8	50	20	30	100	500	415
1975	9	40	40	4	15	8	8	150	60	9	50	15	10	100	500	
1976	5	30	30	10	9	4	7	150	30	15	60	20	30	100	500	
	(10^3 tonnes S/year).															

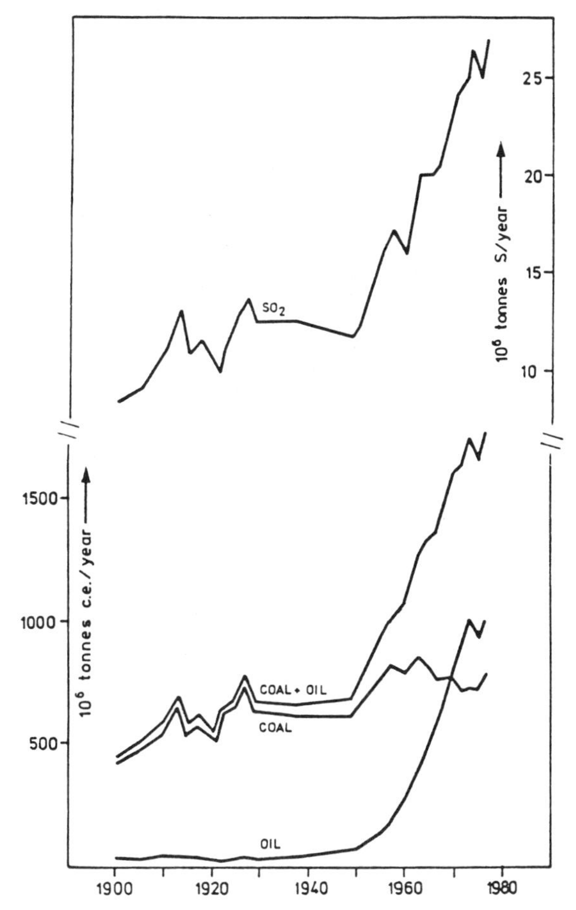

Figure 1. Consumption of coal and oil in Europe and estimates of the resulting sulphur dioxide emissions since year 1900 (5).

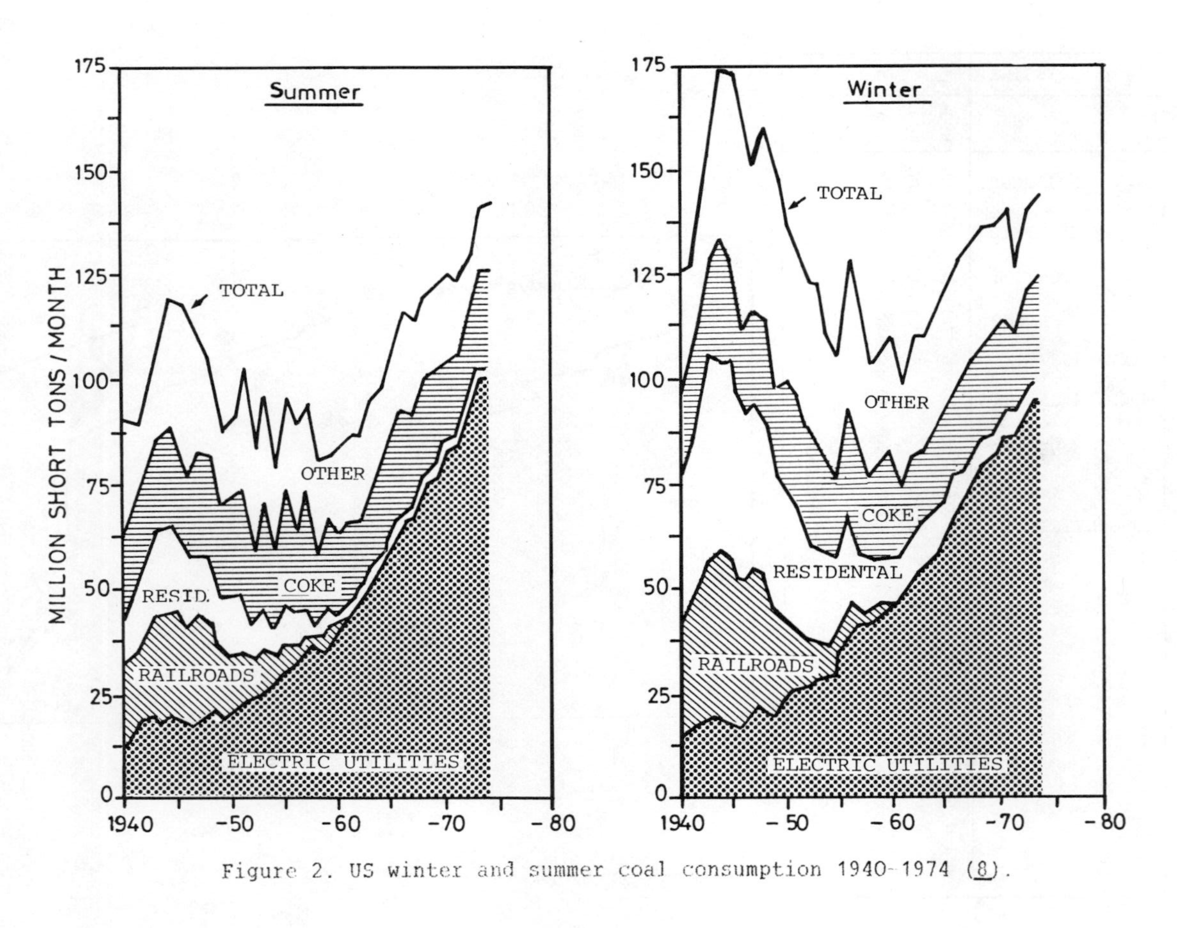

Figure 2. US winter and summer coal consumption 1940–1974 (8).

the highest nitrogen oxide emissions per unit of fuel consumed. Some typical emission factors are given in Table 2.

In the atmosphere the nitrogen oxides give rise to a number of different chemical reactions. Nitric oxide (NO), initially formed in the combustion process, is oxidized to nitrogen dioxide (NO_2). The emission is therefore usually given as the equivalent concentration of NO_2, and designated as NOx. When exposed to sunlight in the atmosphere, NO_2 is decomposed to NO and atomic oxygen, the latter reacting with molecular oxygen to form ozone (O_3). The NO is re-oxidized to NO_2 by O_3. (There is always a natural background of 20-30 ppb O_3 in the atmosphere, which is due to intrusions from the stratosphere.) In this way a photostationary equilibrium is established. When organic components are added, a net production of O_3 results (11). The mixture of oxygen atoms and free radicals leads also to the formation of various highly reactive organic peroxides and nitrates, as well as aldehydes and nitric acid. Photochemical reactions also play an important part in the homogeneous gas phase oxidation of sulphur dioxide.

In polluted areas, oxidant production reaches a maximum in the afternoon, and the highest values are often observed some distance away from the main precursor emissions. During night there is no production of oxidants, and in the polluted urban air the oxidants will be rapidly consumed in various oxidation reactions. The final products are nitric acid, various organic nitrates, aldehydes and organic acids. Oxidants, moved out of one polluted area, may the following day enhance the problems in downwind areas more than 500 km away (12,13).

In Europe oxidants were first observed in the Netherlands in the late 1960's (14). Before then many believed that photochemical smog could only be found at lower latitudes with more intense solar radiation. Later experience has shown that if the emissions of nitrogen oxides and organic components are large enough, photochemical oxidants can be produced anywhere during the summer season. A simultaneous increase in the ozone concentrations has on several occasions been observed all over western Europe (15).

The emissions of nitrogen oxides in Europe have been increasing at a faster rate than sulphur dioxide, and are still increasing. One particular reason for this is the increased motor vehicle traffic. There has also been some change-over from gasoline to diesel engines. For the photochemical activity and ozone production the emissions of gaseous organic components are most important, but these emissions are not well quantified in Europe.

Although the use of fossil fuels in Europe is higher in winter than in summer, the concentration of nitrate in the precipitation is higher in summer, because of the higher photochemical activity. The effect is less pronounced for sulphates, because sulphur dioxide also can be oxidized, after absorption in liquid droplets, catalytically by iron or mangenese ions, or by dissolved ozone and hydrogen peroxide. The limiting factor is that the absorption stops, when the acidity of the droplet approaches pH 3.

Geographical Distribution

The modelling of long range transport of sulphur pollutants in Europe is based on an emission survey in a grid of 150 km x 150 km.

Table II. Fuel consumption and estimated No_x emission within OECD Europe in 1975. Reprinted with permission from Ref. 18.

	Emission-factor kg NO_2/tonne fuel	Fuel consumption Tg	NOx-emission Tg NO_2
Hard coal			
Power plants	9	133	1.2
Industry	6	22	0.1
Other	2	24	0.05
Brown coal			
Power plants	4	137	0.5
Residual fuel oil			
Power plants	12	69	0.8
Refineries	8	19	0.15
Industry	8	95	0.75
Other	6	27	0.16
Gas/diesel oil			
Industry	8	24	0.2
Other	4	121	0.5
Transport	36	46	1.7
Motor gas			
Transport	25	90	2.2
Natural gas			
Power plants	1	336	0.3
Industry	0.3	642	0.2
Other	0.2	554	0.1
			9.0

For sulphur dioxide, the survey has been worked out in co-operation with the participating countries. For countries which have only been able to provide information on the total emissions for different industrial sectors, the survey was based on national fuel consumption statistics from OECD and ECE sources, emission factors, and population densities. For some countries the accuracy is within ±10-15%; in other cases the data are less accurate. The major emission areas are, however, sufficiently well defined for model calculations (10).

The geographical distribution of sulphur dioxide emissions in Europe is shown in Figure 3 (16,17). For larger areas the sulphur dioxide emissions are approximately proportional to the population density, are higher in areas with a particularly high degree of industrialization. The emissions of nitrogen oxides largely follow the same pattern (18). About half of the emissions originate from motor vehicle traffic, but because traffic density depends on other activities and the population density, an approximate proportionality is obtained for sulphur dioxide and the nitrogen oxide emissions in the larger areas. On a similar basis approximate emission surveys have also been worked out for heavy metals (19) and hydrocarbons (20).

The geographical distribution of sulphur dioxide emissions in North America in Figure 4 (21) shows the eastern part of the North American continent as the main emission area. Emission surveys for nitrogen oxides and hydrocarbons show similar distributions (21).

In central Europe the annual mean concentration of sulphur dioxide is about 20 $\mu g/m^3$ (see Figure 5). The annual concentration pattern of sulphate particles is similar, but because of the time required for sulphur dioxide to be transformed into sulphate particles the annual mean concentration level is lower. The maximum values of about 10 $\mu g/m^3$ (see Figure 5) are slightly shifted to the north-east due to the predominant westerly winds.

Dry deposition of sulphur dioxide is a significant factor in the central part of the area, and is responsible for the removal of about 50% of the total emissions. Compared to this, the dry deposition of sulphate is of lesser significance. About 30% of the total sulphur emission is removed by precipitation scavenging. Maximum wet deposition is found in orographic precipitation areas frequently exposed to polluted air masses, such as the Scandinavian mountains and the Alps, as seen in Figure 6 (2).

Measurements during recent years have shown that much of the remaining pollutants move into the Arctic. As seen in Figure 7 (22), the main transport takes place in the winter across the northern shoreline of the USSR. In summer this pathway is largely blocked by the polar front, and transport from western Europe becomes more important.

The day to day situation is very different from this average picture. With southerly winds, concentrations of 20-30 $\mu g/m^3$ of sulphur dioxide and sulphate particles are frequently observed in southern Scandinavia. For this area, more than 90% of the annually deposited sulphate comes from other countries. Half of the deposition may be due to 10 or so episodes with highly polluted precipitation. A similar situation is observed in other remote areas.

The main acidic component is sulphuric acid, with an addition of 20-50% of nitrate and ammonium ions on an equivalent basis. The

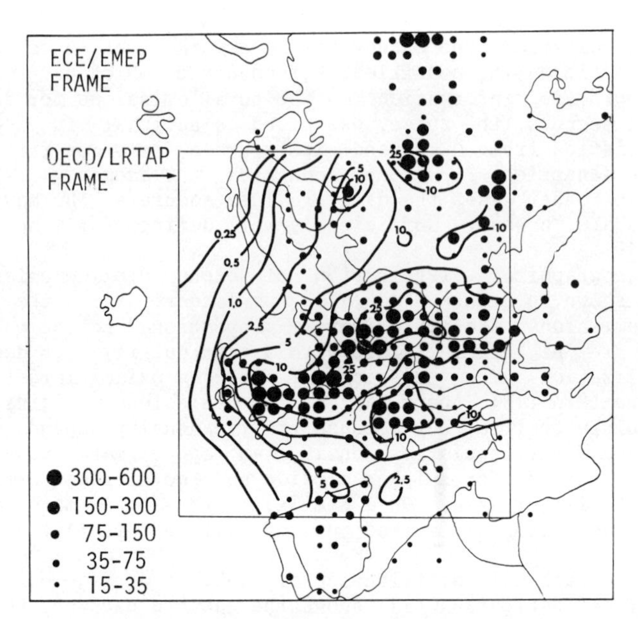

Figure 3. Estimated sulphur dioxide emission in 150 km x 150 km grid squares for Europe 1978 (kilotonnes S/year), and annual mean concentrations of sulphur dioxide (1974) ($\mu g SO_2/m^3$), (16, 17).

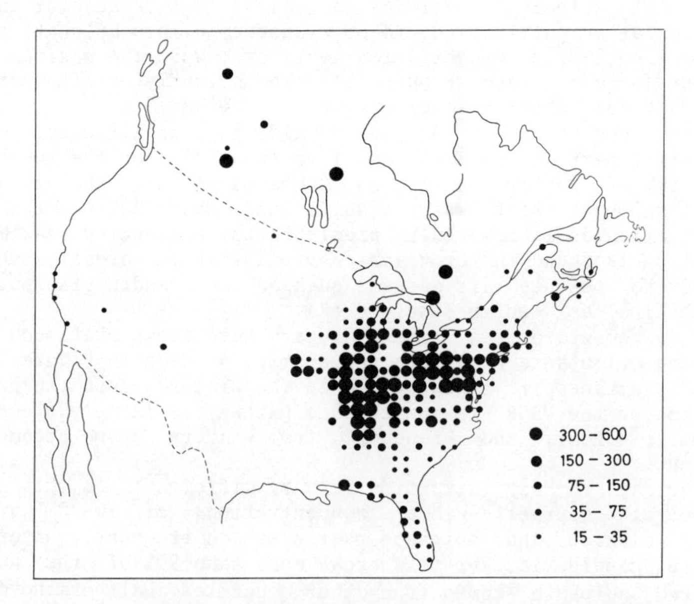

Figure 4. Sulphur dioxide emissions in North America, 1970-75, (21). Unit: kilotonnes SO_2/year.

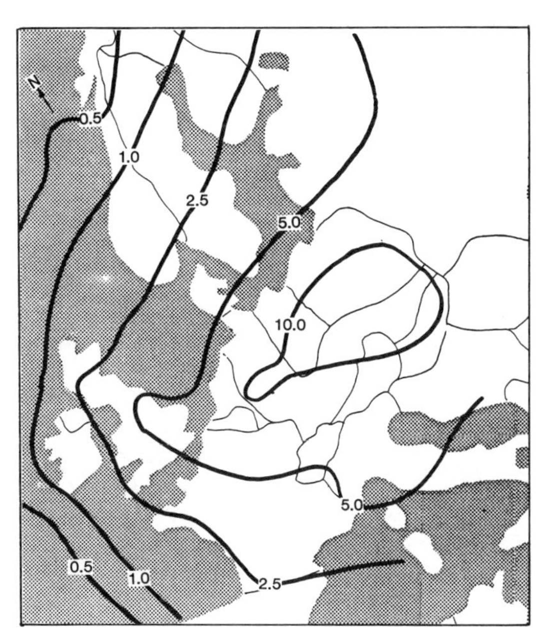

Figure 5. Estimated mean concentration field for particulate sulphate for 1974. Observed mean concentrations given by Italic numbers. Unit: μg SO_4/m^3 (2).

Figure 6. Estimated sulphur wet deposition patterns for 1974. Unit: g S/m^2 (2).

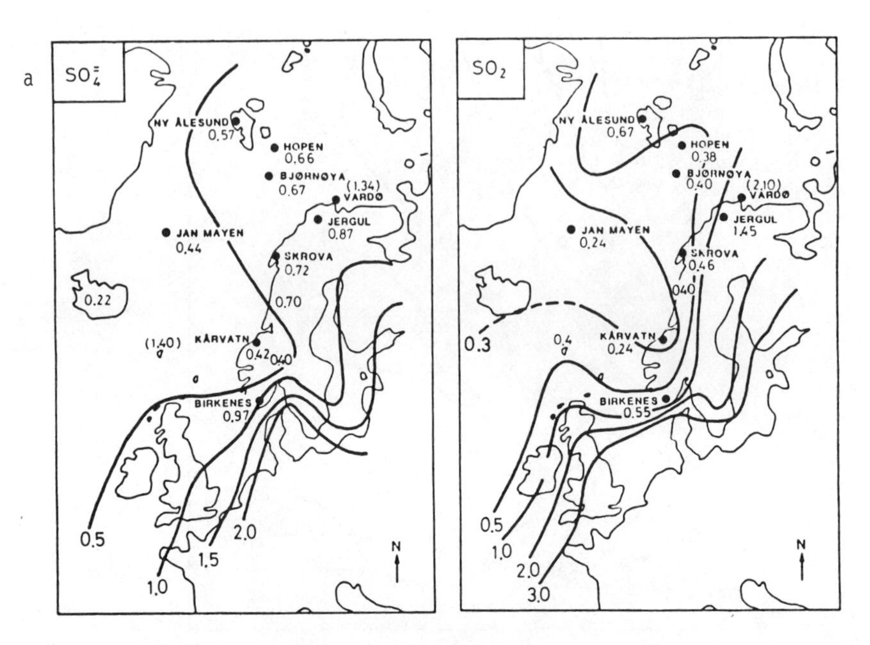

Figure 7a. Average winter/spring concentrations of SO_2 and SO_4 in air in the Norwegian sector of the Arctic. December 1982 May 1983. Unit: μg S/m^3.

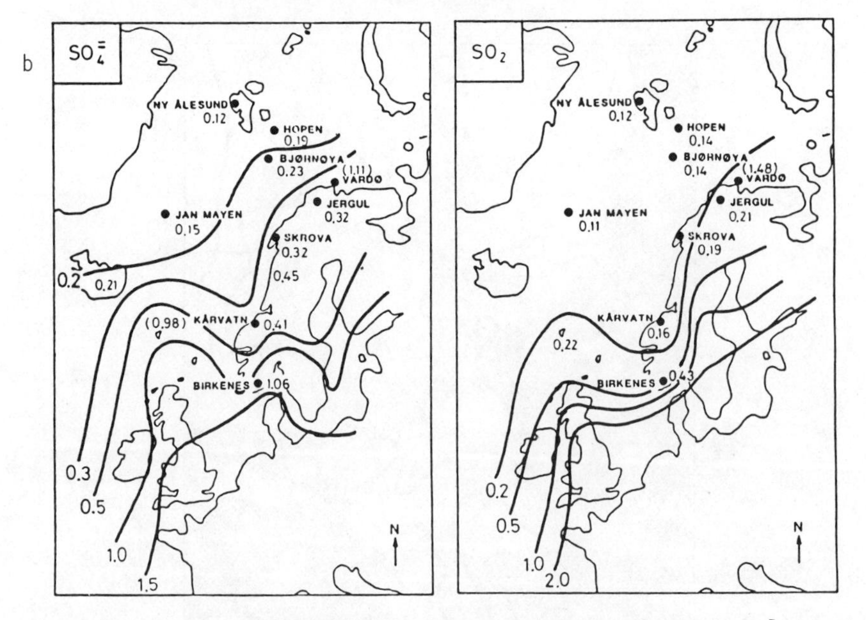

Figure 7b. Average summer/fall concentrations of SO_2 and SO_4^- in air in the Norwegian sector of the Arctic. June - November 1982. Unit: μg S/m^3.

annual mean concentrations of sulphate in precipitation in Europe from World Meteorological Organization data for the period 1972-76 (23) are shown in Figure 8.

Episodes with maximum acidity are often observed when highly acidic particles are formed in air, which has remained over the sea for several days (with no ammonia emission), and later are scavenged by orographic precipitation. In 1978 an exceptional case of 10 mm precipitation with a pH of 2.5 was reported by the Meteorological Service of Iceland. Cases of pH 2.7 have been observed in Scotland and on the west coast of Norway.

The mean acidity of precipitation in North America for the period 1976-79 (24), is shown in Figure 9. The maximum values (i.e., minimum pH) are associated with the eastern part of the continent, and are closely related to the major emissions in the Ohio valley. In northeastern USA and Canada, both deposition and sulphate concentrations are at a maximum in summer.

In Scandinavia the concentrations of sulphate in precipitation are generally highest during the spring, while the emissions of sulphur dioxide in Europe have a maximum in January (about 2 times the emissions in summer). This delay can be attributed to a precipitation minimum in western Europe during the early spring, and more rapid conversion of sulphur dioxide to sulphate with increased solar radiation. The seasonal variation of the concentration of nitrate in precipitation is similar, but with a longer maximum period (25).

This reflects the different climates and fuel consumption patterns in Europe and North America. In Norway the precipitation at the present contains about equivalent amounts of nitrate and ammonium ions. In the 1950's this ratio was also very constant, but the concentration of nitrate ions was only 1/2 of the ammonium ions. The basic reason for this constant ratio seems to be that in northwestern Europe the emissions of nitrogen oxides from industry and motorized traffic largely take place in regions of major agricultural activities. The higher ratio today indicates that the nitrogen oxide emissions have increased over the past years. It is interesting to note that in North America the emissions of ammonia come from the mid-western agricultural areas, while most of the nitrogen oxide emissions come from areas further east. As a result the nitrate to ammonium ions ratio in precipitation is much more variable.

The acidity of precipitation is mainly governed by its content of sulphate, nitrate and ammonium ions. It is closely related to the chmical composition of the aerosols, and may to a large extent depend on the pathway of the polluted air masses. The effects of the acid precipitation are not simply related to the acidity of the precipitation, but are the result of complex interactions in which all the major ions in precipitation are of significance.

The chemical composition of aerosols is influenced by other substances present in the air. An equilibrium is rapidly established between ionic components in the particles or droplets and the corresponding gaseous components, i.e., NH_4^+/NH_3, NO_3^-/HNO_3, Cl^-/HCl, SO_4^{--}/SO_2. Particles formed over the sea are often particularly acidic, because of the low partial pressure of ammonia. If later these particles pass over agricultural areas, they are rapidly neutralized by ammonia, at least at ground level. On the other hand, ammonium sulphate particles which pass over the acidic forest soils in Scandinavia, have been shown to give off ammonia (26).

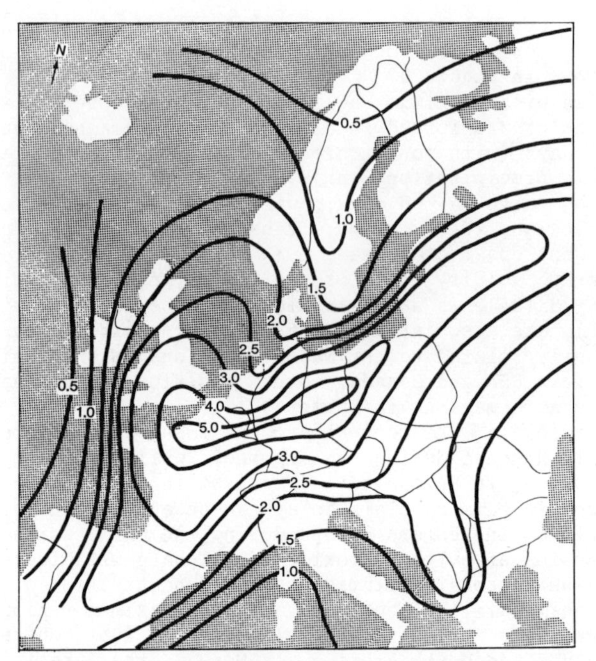

Figure 8. Annual mean concentration of sulphate in precipitation mg $SO_4^{=}$/l 1972-76 (23).

Figure 9. Weighted mean pH of precipitation for North America for the period 1976-1979. Dashed lines indicate where data are sparse and thus only the general pattern is indicated (24).

Smaller acidic sulphate particles may lose chloride and nitrate ions in the form of gaseous hydrochloric and nitric acid. Thus, the chloride in airborne sea salt may be driven off as hydrochloric acid, which may be subsequently absorbed by larger, less acidic particles. Similar chemical reactions can also take place in samples of particles collected on filters, particularly if the coarse and fine particles are not separated. The pressure drop across the filter may also cause evaporation of the more volatile components. The chemical analysis of the collected particles may then give a distorted picture of the true airborne composition of the aerosol.

Aerosol particles serve as condensation nuclei for cloud droplets and precipitations. As a rule of thumb, about 1 ml H_2O is precipitated from each m^3 of air in a cloud. With a 100% scavenging by precipitation, 1 $\mu g/m^3$ of a substance should give 1 mg/l in the precipitation. Some typical analyses of polluted air masses, which have been transported to Norway across the North Sea, are shown in Figure 10 (2). The high concentrations of nitrate and ammonium in the precipitation, relative to those in the aerosols, indicate that absorption of sulphur dioxide and gaseous nitric acid has taken place.

Effects

Acidification of rivers and lakes. Loss of fish populations due to acidification of rivers and lakes was the first recognized sign of ecological damage due to acid precipitation. Although problems with acid water and salmon and trout stocks in Southern Norway were noted in the 1920's (27), the fish decline has escalated since 1950 and today it affects many thousand lakes in Norway and Sweden (Figure 11).

Surveys have shown a general correlation between the pH and fish status in lakes, and the physiological response of fish to acidic water is now well known (28). Acid water affects the uptake of sodium and chloride ions through the cell membranes in the gills, and leads to a disturbance of the electrolyte balance (29). The concentration of reactive (uncomplexed) aluminium ions in acidified waters is particularly detrimental (30), apparently because this interferes with the metabolic uptake of electrolytes from the water, while calcium ions have an ameliorating effect (31).

Generally, the problems with acid water occur mostly in areas with granitic or similarly resistive bedrock, and with sparse soil cover derived from the same parent material. Calcium concentration levels in surface waters are generally low, typically less than 2 mg/l. This explains why the problem so far has mainly surfaced in Scandinavia, and in areas of similar geological formations on the Canadian shield and in the Adirondack Mountains. Some acidification of lakes and water courses has also occurred in England and Scotland (32,33), and in the Erzgebirge between the German Democratic Republic and Czechoslovakia (34,35).

The interactions between soluble ions in precipitation, soil, and vegetation, and the effect of these interactions on runoff water quality are complex. The soil cover which was formed in Scandinavia when glaciation retreated some 9000 years ago, largely consists of granitic sand, covered with a top layer rich in humus (podsol).

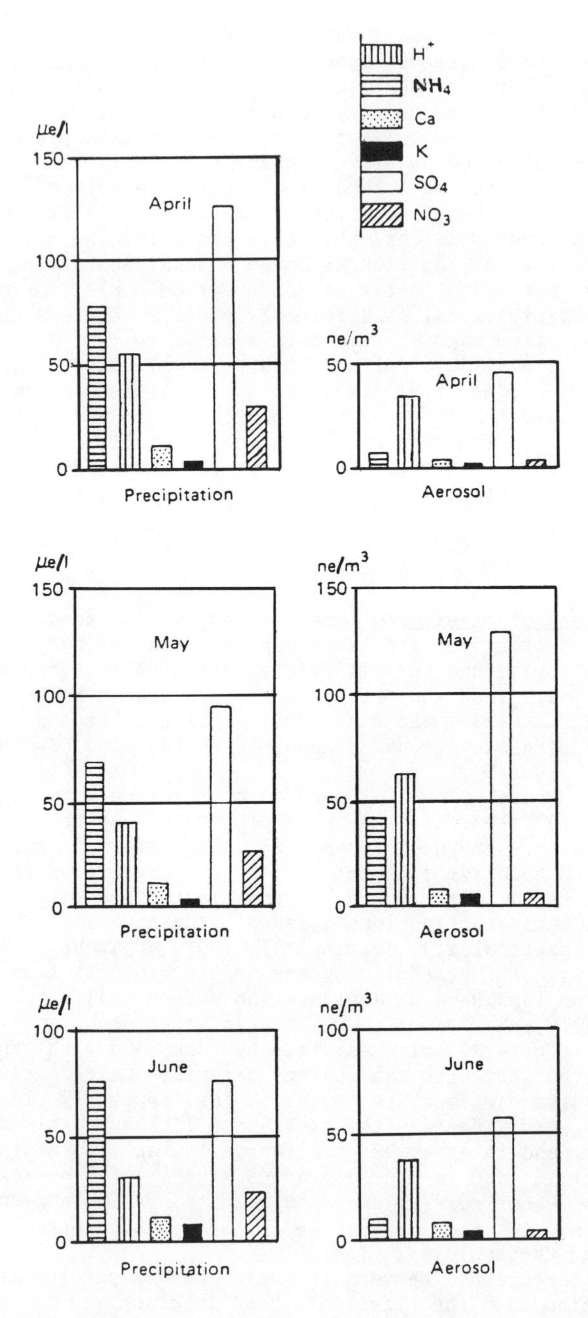

Figure 10. Concentrations of water soluble compounds in precipitation and aerosols at Birkenes, April - June 1975 (2).

The sparce amount on nutrient elements primarily originate from the vitrification of the feltspar minerals. The hydrogen ions, required by the vitrification process, are available from the carbonic acid dissolved in the water or from organic acids produced by the roots of the plants. In the top soil, roots and dead plant material, in the form of humus, function as a cation exchanger where dissolved metal ions as calcium, magnesium and potassium from the vitrification of minerals, precipitation and organic litter are held back in exchange for hydrogen ions.

In passing this layer, the precipitation water becomes more acid, and iron and aluminium are dissolved in the upper part of the sand layer. Further down, the acidity is reduced by the dissolution of calcium, magnesium and potassium from the minerals, and iron and aluminium are redeposited as hydroxides (see Figure 12). This is a natural process which during some thousand years has led to formation of the characteristic podsol profile: on the top a layer of black humus, then a nearly white layer of leached silicates, and underneath a deposition layer, which takes on first a yellow and then a red colour as the iron is precipitated out.

In this system, the acidity of the soil liquid is limited by the amount of mobile anions. The more important are the sulphate, nitrate and chloride ions, all of which originate from the precipitation. The chloride ions are usually accompanied by an equivalent amount of sodium ions, which are not taken up by the plants and retained by the humus to any significant degree. Most of the nitrates are consumed by the vegetation.

The acidity of the soil may conveniently be characterized by the content of metal ions, relative to the total cation exchange capacity of the soil. In the acid types of soil considered here, this base saturation degree is usually below 10%. The degree of base saturation will be reduced when (1) the roots take up exchangeable cations from the soil, and (2) when accumulation of dead plant material increases the amount of humus, and thereby the cation exchange capacity. To a certain degree, both of these processes are reversible, but if plant products are removed from the area without application of fertilizers, manure or lime, this represents an acidification by reducing the available supply of cations.

The acidifying effect of the vegetation is therefore limited to a regulation of the wash-out of the basic components formed by vitrification of the minerals. In addition to this, humus components (fulvic acid), and mineralization of reduced sulphur and nitrogen compounds may to some extent contribute to the acidity and the anion concentration in the drainage water. But this seems to be of limited significance for the "regional acidification problem".

The acidification of the drainage water is particularly related to sulphate ions. Along their route in the soil, the sulphate ions must bring with them an equivalent amount of hydrogen ions or other cations. The humus ion exchanger does not willingly give away cations; on the contrary, it removes metal ions from the precipitation. Some cations, however, are released all the time by vitrification.

Aluminium hydroxide which is accumulated in the deposition layer, acts as a weak basic anion exchanger, and may hold back significant amounts of the sulphate ions in the water. In southern Norway the total amount of sulphate retained may correspond to

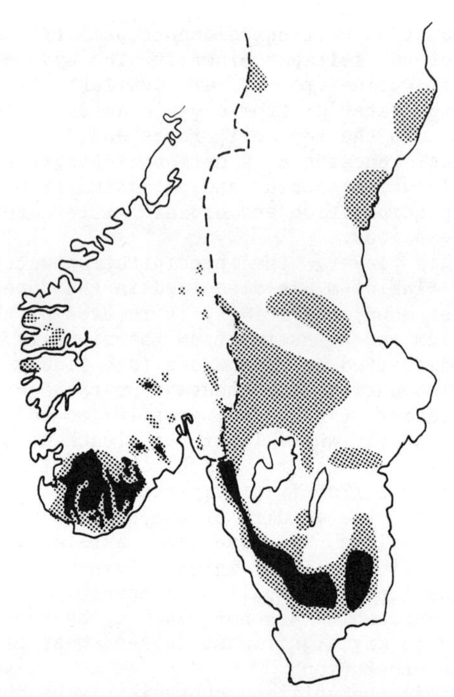

Figure 11. The right side shows areas in Sweden with lake water pH 5.0 (black) all year round in at least one third of the lakes and areas with pH 5.5 (gray) some time during the year in at least half of the lakes (SNV, 1981). The left side shows areas in Norway when the fish is virtually extinct (black) and areas where the fish population is strongly affected (gray) (Munitz and Leivestad, 1980).

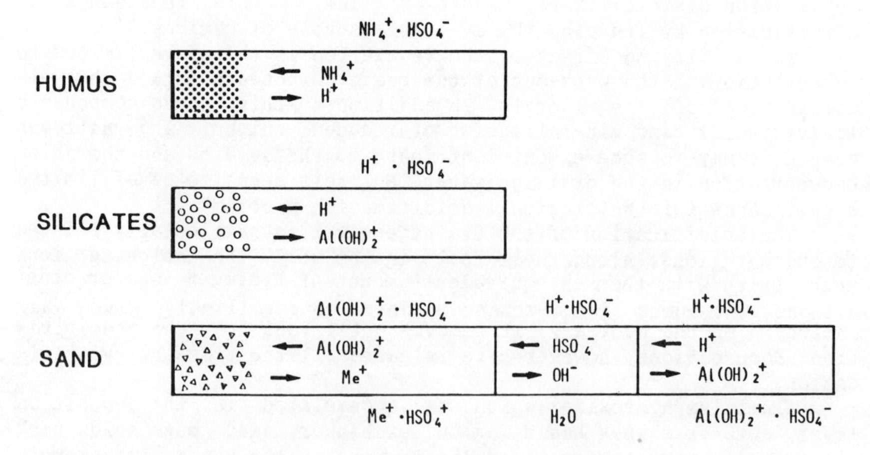

Figure 12. Chemical reactions of acid rain in podsol profiles.

several years of precipitation. At present the amount of sulphate in the annual precipitation is approximately equal to the amount in drainage waters. The accumulation of sulphate in the soil must therefore be a process which has taken place over an extended time period.

When the soil water passes through this buffer of aluminium and iron hydroxides, sulphate ions may be given off or taken up, depending on the initial concentrations in the precipitation. However, as a result of the increasing concentration of sulphate in the precipitation during the later decades, the hydroxide layer is now gradually being saturated with sulphate ions. Consequently the soil water becomes more acid and starts to dissolve aluminium hydroxy-ions from the clay minerals in the silicate layer and from the deposition layer. In mineral soils the ratio between hydrogen and aluminium ions in the soil water is now close to the theoretical ratio for equilibrium with aluminium hydroxide in the form of the mineral gibsite. However, the aluminium ion concentration may be less in the drainage water from predominantly organogenic soils, such as peatbogs.

Forest damage. After the weekly magazine Der Spiegel had published in 1981 a series of articles, where it was shown that the forests in the Federal Republic of Germany were dying because of air pollution, regional air pollution problems became almost overnight an international issue of first priority. Intensive studies have since shown that forest damage occurs in all the central European countries, while Scandinavia and England so far seem to be less affected (36,37).

Initially forest damage was believed to be another effect of acid precipitation: toxic aluminium ions released from the soil minerals were poisoning the root system of the trees. But the forest damage also occurred in areas with soils rich in carbonates, and in valleys it often appeared at certain height levels where fog was frequently observed. This pointed to air pollution as a direct main cause for the damage.

The symptoms of this forest damage are not specific for air pollution, and several factors may contribute to a general "stress" situation. The acidity of dew and fog can be much higher than in regular precipitation. The acidification of the soil may lead to increased dissolution of heavy metals, naturally present or accumulated from air pollutant deposition over a long period of time. During recent years, the emission and deposition of nitrates and ammonium have increased steadily. Agricultural experience has shown that this can prolong the period of active growth of the trees, which makes them less resistant when the cold winter comes.

Today it is generally believed that photochemical oxidants may be a most important contributing factor. In Europe the ozone levels in summer frequently exceed levels known to cause forest damage in North-America, and have led to large scale attacks by bark beatles. However, in the forests of FRG and other central European countries, damaged trees are quickly removed for productivity reasons, and this is exactly the way Canada has prevented large scale bark beatle attacks on the northern shores of the Great Lakes. In the Mediterranean area extensive insect attacks on the pine forest along the coast of Italy, France, Spain and North Africa, in a zone of about 2 km along the coast have been observed. No detailed examination of the damage has so far been carried out. It is, however, well known that nitrogen oxides and hydrocarbons, transported out to the sea with the land breeze during the night, may return as photochemical oxidants in the morning and cause such damage by fumigation near the shore line.

At the moment one has no comprehensive survey of the photochemical oxidant situation in Europe. It has been pointed out many times during the past years that such survey is needed, but so far the affected countries have been reluctant to spend money on a case for which causes have as yet not been proven.

Conclusions

The present experience indicates that the regional air pollution situation in Europe and in North-America has gradually become more serious. In general terms, a change of our chemical environment has been taking place. The atmosphere has become more acidified and the oxidation potential has increased.

If this development continues, there is little doubt that more regional problems will appear. Incidents of "red tide" are reported from the North Sea area. This is an over-growth of toxic algae which follows excessive deposition of nitrates by precipitation and kills the fish. Watersoluble mercury compounds in the atmosphere, partly due to emissions from coal-fired power plants and partly due to an oxidation of the natural mercury vapour, have resulted in a serious contamination of freshwater fish in Scandinavia. Already fish from many lakes cannot be marketed in Sweden, because the mercury content exceeds 1 mg/kg.

With respect to corrosion, the conventional classification of climates in marine, inland, industrial , etc. types is not sufficient. It should now be specified with respect to the actual chemical components in the atmosphere, as well as humidity and other factors. Recent research in this field has led to much more precise methods for estimating corrosion rates in polluted atmospheres (38). Economically, perhaps even more important problems are caused by the increased corrosion of water supply pipelines. Not only copper is dissolved, but also cadmium from soldered joints, and larger steel and cement pipelines may also be affected.

So far it has more or less been taken for granted that this general development can be reversed by reducing pollutant emissions, if the situation becomes intolerable. In principle this may in many cases be correct, but it may take a long time. In some areas the aluminium deposited in the podsol profiles is leached out by acid precipitation and brought into the ground water. It took a long time to build up these deposits by vitrification of soil minerals. The rapid deterioration of monuments and historical buildings can be stopped, but not reversed. It is therefore important to try and repair the general situation before this acidification of our environment in a wider sense has gone too far.

Literature Cited

1. Odén, S. Statens Naturvetenskapeliga Forskningsråd, Ekologikommitéen , Bull. No. 1, Stockholm, 1968.
2. "The OECD programme on long range transport of air pollutants", OECD, Paris, 1977, 2nd ed.
3. Ottar, B. Atmos. Environ. 1978, 12, 445-454.
4. Eliassen, A. Atmos. Environ. 1978, 12, 479-487.
5. Semb, A. Atmos. Environ. 1978, 12, 455-460.
6. "Convention on long-range transboundary air pollution", United Nations Economic Commission for Europe, Geneve, 1979.
7. Eliassen, A.; Saltbones, J. Atmos. Environ. 1983, 17, 1457-1473.
8. Husar, R.B.; Patterson, D.W.; Halloway, J.H.; Wilson, W.E.; Ellestad, T.G. Proc. 4th Symp. Atmos. Turb. Diffusion and Air Pollution, Am. Meteorol. Soc., Boston, MA, 1979, p. 249-256.
9. Cullis, C.F.; Hirschler, M.M. Proc. Int. Symp. Sulphur Emissions and the Environment. The Chemical Society, London, 1974, p. 1-24.
10. Semb, A. Proc. WMO symp. long-range transport of pollutants and its relation to general circulation including stratospheric/tropospheric exchange process, WMO No. 538, pp. la-lm, Geneva, 1979.
11. Atkins, D.H.F.; Cox, R.A.; Eggleton, A.J.E. Nature 1972, 235, 1391-1401.
12. "Ozone and other photochemical oxidants", National Academy of Science, Washington D.C., USA, 1977.
13. "Photochemical oxidants and their precursors in the atmosphere". OECD, Paris, 1978.
14. Guicherit, R.J.; van Dop, H. Atmos. Environ. 1977, 11, 145-156.
15. Schjoldager, J.; Dovland, H.; Grennfelt, P.; Saltbones, J. Norwegian Institute for Air Research, Report OR 19/81, Lillestrøm, Norway, 1981.
16. Dovland, H.; Saltbones, J. Norwegian Institute for Air Research, EMEP/CCC-report 2/79, Lillestrøm, Norway, 1979.
17. Amble, E. Norwegian Institute for Air Research, Report TR 8/81, Lillestrøm, Norway, 1981.
18. Semb, A.; Amble, E. Norwegian Institute for Air Research, Report TR 13/81, Lillestrøm, Norway, 1981.
19. Pacyna, J.M. Atmos. Environ. 1984, 18, 41-50.
20. Eliassen, A.; Hov, Ø.; Isaksen, I.S.A.; Saltbones, J.; Stordal, F. J. Appl. Met. 1982, 21, 1645-1661.
21. "Acidification in the Canadian aquatic environment", National Research Council Canada, Report NRCC No 18475, Ottawa, 1981.
22. Joranger, E.; Ottar, B. Geophys. Res. Letter, 1984, 11, 365-368.
23. Wallén, C.C. Monitoring and Assessment Research Centre, Report No. 22, Chelsea College, University of London, 1980.
24. "Report of the United States - Canada Research Consultation Group on the Long-Range Transport of Air Pollutants". Altshuller, A.P. and McBean, G.A. eds. U.S. State Department, Washington DC, and Canada Department of External Affairs, Ottawa, 1980.
25. Joranger, E.; Schaug, J.; Semb, A. Proc. Int. Conf. Ecological Impact of Acid Precipitation, Drabløs, D. and Tollan, A. eds., Norwegian Institute for Water Research, Oslo, Norway, 1980, p. 120-121.

26. Brosset, C. Ann. N.Y. Acad. Sci. 1980, 338, 389-398.
27. Dahl, K. Salm. Trout. Mag. 1927, 46, 35-43.
28. Overrein, L.N.; Seip, H.M.; Tollan, A. Norwegian Institute for Water Research, Report SNSF FR 19/80, Oslo, Norway, 1980.
29. Leivestad, H.; Muniz, I.P. Nature 1976, 259, 391-392.
30. Driscoll, C.T.; Baker, J.P.; Bisogni, J.J.; Schofield, C.L. Nature, 1980, 284, 161-164.
31. Brown, D.J.A. Bull. Environ. Contam. Toxicol. 1983, 30, 382-87.
32. Sutcliffe, D.W.; Carrick, T.R. Freshwater Biol. 1973, 3, 437-462.
33. Wright, R.F.; Harriman, R.; Henriksen, A.; Morrison, B.; Caines, L.A. In "Ecological effects of acid precipitation", Drabløs, D. and Tollan, A. eds., p. 248-249. Norwegian Institute for Water Research, Oslo, Norway, 1980.
34. Hanuska, L. Werk. Internat. Verein. Limnol. 1972, 18, 968-980.
35. Peukert, V. Wasserwirtschaft-Wassertechnik, 1974, 24, 158-161.
36. Ulrich B.; Pankrath, J. eds. "Effects of Accumulation of Air Pollutants in Forest Ecosystems". D. Reidel Publ. Comp., Dordrecht, Holland, 1983, 127-146.
37. Ulmer, G.A., "Unser Wald darf nicht sterben". Günter Albert Ulmer Verlag. Schwabenstrasse 31, 7036 Schönaich, FRG, 1983.
38. Haagenrud, S.E.; Henriksen, J.F.; Wycisk, R. Proc. "Corrosion 85", National Association of Corrosion Engineers, Houston, TX, 1985.

RECEIVED February 5, 1986

2

Wet Deposition Chemistry

Richard G. Semonin

Illinois State Water Survey, 2204 Griffith Drive, Champaign, IL 61820

Interest in precipitation chemistry dates back to the 18th century, but only since 1977 has the U.S. established a national network for monitoring at a level sufficient to describe the chemical climate. The siting criteria, quality assurance, and standardized equipment and analytical methodology are addressed to insure the highest quality data for interpretative research. The national distribution of concentration and deposition for calcium, ammonium, chloride, nitrate, sulfate, and pH are shown in a series of maps. The maps show high concentrations of calcium and ammonium in the Great Plains, high concentrations of nitrate, sulfate, and hydrogen ion in the eastern states, and high concentrations of chloride along the coastal states. The primary conclusions are 1) the greatest deposition is directly associated with the largest emission regions, and 2) dry deposits are largely unknown but clearly must be measured before the biochemical and geochemical cycles will be fully understood.

We have often heard the saying, "What goes up must come down", but all too frequently have not given the concept too much thought. It now has become a very important issue not only in this country, but everywhere on the planet earth and is the center focus of the acid rain debate. The debate today does not question the statement given above, but rather addresses the tougher problem of "When something goes up how, when and where does it come down?" A related question might be "When it comes down, is it beneficial or harmful to its final resting place?" A further important question is "Is the acidity of precipitation increasing and is it spreading to larger areas in the U.S. and beyond its borders?" The answers to these questions are central to the establishment of policy regarding how to deal with the national and international concerns about acidic deposition.

To begin to address these and other questions, it is appropriate to very briefly examine what little is known about previous studies of the chemistry of precipitation. It must be borne in mind,

0097-6156/86/0318-0023$06.00/0

however, that while there are some indications of great interest in atmospheric chemistry over many years, the technology for providing analyses of both the atmosphere and its precipitation is still improving. An implication of this is that it is most difficult to compare more recently acquired data with those of the relatively distant past to determine the extent and trend of changing precipitation chemistry.

It seems central to the on-going debates, both scientific and political, that the local and regional trend of important chemical ions in precipitation be determined in order that their contribution to the chemical cycle of local ecosystems can be more accurately assessed. Equally important to such an assessment is the quantification of the magnitude of the wet and dry deposition from the atmosphere and their ultimate disposition in the identified cycles.

In this brief paper, we will not attempt to address all of the complex biochemical and geochemical sciences, but we will attempt to describe the chemistry of precipitation as a single, and frequently minor, input to those chemical budgets of importance to the environment. We will also not attempt a comprehensive description of dry deposition as there is no single agreed-upon method for its measurement. Experience tells us that it does not precipitate all of the time, and in the eastern mid-latitude of the U.S. no precipitation is observed about 90% of the time during any given year. However, experience also tells us that material from the atmosphere is continuously returning to the earth in dry form. Some preliminary data obtained from exposed bucket collections will be discussed as a "poor man's" estimate of a portion of the dry deposition.

Precipitation Chemistry Background

There is early documentation of interest in the chemical composition of rain dating back to 1727, followed by a fairly thorough study of the subject in the mid-1800's (1). These early efforts focused on sulfur in rain, snow, and dew, and some of the perceived impacts on health and agriculture. The modern interest began in the post-World War II years when some monitoring of precipitation chemistry was initiated in the Scandinavian countries (2). A particular effort was put forth in Sweden with agriculturalists and meteorologists expressing equal interest in the chemistry of precipitation, but for entirely different reasons. The agricultural interests focused on the quantity of nutrients being deposited by precipitation as an aid to plant growth while the atmospheric scientists were attempting to use the chemistry to further their knowledge of the origin of precipitation water and to trace atmospheric motions. The national commitment of Sweden to continue the network operation placed them in the forefront of the emerging issues related to acid rain.

U.S. Sampling Activities. During the early quarter of this century, agriculturalists in the U.S. showed interest in the amount of nutrient falling on productive soils as evidenced by the literature in the 1920's (3,4). Following the lead of Sweden, some interest in the chemistry of precipitation was expressed by meteorologists in the U.S. which culminated in the establishment of a national network in 1955 (5,6). This network was operated for a single year (1955-1956) and the data have been used by some as the baseline data for demonstrating that acid rain is worsening and spreading in the eastern U.S. (7-9).

A short time after the demise of this initial network a similar one was begun by the U.S. Public Health Service (PHS) in 1960. This network was maintained until 1966 with management responsibilities eventually falling upon the National Center for Atmospheric Research (NCAR) (10). No national network was operated after the closing of the PHS/NCAR stations in the 1960's although several isolated measurement programs were carried out in various parts of the country (11,12). An excellent summary of sampling activities in North America is available for additional information (13).

The First International Symposium on Acid Precipitation and the Forest Ecosystem in 1975 concluded with workshops addressing various needs to properly research the topic of acidic deposition in the U.S. (14). Among the recommendations from one of the workshops was that a network be established for the purpose of long-term monitoring of the precipitation chemistry across the U.S. and its territories. This currently operating network extends from Alaska to Puerto Rico, and from Maine to American Samoa with 190 sampling sites. Many of the stations have been identified as National Trends Network (NTN) sites - the monitoring network of the National Acid Precipitation Assessment Program (NAPAP) (15).

Precipitation Chemistry Character

The chemistry of precipitation is characterized by trace quantities of most substances found in the atmosphere. Concentrations are typically measured in parts per million, parts per billion, and even parts per trillion. When considered as an ionic solution, about 95% of the total ionic strength is accounted for by the analysis of calcium, magnesium, ammonium, sodium, potassium, hydrogen, chloride, nitrate, and sulfate. The hydrogen ion is usually determined from measurements of the sample pH.

The acidity of precipitation is commonly presented as pH. However, it is important to also examine all of the other ions in a sample to understand the effects on the environment. There has been a suggestion put forth to limit the sulfate deposition without regard to the precipitation acidity. It is presumed that a sulfate reduction will automatically result in higher pH values, and, thus, a double environment benefit will result. But without better quantification of all sources of sulfate and their relation to acidity, any reduction strategy may not bring about the desired result.

Precipitation Collection for Chemical Analysis

A discussion of the quality of precipitation chemistry must include the analytical methods used as well as a description of the device used to collect samples. Most importantly, the quality of the data must be assured when assessing the impacts on the precipitation-receiving environmental system. Students of precipitation chemistry have used everything from glass bottles to baby bottle plastic liners to collect samples for analysis. Laundry baskets, staked to the ground, as well as fence-post mounted plastic bottles have also been used in network operations (19). One of the most interesting collectors was the entire roof of a wood frame building convered with polyethylene sheeting and special guttering. A very large sample could be collected in a very short period of time even during the lightest rainfalls.

A number of other devices have been built directed toward acquiring one sample after another during a single rain event to obtain fine detail of the chemical structure of precipitation (20,21). There are various means used to control the sample collection by either the volume per sample or time interval between sample collections. Owing to the rapid collection of numbers of samples in a very short period of time, none of these sequential collection devices have been used in a regional network.

The most widely used sampler in the United States networks today is comprised of two buckets, and a rain-activated switch to operate a movable cover. During non-precipitating periods, the cover remains tightly sealed on one bucket. Precipitation falling on the sensitive switch completes an electrical circuit activating a motor which lifts the cover from one side and places it on the opposite bucket. The sampler thus provides a dry sample as well as a wet sample. This is the standard instrument used throughout the NADP/NTN network.

Anyone wishing to collect rain or snow for chemical analysis is cautioned to first check the collection vessel for the chemicals of interest to see if, in fact, the analysis will be contaminated. For example, it would be unwise to collect samples in a weighing-bucket raingage for zinc analysis when the bucket is zinc-coated and leaches into the sample.

A second serious consideration is whether one wishes to collect a bulk sample as opposed to a wet-only sample. A bulk sample is one which is directly exposed to the atmosphere and remains open throughout a prescribed interval of time. This is not a very satisfactory way of collecting precipitation samples because of the natural tendency of birds to perch on the rim of the collector always facing outward contributing to the debris deposited inside the container. Equally important, dust, leaves, and other natural wind-blown materials are likely to enter the sampler and contaminate the precipitation in an unpredictable manner.

The interval between the collection of samples is largely determined by the goals for the sampling program. If one wants to study the effects of precipitation chemistry on the forest, for example, it highly unlikely that it is necessary to collect samples on intervals of anything less than a one-week period and perhaps even one month may suffice for the majority of biological effects studies. On the other hand, if one wants to study the variability of precipitation chemistry in convective storms during the warm season, a sequential sampler may be necessary to obtain samples as frequently as one or more per minute (20). So in establishing a sampling program, it is most important to carefully consider the goal of that program and then determine the need for event sampling, as opposed to less costly longer periods, to achieve that goal. The NADP/NTN weekly collection network is an arbitration between event samples and monthly samples, but was chosen to address the program goals of determining 1) the long-term trend of precipitation chemistry and 2) atmospheric deposition effects on the environment.

Precipitation Chemistry Quality Assurance

Once a sample has been confined within the collecting vessel, the safest thing is to immediately seal that vessel and carry it or ship it to the analytical laboratory. However, it is a practice in some operations to allow prior handling of the sample such as withdrawal of aliquots for the local determination of a particular parameter. For example, the NADP/NTN allows extraction of a few milliliters for the field determination of pH and conductivity. Immediately after the aliquot has been withdrawn, the sample is sealed and then shipped to a central laboratory for further chemical analysis. Shipment of the sample is an important consideration for any type of sampling program since one must be sure that the collecting vessel does not leak in transit.

Written documentation of everything concerning the sample up to this point should be provided for the laboratory staff as the analysis of precipitation chemistry proceeds. Certainly, any laboratory, whether it is adjacent to the sampling site or several thousands of kilometers distant, should have certain analytical capabilities for the determination of trace materials in precipitation. The analysts must be trained to recognize the expected concentrations in precipitation and detect contamination in a sample. Contamination can originate from either natural causes or handling of the sample.

Finally, one must be alert that even though a determination may be perfectly accurate and within statistically allowable errors of the instrumentation, the value may, in fact, be excluded from a data set for other reasons. For example, a loose covering over the collection vessel can allow crustal dust to enter into the collector during non-precipitating intervals and can artificially raise the concentrations of those materials. A "leaky" seal results in values that are not representative of precipitation but are more representative of a bulk sample. The major point is that the sample quality control does not begin or end in the laboratory, but must be extended to include everything from the sample collection in the field to the point of preparing the data for dissemination or further interpretation and archiving (22).

Concern has been expressed about the chemical integrity of samples collected less frequently than the duration of a single storm. There is reason for some scientific inquiry on this matter, but the available data suggest that any chemical changes in a sample will occur in a relatively brief period after the precipitation has ended (23). However, event samples may not be any more stable than weekly samples if the delay between collection of the sample and its analysis is of the order of one or more days. Consequently, until real-time chemical analysis can be performed in the field, all currently available data contain largely unknown errors from chemical changes that occur between the end of an event and the analysis.

Selected Interpretative Analyses

There are at least two obvious ways of viewing the chemistry data from precipitation samples. The first is the concentration of samples and the second is the deposition (or loading) of ions of interest to the surface. From a simple perspective, the concentration is of interest to atmospheric chemists while the deposition is of interest to effects research scientists. It should be kept in mind that the concentration and precipitation are the observables and the deposition is a derived quantity. The deposition is calculated by multiplying the observed concentration by the amount of precipitation associated with the sample thereby obtaining a value of the mass deposited per unit area. In the following discussion, both the concentration and deposition will be shown and described. Only the major ions calcium, ammonium, chloride, nitrate, sulfate, and hydrogen ion will be shown. There are many ways that the data can be displayed, but for this discussion median values from the entire data set were selected for each site and maps were hand-drawn to illustrate national patterns for each ion.

Calcium. The calcium contribution to the ionic strength of a sample is thought to be mainly due to the incorporation of soil aerosol into the precipitation before it is collected. Owing to agricultural practices and the semi-aridness of the Great Plains, it is not surprising to find the highest concentrations in that region (Figure 1). A large area of relatively high concentration is seen extending from Montana-North Dakota south through central and western Texas. A secondary area is seen extending northwest into southern Idaho from the Four Corners area where Utah, Colorado, New Mexico and Arizona join. The coast lines of the U.S. are areas of relatively low concentration with values 3 to 6 times less than in the interior continent.

The deposition, expressed in kilograms per hectare per year, also shown in Figure 1, maximizes slightly to the southeast from the center of maximum concentration and is found over the southeast South Dakota, southwest Minnesota, northwest Iowa, and northeast Nebraska area. This slight shift is due to the somewhat greater precipitation toward the southeast while maintaining high concentrations. The isolated maximum in the Four Corners area is a direct result of an isolated maximum in precipitation and the maximum concentration. The same is true of the maximum in southeastern Louisiana. Note the excursions of high deposition into northern Illinois, western Kentucky, and even a small maximum in northern West Virginia. During periods of drought and attendant dust storms, such excursions can be more severe and cover much larger areas leading to misinterpretation of the meaning of the chemistry regarding trends (24).

Ammonium. The calcium likely appears in precipitation as the result of natural causes. It is in relative abundance in the natural sources of the oceans and the earth's crust. The source strength has not been quantified although some have attempted adjustments for seawater contributions to observed concentrations (25).

The ammonium ion source is also largely undetermined. The concentration pattern shown in Figure 2 suggests a relationship to the large feed lots associated with the cattle industry of the central Plains region. However, this possible source has not been measured, but only surmised from the geographical relationship between the maximum concentration and the known feed lot distribution. In one sense, the distribution of ammonium in precipitation has a natural source, but somewhat controlled by man.

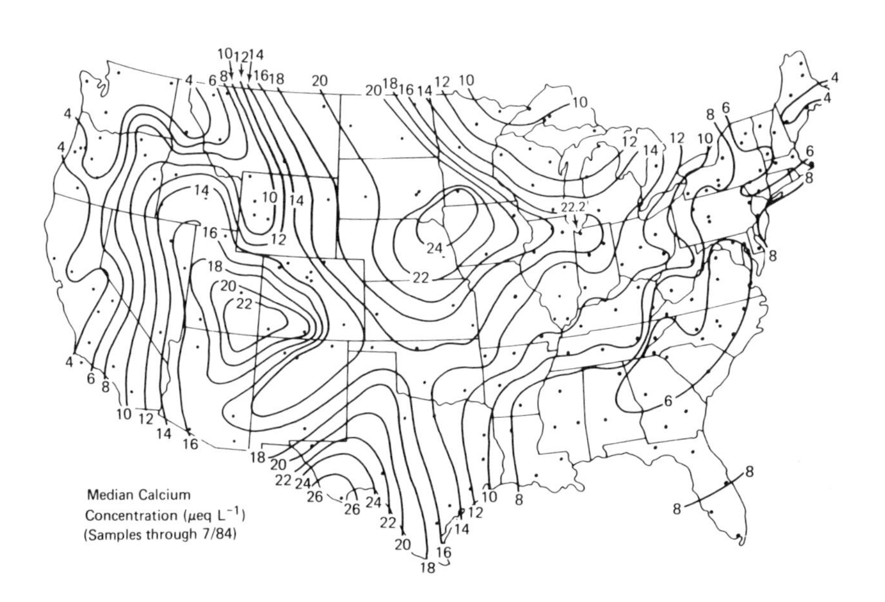

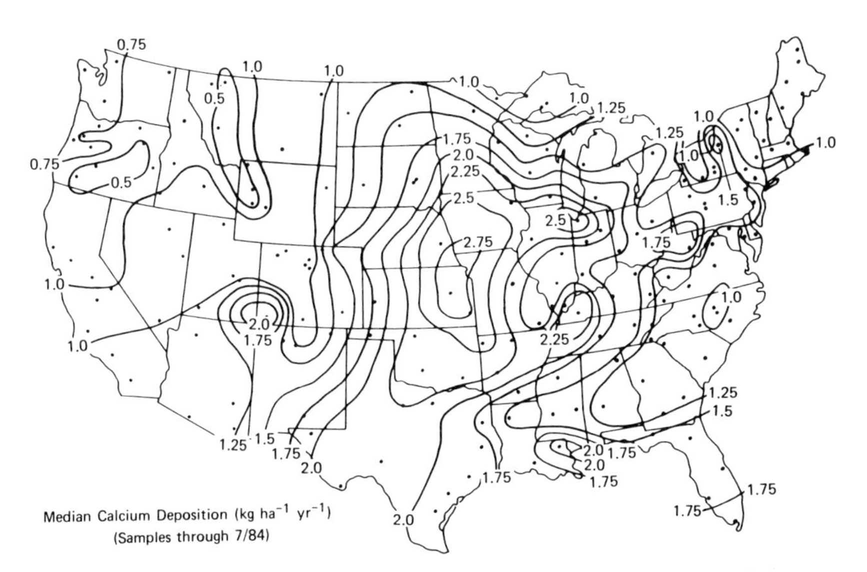

Figure 1. Top: the median calcium concentration (μeq L^{-1}), samples through 7/84; bottom: the median calcium deposition (kg ha^{-1} yr^{-1}), samples through 7/84.

Unlike the calcium ion, the ammonium maximum is over the center of the interior U.S. decreasing outward in all directions with a few small, isolated high values scattered in other areas. For the most part, the coastline precipitation contains the lowest observed ammonium concentrations.

The deposition of ammonium is dominated by the concentration pattern as modulated slightly by the precipitation (Figure 2). The maximum deposition of greater than 4 kilograms per hectare per year is centered over the identical area of the maximum concentration. Since the deposition is influenced greatly by the regional weather (winds, storm systems and movement), it is interesting to observe the close relationship between the presumed source, the concentration, and the deposition. This distributional relationship certainly suggests a region for the testing of long-range transport and transformation models or it suggests something about the atmospheric chemistry of ammonium.

Chloride. The distribution of this ion is acknowledged to be controlled by the proximity of a sampling site to the oceans. This is borne out by the concentration pattern shown in Figure 3. The ratio of sodium to chloride at many of the coastal sites is very close to that for seawater of 0.86.

There are two features of the concentration pattern that is interesting and cause for some speculation. Chloride shows a relative maximum extending from the Gulf coast northward across Texas into the upper Great Plains states. The seawater ratio, however, does not hold beyond northern Texas and the ratio becomes one or greater further north. This observation suggests either an inland source of sodium (soil?) or a selective decrease of chloride during the precipitation process. There is also a low concentration area observed over the Smokey Mountains, but the seawater ratio seems to be sustained. One could interpret this observation as due to the simple reduction of the seasalt component in the atmosphere with increased altitude and distance from the coast. There is also a tongue of high concentration extending from the central Gulf coast northeastward to Ohio with a strong possibility of a seawater influence as evidenced by the nearness to that ratio. This may reflect the meteorologists notion that the source of atmospheric water vapor for precipitation in the Midwest originates in the Gulf of Mexico, particularly during the warm season.

The deposition of chloride, obviously, is a mirror image of the precipitation pattern showing decreasing values with distance from the U.S. coastline (lower half of Figure 3). This ion, along with sodium, are worthy of additional study because of their domination by natural sources, and their obvious relationship to coastal influences.

Nitrate. The nitrate concentration in Figure 4 shows the highest values over central New York and Pennsylvania and an equal maximum over southwestern Michigan. The extension of these isolated maxima to the west as far as South Dakota and Nebraska appears to be associated with the ammonium concentration maximum in the same area indicating the possible presence of ammonium nitrate.

The major source for atmospheric nitrate is attributed to vehicular traffic and certainly the small center of high concentration

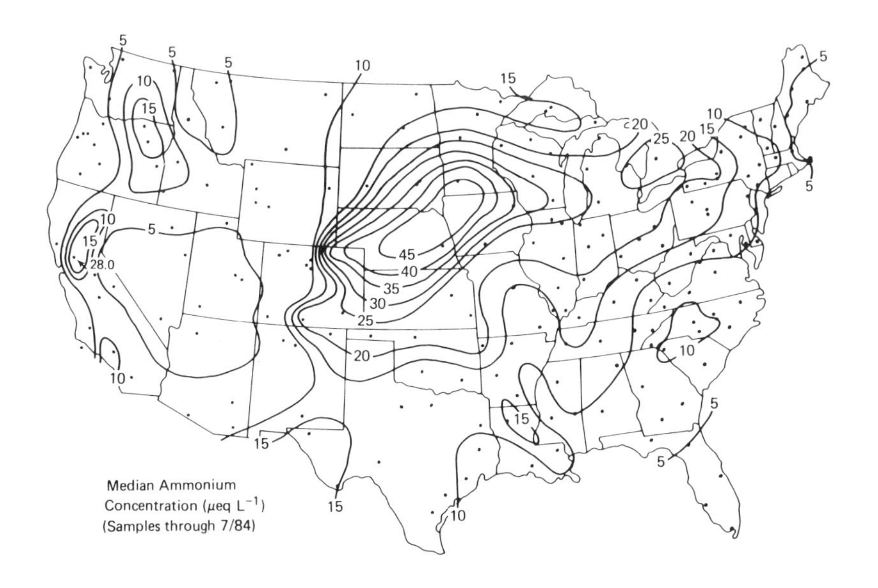

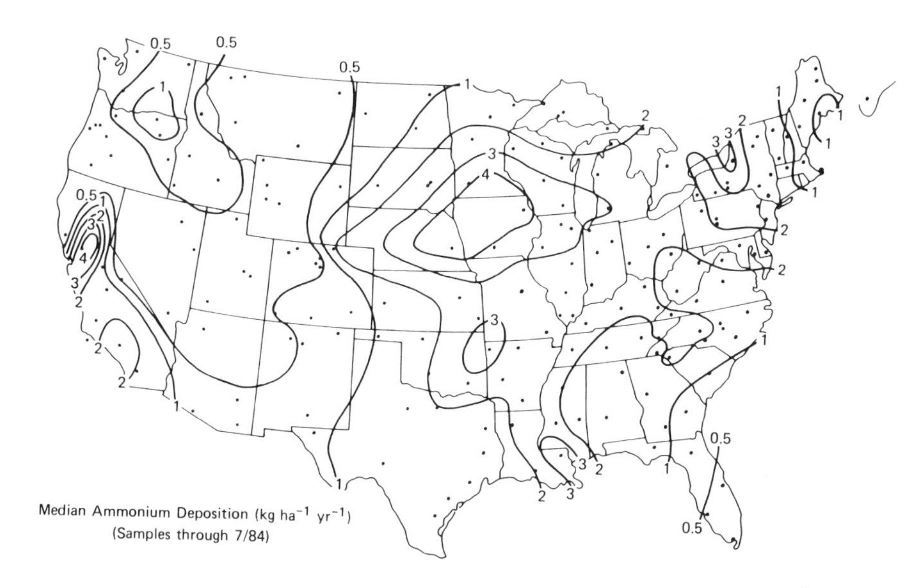

Figure 2. Top: the median ammonium concentration (μeq L^{-1}), samples through 7/84; bottom: the median ammonium deposition (kg ha^{-1} yr^{-1}), samples through 7/84.

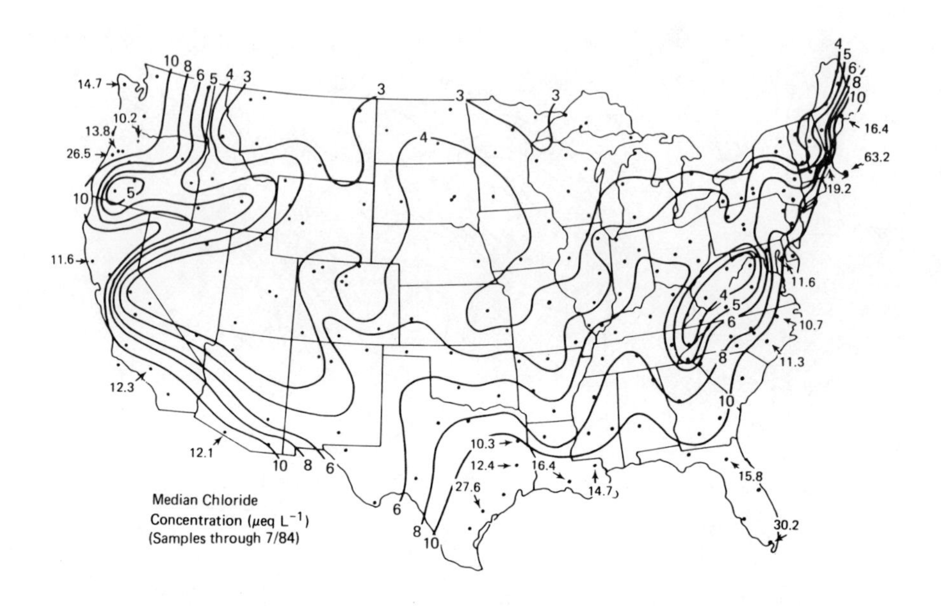

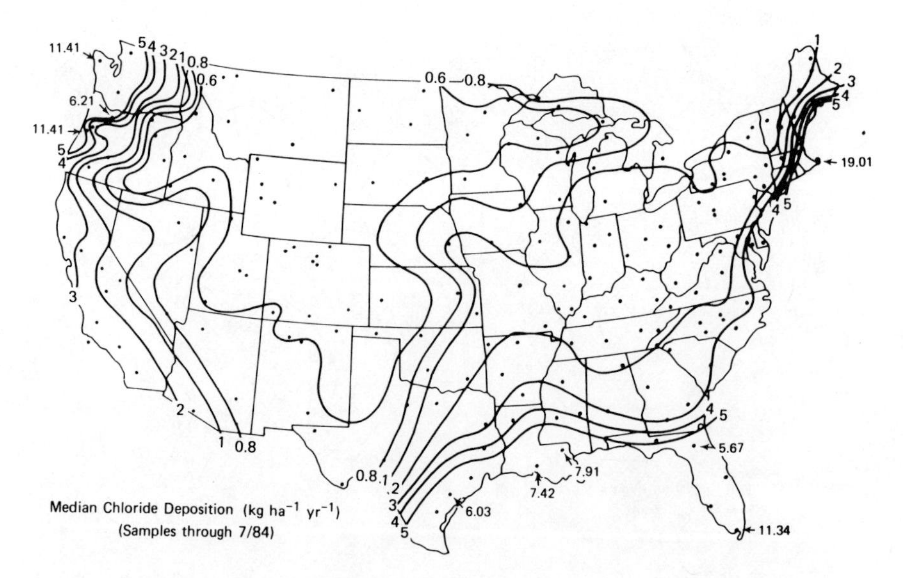

Figure 3. Top: the median chloride concentration ($\mu eq\ L^{-1}$), samples through 7/84; bottom: the median chloride deposition ($kg\ ha^{-1}\ yr^{-1}$), samples through 7/84.

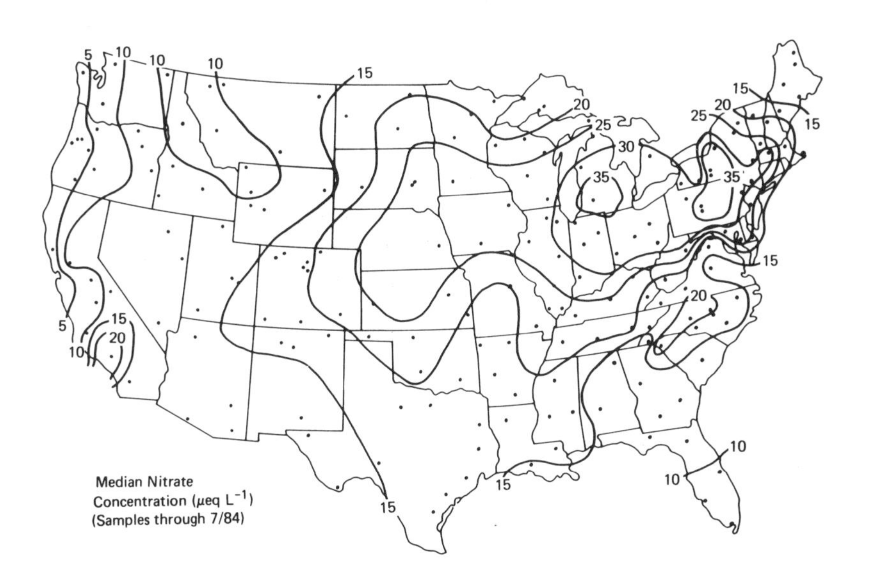

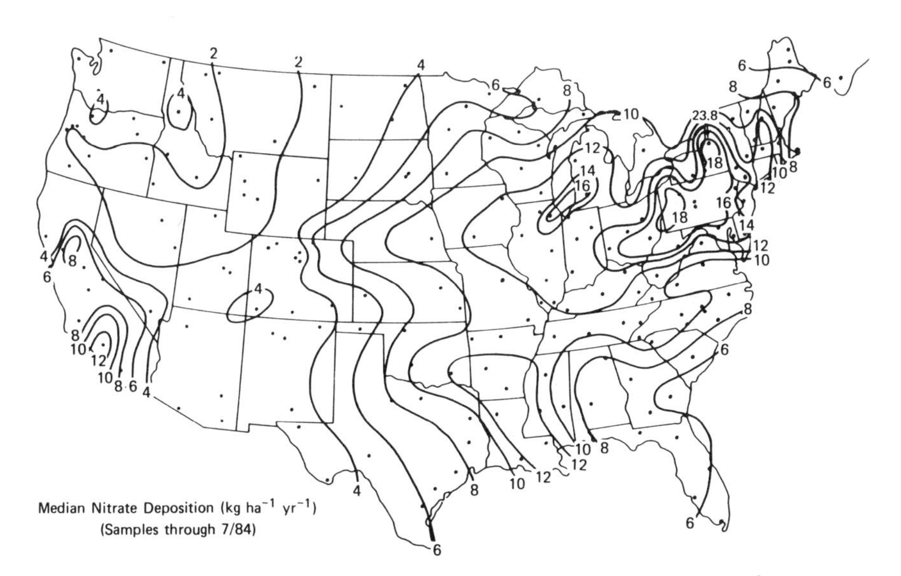

Figure 4. Top: the median nitrate concentration (μeq L^{-1}), samples through 7/84; bottom: the median nitrate deposition (kg ha^{-1} yr^{-1}), samples through 7/84.

over southwest California seems to confirm that relationship. It is not so easy to relate the other areas of high concentration to such high density mobile sources. This pattern is not easily traceable to the currently identified sources and considerable research is needed to explain the distribution of nitrate in precipitation.

The deposition pattern in Figure 4 is almost identical to the concentrations. The one exception to the simple correlation between concentration and deposition is in southeast Louisiana. The heavy precipitation in that area explains the relative maximum in deposition extending from the Delta region northward into central Arkansas.

Sulfate. The sulfate concentration is shown in Figure 5. The most obvious feature in this figure is the large area of high concentration in the eastern U.S. The values decrease outward from maxima over southwestern Pennsylvania and south-central New York. It is interesting to note that the east-west axis of the maximum lies to the north of the Ohio River frequently presumed to be the major source region for sulfur dioxide in the east.

The west coast population centers of San Francisco and Los Angeles appear with minor maxima associated with them. The other two small maxima over the central Washington and Oregon border and over the southern Arizona and New Mexico border are difficult to explain. Equally noticeable is the lack of high concentrations in the Four Corners region.

The deposition shown in Figure 5 is a good visual representation of the product of the concentration and precipitation. The axis of the major high deposition area is shifted to the south and oriented northeast-southwest. This slightly shifted pattern from the concentration maximum is due to the gradual increase of precipitation from Illinois southeastward. The pattern, then, gives the appearance of little long-range transport from the primary source region, that is, the heaviest deposition occurs directly over the highest sulfur dioxide emission area.

pH and Hydrogen Ion. The median pH distribution shown in Figure 6 reveals values less than 5.0 over almost the entire eastern half of the U.S. It has been argued that this pattern has not changed significantly since prior to the early 1950's (24). The major features to be noted are the low pH values in the east, and high values in the Great Plains from the Canadian border south to the Texas border, and the more variable pattern in the mountainous west.

The hydrogen ion deposition, also shown in Figure 6, exceeds 20 grams per hectare per year over most of Michigan, Illinois, eastern Missouri and Arkansas, Mississippi, Alabama, and the northern half of Georgia. The greatest deposition of over 60 grams per hectare per year is observed over western Pennsylvania.

Summary and Discussion

The author has briefly recounted the history of scientific interest in precipitation chemistry and focused attention on some of the important problems dealing with the quality assurance of analyzed samples. These considerations lead the author to conclude that comparisons between data collected prior to the late 1970's and those

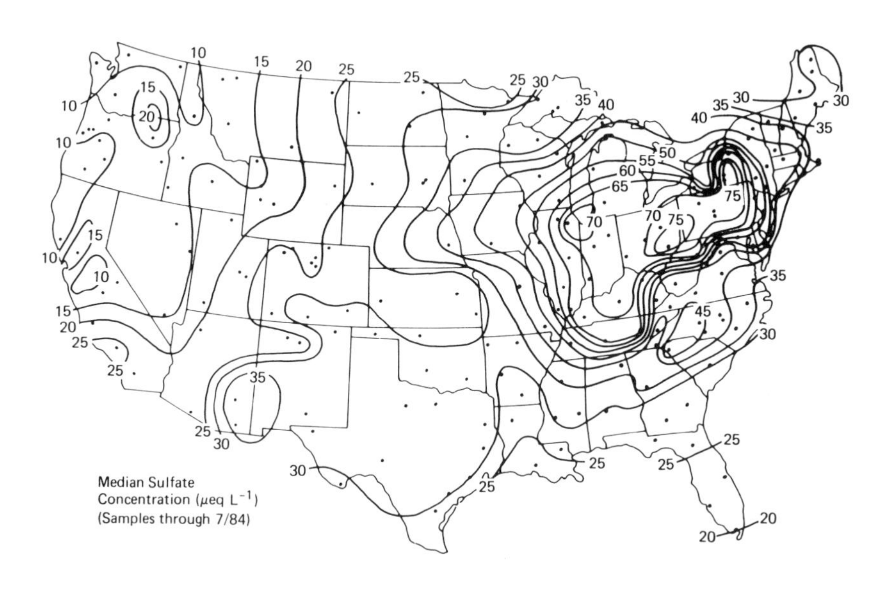

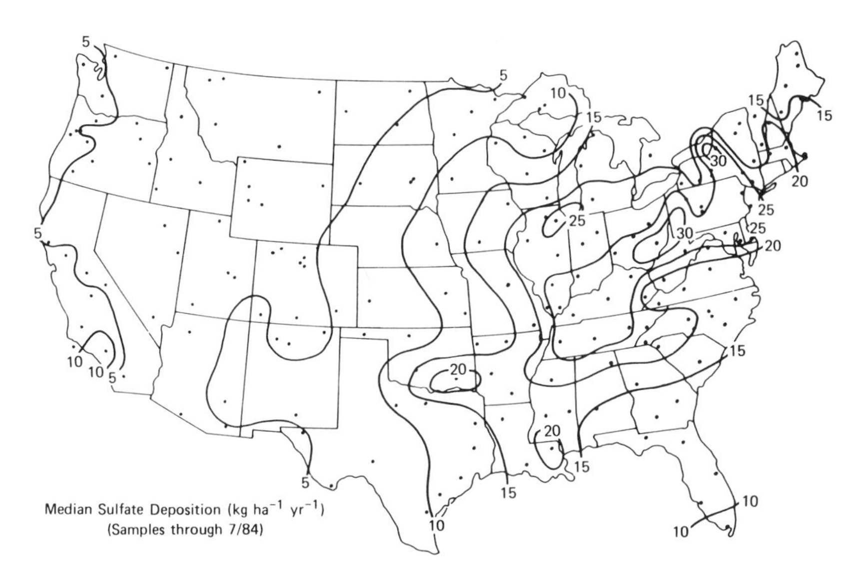

Figure 5. Top: the median sulfate concentration (μeq L^{-1}), samples through 7/84; bottom: the median sulfate deposition (kg ha^{-1} yr^{-1}), samples through 7/84.

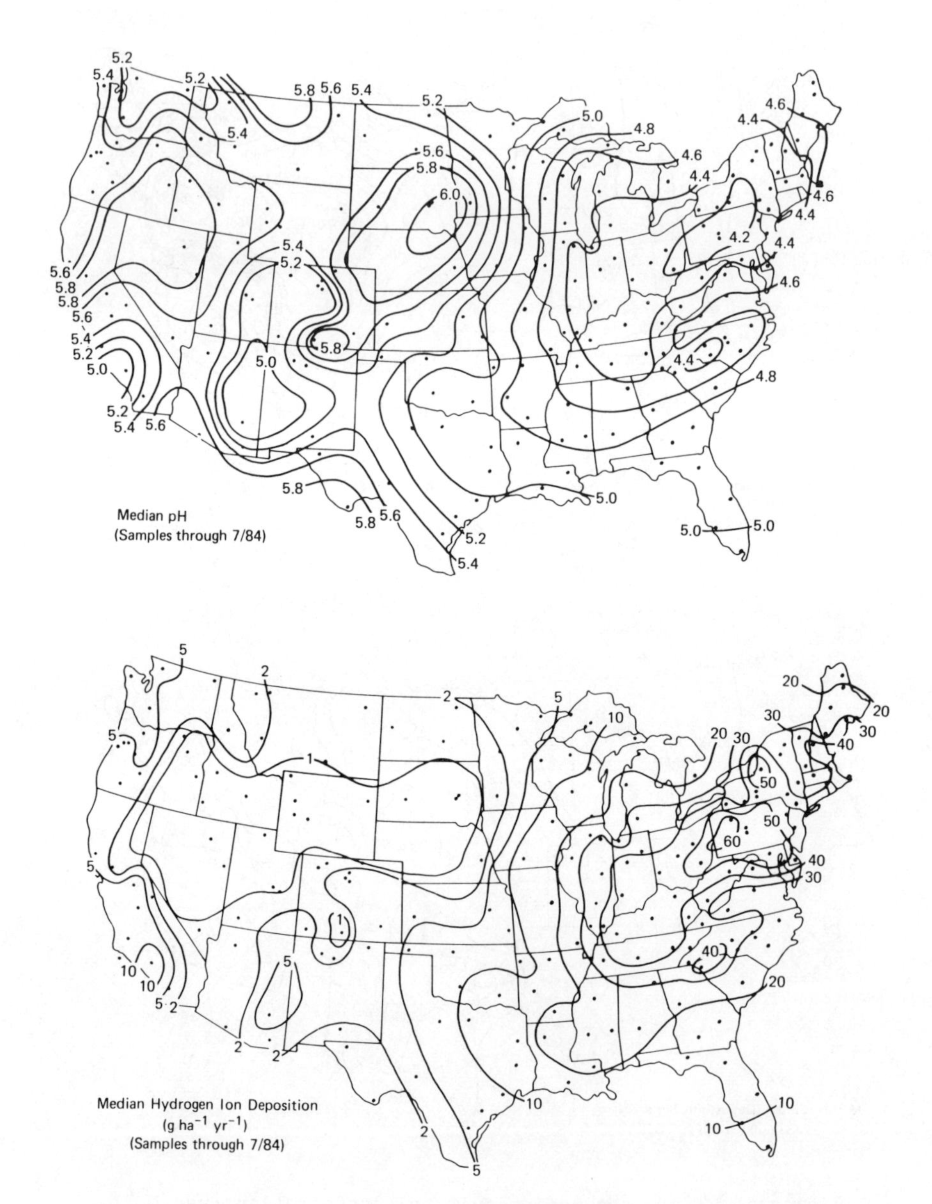

Figure 6. Top: the median pH (samples through 7/84); bottom: median hydrogen ion deposition (g ha^{-1} yr^{-1}), samples through 7/84.

more recently acquired should not be used to establish trends. The variety of sampling methods used, frequently unknown analytical procedures used, and the absence of measured key chemistry variables do not permit objective interpretation of the older data. These inadequacies of the available data prior to the establishment of the NADP/NTN network gave rise to controversy concerning the reality of a presumed trend toward greater precipitation acidity in the Northeast U.S. and areal spreading to the Midwest and Southeast (9,24). The trend issue is of importance as national policy is gradually emerging from the work of the National Acid Precipitation Assessment Program (NAPAP).

An example of the variability of pH and sulfate is shown in Figure 7. The data shown are from a site in rural, east-central Illinois. The individual weekly samples were used to generate 12 week moving averages which emphasizes the seasonality of the sulfate in precipitation. A linear fit to the concentration data shows a marked decline of sulfate over the period of record (not shown on the figure). The peak in concentration during the summer months is readily seen in this figure although there are year-to-year differences in the maximum value. The middle part of Figure 7 depicts the precipitation moving average. The seasonality of the Illinois precipitation is also quite noticeable with summertime peaks. The next curve, moving up on the figure, is the sulfate deposition. The seasonal peaks are emphasized for the deposition because the concentration and the precipitation are in phase. Note that the deposition, however, is dominated by the precipitation. For example, the greatest concentration was observed in the 1980 summer, but the greatest deposition was observed in the 1981 summer due to the higher summer rainfall in that year. Finally, the pH moving average short-term trend is shown at the top of Figure 7. While there appears to be a direct correlation between the sulfate concentration and the pH, it is by no means perfect. One of the most obvious discrepancies appears in 1984 where the sulfate peak concentration occurs with a relative maximum pH although there is a decline noticeable shortly thereafter.

One final point can be made from the data in Figure 7. It we assume that a reduction by 50 percent of the peak concentration (90 microequivalents per liter) in the 1983 summer, the resulting concentration would be approximately that observed in the previous winter. Yet, the pH change over those two seasons was observed to be from about 4.3 to 4.5 or 0.2 pH units. Clearly, the acid rain issue is a complex one with no easy solutions.

It was stated at the outset that no dry deposition monitoring method has been decided and approved, but the dry bucket data from the precipitation network are available and, perhaps, are a source of some information. The total deposition was calculated at five sites in the NADP/NTN network and the percentage that was observed wet and dry was determined. The sites are located in east-central Illinois, northeast Ohio, southeast New York, central North Carolina, and extreme southwest North Carolina. All sites are rural in character with the southwest North Carolina site standing in a forest clearcut area. The data are shown in Figure 8. Recall the point made earlier that the first three ions shown in the figure are related to crustal dust. At all but one of the sites, these ions are

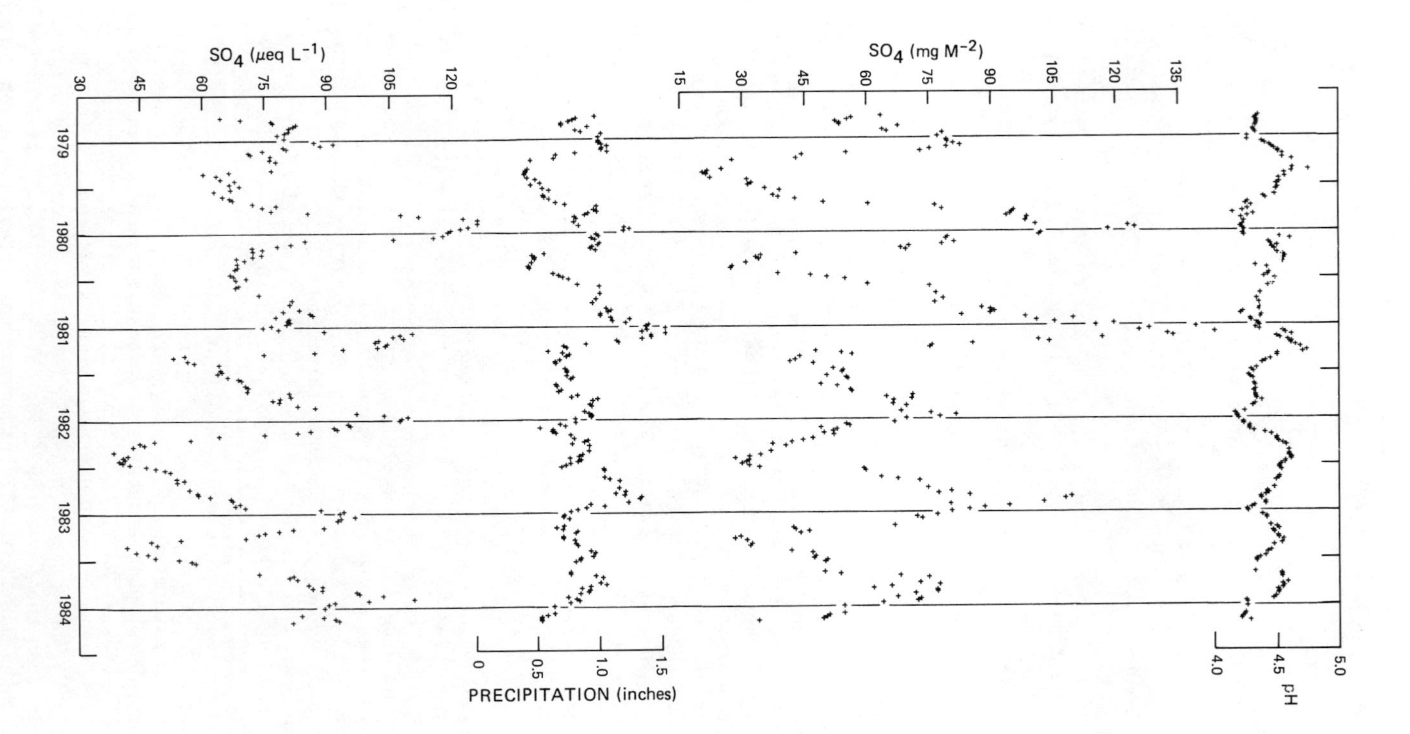

Figure 7. The pH (top), sulfate deposition (first down from the top), precipitation (first up from the bottom), and the sulfate concentration (bottom) from the Bondville, Illinois NADP/NTN network site.

		(mg M^{-2})								
		Ca	Mg	K	Na	NH_4	NO_3	Cl	SO_4	Total
IL 11 (17)	T	11.2	1.7	1.3	1.0	6.6	35.2	2.5	62.4	121.8
	%D	69.2	70.5	74.2	25.4	17.7	26.0	24.0	27.9	31.6
	%W	30.8	29.5	25.8	74.6	82.3	74.0	76.0	72.1	68.4
OH 71 (14)	T	13.1	1.9	1.4	1.6	5.2	32.0	3.5	71.0	129.7
	%D	78.5	67.5	77.7	31.5	10.9	20.7	31.5	30.9	33.4
	%W	21.5	32.5	22.3	68.5	89.1	79.3	68.5	69.1	66.6
NY 51 (13)	T	4.8	1.9	0.6	3.7	3.0	31.4	4.4	69.5	109.4
	%D	55.5	58.9	38.8	36.5	11.0	10.3	16.7	17.2	18.3
	%W	44.5	41.1	61.2	63.5	89.0	89.7	83.3	82.8	81.7
NC 41 (12)	T	3.8	1.2	1.4	3.8	6.7	22.8	5.5	44.8	89.9
	%D	58.6	53.9	73.1	40.4	53.0	23.0	32.1	30.8	33.1
	%W	41.4	46.1	26.9	59.6	47.0	77.0	67.9	69.2	66.9
NC 25 (10)	T	2.5	0.7	0.9	2.6	3.2	22.1	3.6	45.6	81.1
	%D	18.7	16.0	52.5	12.7	9.3	1.7	3.2	7.5	6.9
	%W	81.3	84.0	47.5	87.3	90.7	98.3	96.8	92.5	93.1

Figure 8. The total and percent wet and dry deposition at selected NADP/NTN network sites.

deposited primarily dry while those associated with the oceans and anthropogenic sources are deposited wet. The one exception is the forest clearcut site in North Carolina where wet deposition accounts for almost all of the deposition for all ions but potassium.

In conclusion, as the data base for precipitation chemistry grows so does our knowledge of its variability and trend. It seems rather clear, that there is not a rapidly declining quality of precipitation and any changes in recent years are going to be difficult to quantify in the presence of the observed great variability (26). It is equally clear, that to try to estimate a trend using data prior to the implementation of the NADP/NTN network opens the door to controversy since those data were not collected for trend analysis and did not include some of the important ions necessary to address the precipitation acidity issue. The total deposition must be measured if the insult to the biosphere is to be fully assessed. In the absence of a dry deposition monitoring network, the bucket estimates suggest that for some ions dry deposition is most important while for others wet deposition dominates. In any event, there does not appear to be a 50-50 split between the two forms. Damage mitigation strategies must consider this embalance between wet and dry deposition and its regionality. It must be obvious, from the foregoing data and discussion, that additional work needs to be accomplished before answers to many key questions will be forthcoming.

Acknowledgment

This work was supported by the National Acid Precipitation Assessment Program under USDOE Office of Health and Environmental Research Contract DE-AC02-76EV01199.

Literature Cited

1. Cowling, E.B. Environ. Sci. Technol., 1982, 16, 110A-123A.
2. Rossby, C.G.; Egner, H. Tellus, 1955, 7, 118-133.
3. MacIntire, W.H.; Young, J.B. Soil Sci., 1923, 15, 205-227.
4. Stewart, R. Univ. of Illinois Ag. Exp. Station Bulletin 227, 1920, 99-108.
5. Junge, C.E.; Gustafson, P.E. Bull. Am. Meteor. Soc., 1956, 37, 344.
6. Junge, C.E.; Werby, R. J. Meteorol., 1958, 15, 417-425.
7. Cogbill, C.V.; Likens, G.E. Water Resources Res., 1974, 10, 1133-1137.
8. Cogbill, C.V. Water, Air, and Soil Pollution, 1976, 6, 407-413.
9. Likens, G.E.; Butler, T.J. Atmos Environ., 1981, 15, 1103-1109.
10. Lodge, J.P., Jr.; Pate, J.B.; Basbergill, W.; Swanson, S.G.; Hill, K.C.; Lorange, E.; Lazrus, A.L. "Chemistry of United States Precipitation. Final Report on the National Precipitation Sampling Network", National Center for Atmospheric Research, Boulder, CO, 1968.
11. "The MAP3S Precipitation Chemistry Network: First Periodic Summary Report (September 1976-June 1977)," Battelle Pacific Northwest Laboratory Reprot PNL-2402, 1977.
12. Semonin, R.G. Water, Air, and Soil Pollution, 1976, 6, 395-406.

13. Wisniewski, J.; Kinsman, J.D. Bull. Am. Meteor. Soc., 1982, 63, 598-618.
14. Dochinger, L.S.; Seliga, T.A. First Intern. Symp. on Acid Precip. and Forest Ecosystems, 1976.
15. "National Acid Precipitation Assessment Plan", Interagency Task Force on Acid Precipitation, 1982.
16. Peden, M.E.; Lockard, J.M. In "Study of Atmospheric Pollution Scavenging, 20th Progress Report", Illinois State Water Survey; Champaign, IL, 1984, 75-84.
17. Bachman, S.R.; Lockard, J.M.; Peden, M.E., In "Study of Atmospheric Pollution Scavenging, 20th Progress Report", Illinois State Water Survey; Champaign, IL, 1984, 233-250.
18. Galloway, J.N.; Likens, G.E. Tellus, 1978, 30, 71-82.
19. Semonin, R.G. Proc. of the 12th Annual ENR Conference, 1983, 182-217.
20. Semonin, R.G. Proc. of the Conf. on Metropolitan Physical Environ., 1977, 53-61.
21. Miller, J.M.; Stensland, G.J.; Semonin, R.G. "The Chemistry of Precipitation for the Island of Hawaii During the HAMEC Project", NOAA/ARL; Boulder, CO, 1984; 3-9.
22. "NADP Quality Assurance Plan for Deposition Monitoring", National Atmospheric Deposition Program, 1984.
23. Peden, M.E.; Skowron, L.M. Atmos. Environ., 1978, 12, 2343-2349.
24. Stensland, G.J.; Semonin, R.G. Bull. Am. Meteor. Soc., 1982, 63, 1277-1284.
25. Granat, L. Tellus, 24, 550-560, 1972.
26. Semonin, R.G.; Stensland, G.J. Weatherwise, 1984, 37, 250-251.

RECEIVED January 2, 1986

3

Measurement of Dry Deposition onto Surrogate Surfaces: A Review

James S. Gamble and Cliff I. Davidson

Department of Civil Engineering, Carnegie-Mellon University, Pittsburgh, PA 15213

A review of dry deposition measurement techniques is presented, focusing on surrogate surfaces. Detailed field data from the literature are available for three types of collectors: filter paper with rainshields positioned overhead, Petri dishes, and flat Teflon or polyethylene plates. The data suggest that deposition velocities of submicron particles increase in the order flat plates $<$ Petri dishes $<$ filter paper. For supermicron particles, the order is filter paper $<$ Petri dishes $<$ flat plates. These results are interpreted in terms of the geometry of the collector, surface roughness, and peripheral shielding of the surface.

Recent concerns over acid deposition have enhanced interest in the study of pollutant transport from the atmosphere. Knowledge of such transport is important for several reasons:

1. Pollutant mass balances on regional and global scales require accurate deposition rates as inputs.

2. Ecosystem effects are related to the amounts of material depositing and to the specific deposition processes.

3. Reductions in air pollutant concentrations downwind of sources depend on loss of material from the atmosphere.

Atmospheric transport to surfaces involves wet and dry deposition processes. Wet deposition refers to scavenging during precipitation, and may include in-cloud processes as well as uptake of pollutants by falling raindrops and snowflakes. For particles, wet deposition often involves scavenging during formation of cloud condensation nuclei or ice nuclei. Collection of particles by existing hydrometeors, as polluted air masses sweep through clouds, may also be important. For wet deposition of gases, diffusion to cloud droplets or ice crystals, with subsequent chemical reactions, is an important process (1).

Dry deposition refers to transport between and during precipitation events. Particles may deposit by sedimentation, inertial impaction, interception, diffusion,

0097-6156/86/0318-0042$06.25/0

or a combination of these mechanisms (2,3). Gases generally deposit by diffusion from the atmosphere onto surfaces, with subsequent chemical reactions (4).

Besides wet and dry deposition, recent research has investigated the scavenging of particles and gases by fog droplets. The low pH of fog in some areas suggests that this mechanism may be important for removing acid species from the atmosphere (5–7).

Several methods have been employed to measure atmospheric deposition. For example, precipitation sampling buckets have been commonly used in wet deposition monitoring programs (8). Fog has been sampled with rotating arm collectors (9). Although these techniques have received some criticism, the methods are generally regarded as adequate by many research groups.

Attempts to routinely measure dry deposition, on the other hand, have encountered severe difficulties. Table 1 summarizes several measurement techniques currently available. Note that each method has important disadvantages which limit the utility for routine monitoring (10).

One of the simplest and most straightforward methods of measuring dry deposition is with surrogate surfaces. Table 1 identifies the primary disadvantage of this technique: deposition on artificial collectors may be quite different from that on natural surfaces of interest. Furthermore, it may be difficult to calibrate surrogate surface fluxes with those on natural surfaces, since deposition rates vary with meteorological and surface parameters in complex ways. Nevertheless, development of surrogate surfaces for dry deposition monitoring may be valuable for several potential uses:

1. Long-term trends in dry deposition onto natural surfaces may be reflected, at least semi-quantitatively, to similar trends measured with surrogate surfaces.

2. Certain pollutant species may have fluxes onto natural surfaces which are rate-limited by delivery from the atmosphere, and thus have deposition rates which are relatively independent of surface characteristics. Examples include large particles, which deposit primarily by sedimentation, and nitric acid vapor which is highly reactive with some natural and surrogate surfaces.

3. Many specific processes of dry deposition in the field are best studied with surrogate surfaces which can be modified to suit the design of the experiment. Examples include investigations into the effects of obtacle geometry and surface roughness.

4. Relating fluxes on surrogate surfaces to those on some natural surfaces may be feasible for certain designs of collectors. Research to assess this possibility is needed (11).

This paper addresses the use of surrogate surfaces to assess dry deposition of atmospheric pollutants. Several of the designs of artificial collectors reported in the literature are reviewed. Published data obtained with these collectors are then summarized and interpreted.

Table I: Examples of techniques for assessing dry deposition under field conditions after Hicks et al., 1980 (10)

Category: Flux Measurements

These methods use surrogate or natural surfaces to directly measure the flux of the depositing material on a small surface area.

	Surrogate Surfaces	Vegetation Washing	Snow Sampling
Description:	-Artificial collectors (e.g. buckets, funnels, Petri dishes, filter papers, and flat plates) set up in the field and exposed for known time periods. -Rainshields occasionally used to distinguish between wet and dry deposition. -Sticky films or other adhesive material sometimes used to minimize particle resuspension.	-Requires detailed rain chemistry studies including rainfall and vegetation throughfall measurements. -Typically uses washing of individual leaves and measurement of species concentration in wash fluid.	-Uses comparison of species concentrations in fresh and older snow. -Dry deposition inferred by difference.
Advantages:	-Simple and inexpensive. -Permits investigation of individual deposition processes. -Permits chemical analysis.	-Direct measurement of fluxes. -Possible to extend results to entire foliage. -Permits chemical analysis.	-Direct measurements of fluxes. -Particularly useful during periods of low deposition fluxes. -Permits chemical analysis.
Disadvantages:	-Relation to deposition on natural surfaces unknown. -Design and use of artificial collectors not standardized.	-Large spatial variations in deposition within foliage. -Occasionally difficult to isolate deposition on surface from internal plant material.	-Snow sublimation, migration of contaminants, and blowing snow may cause inaccuracies. -May have small differences between large numbers.

Table I: Examples of techniques for assessing dry deposition under field conditions (continued)

Category: Estimates of Regional Accumulation

This group of dry deposition monitoring techniques attempts to measure the accumulation of a species in a region of known size over a specified period of time.

	Atmospheric Radioactivity	Mass Balance	Tracers
Description:	-Uses existing atmospheric radioactive fallout as a tracer of opportunity.	-Used in watershed monitoring. -Uses a material budget model formulation. -Measurements made of wet deposition and weathering (as regional inflows) and runoff (as regional outflows) for given species. -The amount of dry deposition is not measured but is inferred by the difference of measured inflows and outflows.	-Tagged materials (e.g. radioactive or fluorescent) released over natural surfaces. -Samples of surface are analyzed for concentrations of tracer. -airborne tracer concentrations above surface also measured.
Advantages:	-Permits evaluation of net uptake rates of species for well-defined ecosystems. -Long term integrations possible.	-Net dry deposition inferred for large area.	-Dry deposition is measured directly on natural surfaces.
Disadvantages:	-Detailed investigation of specific deposition processes generally not possible. -Distinction between wet and dry deposition difficult. -Differences in size distributions of a monitored species and background radiation may invalidate any comparisons made using this technique.	-Dry deposition not measured directly. -multiple complex variables (e.g. biochemical and geochemical reactions) must be considered in the model development. -Difficult to investigate the influence of surface characteristics.	-Characteristics of tracer may not represent those of the species of interest. -insufficient mixing may invalidate airborne and deposition measurements. -Logistical problems in following tracer plume over large distances.

Continued on next page

Table I: Examples of techniques for assessing dry deposition under field conditions (continued)

Category: Flux Parameterization

These methods involve the use of airborne concentration and related meteorological data to infer the deposition flux, rather than direct measurement of accumulation of a species on surfaces.

	Box Budget	Eddy Correlation	Gradients
Description:	-Airborne Concentrations measured on the boundaries of a given area. -Difference between inflows and outflows, coupled with sources in the region, used to calculate the dry deposition.	-Fast response anemometers measure vertical wind component, w. -Simultaneously, real-time particle or gas analyzers measure concentration, C. -Mean values removed leaving w' and C'. -Product w'C' calculated and time-averaged to yield turbulent flux of monitored species.	-Compare differences between concentrations at two or more heights. -Infer flux as $K \frac{DC}{Dz}$ where: K = Species eddy diffusivity, function of atmospheric stability, surface roughness, and momentum flux.
Advantages:	-Net Surface flux determined for large areas.	-Provides an instantaneous evaluation of dry deposition flux.	-Fast-response sensors not needed. -Conceptually straight-forward and well-studied.
Disadvantages:	-May have small differences between large numbers. -Requires detailed knowledge of all sources within area. -Ignores inhomogeneities of area.	-Requires fast-response sensors for both wind velocity and concentration (1 sec.). -Strict fetch requirements.	-May have small differences between large numbers. -Strict fetch requirements.

Table I: Examples of techniques for assessing dry deposition under field conditions (continued)

Category: Flux Parameterization

These methods involve the use of airborne concentration and related meteorological data to infer the deposition flux, rather than direct measurement of accumulation of a species on surfaces.

	Eddy Accumulation	Variance
Description:	-Compare time-averaged species concentrations in updrafts with those in downdrafts.	-Measure variance of species concentration, and heat (or humidity) simultaneously at a given height and location. -Also simultaneously measure sensible heat flux (or evaporation). -Infer Flux of species, J_c, as: $J_c = J_T \frac{\sigma_c}{\sigma_T}$ or, $J_c = J_H \frac{\sigma_c}{\sigma_H}$
Advantages:	-Permits evaluation of time-averaged fluxes. -Permits chemical analysis for some species.	-No knowledge of surface characteristics or vertical wind velocities necessary. -May use slightly slower response chemical sensors compared with eddy correlation.
Disadvantages:	-Requires very precise chemical analysis techniques (precision = 1%). -Requires sophisticated electro-mechanical systems.	-Temperature method fails for near-neutral conditions. -Humidity sensor unreliable and difficult to maintain. -Strict fetch requirements. -Sign of flux (up or down) is unknown.

Design of Existing Surrogate Surfaces

Many different designs of surrogate surfaces have been used to measure dry deposition in the field. These include smooth flat surfaces, rough flat surfaces, and collectors with complex geometries. Examples of additional design modifications include application of an adhesive coating to minimize particle resuspension, covering the surface with a film of water to study diffusiophoresis, and use of rainshields to prevent contamination and/or washoff by precipitation.

Table 2 describes many of the surrogate surface designs reported in the literature. The descriptions are listed in order of increasing rim height. Unless otherwise noted, these collectors are generally positioned horizontally, facing upward. Note that surface areas range from 19 to about 5800 cm^2, and have been used 0.4 m to 20 m above ground. The dimensions given refer to the horizontal surface alone, excluding vertical walls or rims. Figures 1-6 present examples of some of the surfaces listed in the table.

It is interesting to note that the surface used most widely in the United States is the Aerochem-Metrics bucket. Since 1982, the buckets have been deployed at over 100 sites nationwide as part of the National Atmospheric Deposition Program (NADP) (8). This monitoring program has resulted in a sizeable dry deposition flux data base for numerous trace element and ionic species (12).

Data Presentation and Discussion

The surrogate surfaces summarized in Table 2 have been used to measure dry deposition fluxes for a number of species. In many cases, airborne concentrations have been measured simultaneously with deposition rates, permitting estimation of the dry deposition velocity v_d (cm/sec):

$$v_d = -J/C$$

Where: J = Deposition Rate, $g/cm^2 sec$
C = Airborne Concentration, g/cm^3

Results of those studies reporting deposition velocities are shown in Table 3. For three types of surrogate surfaces, a sufficiently large database exists to permit construction of graphs of v_d versus $\bar{d}p$. In this context, $\bar{d}p$ is the mass median aerodynamic diameter of the airborne species depositing onto the surrogate surfaces.

Previous work has shown that the small fraction of large particles often dominates total mass deposition of a species (2, 32, 38). To the extent that there is a relation between $\bar{d}p$ and the fraction of large particles, these graphs provide a rough indication of the influence of particle size on dry deposition velocity for a particular surrogate surface. Choice of $\bar{d}p$ for most of these species has been taken from Milford and Davidson (39, 40), which includes summaries of some 500 independent sets of size distribution data reported in

Table II: Surrogate Surface Designs

Surface Type	Dimensions	Deposition Surface Material	Height, m (Above ground unless otherwise indicated)	Remarks	Reference
Microscope slides	2.5 cm x 7.6 cm	Glass	0.0	Glass slides coated with silicon grease	Raynor, 1976 (13)
Flat plate	9.4 cm dia.	FEP Teflon	1.5 m above ground level, 1.5 m above building roof	0.3 cm rim on stainless steel holder, top and bottom deposition measurements (Figure 1)	Davidson, 1977; Davidson and Friedlander, 1978; Davidson et al., 1985a (14, 15, 16)
Flat plate	17.6 cm dia.	FEP Teflon	1.5	1.0 cm rim	Elias and Davidson, 1980; Davidson and Elias, 1982 (17, 18)
Flat plate	13.3 cm dia.	FEP Teflon	0.4, 1.5	Mounted flush in aluminum holders, some experiments used fixed rain-shield 30 cm or 45 cm above plate (Figure 2)	Dolske and Gatz, 1982 Davidson et al., 1985b (19, 20)
Flat plate	76 cm x 76 cm	Teflon laminated aluminum sheet	1.0	Rimless	Smith and Friedman, 1982; Dolske and Gatz, 1982 (21, 19)
Flat plate	20 cm dia.	Polyethylene	10,20	Rimless	McDonald et al., 1982 (22)

Continued on next page

Table II: Surrogate Surface Designs (continued)

Surface Type	Dimensions	Deposition Surface Material	Height, m (Above ground unless otherwise indicated)	Remarks	Reference
Filter paper	25 cm x 20 cm	Cellulose (Whatman 541)	1.0	Used a fixed 1 m x 1 m rainshield 12 cm above deposition surface, filter paper mounted on Perspex frame (Figure 3)	Cawse et al., 1973, 1974, 1975, 1976; Peirson et al., 1973; Ibrahim et al., 1983 (23 -28)
Filter paper	25 cm x 20 cm	Cellulose (Whatman 541)	1.0	Used automatic rainshield (Figure 4)	Pattenden et al., 1982 (29)
Filter paper	14.2 cm dia.	Nylon membrane (Nylasorb, Membrana Corp.	17	Mounted on a 15.0 cm dia. Teflon-coated stainless steel plate	Japar et al., 1985 (30)
Filter paper	14.2 cm dia.	PTFE Teflon membrane (Zef-lour, Membrana Corp.)	17	Mounted on a 15.0 cm dia. Teflon-coated stainless steel plate	Japar et al., 1985 (30)
Petri dish	9.5 cm dia. 1.3 cm deep	Polethylene	--	Located in foliage, no rainshield	Lindberg and Harriss, 1981 (31)
Petri dish	9.5 cm dia. 1.3 cm deep	Polycarbon-ate	15-19	Automatic rainshield located in foliage (Figure 6)	Lindberg and Lovett, 1985 (32)

Table II: Surrogate Surface Designs (continued)

Surface Type	Dimensions	Deposition Surface Material	Height, m (Above ground unless otherwise indicated)	Remarks	Reference
Pluviometer (Rain collector)	400 cm^2 area	--	--	Inverted for dry deposition measurements	Servant, 1976 (33)
Sangamo Precipitation Collector	18 cm dia., 10 cm deep cup	--	1.5	Cup suspended in a bucket	Ibrahim et al., 1983 (28)
O.M.E. Collector Funnel	50 cm x 50 cm at top	Teflon-coated collection surface	1.0	--	Ibrahim et al., 1983 (28)
Funnel	26 cm dia. at top, 2.0 cm dia. at bottom, 20 cm height	Polyethylene	0.8	--	Ibrahim et al., 1983 (28)
Aerochem-Metrics and HASL type Dustfall Buckets	25 cm dia., 28 cm deep	Linear Polyethylene	1.5	Automatic rainshield (Figure 5)	Dolske and Gatz, 1982; Semonin et al., 1984; Cadle and Dasch, 1985; Dasch, 1985a, 1985b; Feely et al., 1985; (12, 19, 34-37)

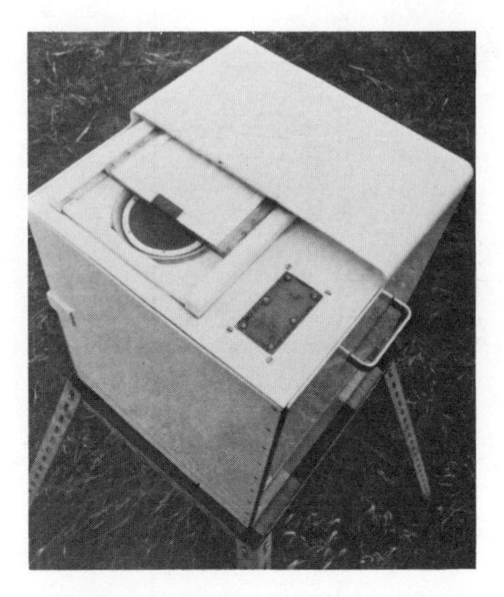

Figure 1. 17.6 cm dia. Flat Teflon Plate (17, 18).

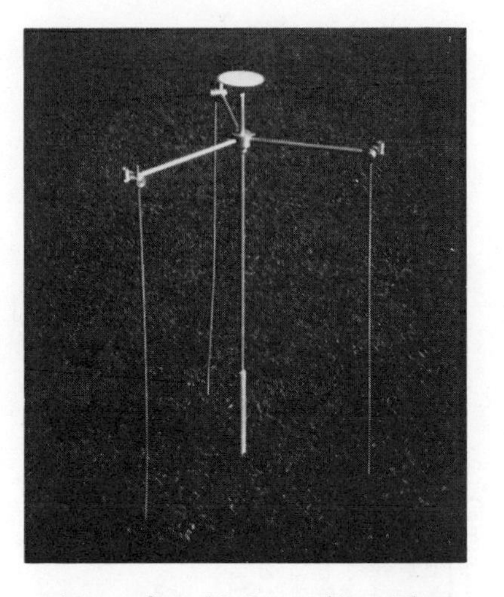

Figure 2. 13.3 cm dia. Flat Teflon Plate shown without rainshield (19, 20).

Figure 3. Field Deposition Collection Apparatus using Whatman 541 filter paper (23-25). Reproduced with permission from Ref. 23. Copyright 1974, United Kingdom Atomic Energy Authority.

Figure 4. Automatic Wet and Dry Deposition Collector using Whatman 541 filter paper. Reproduced with permission from Ref. 29. Copyright 1982, United Kingdom Atomic Energy Authority.

Figure 5. Aerochem-Metrics Dustfall Bucket during wind tunnel flow visualization (12, 19, 34-37).

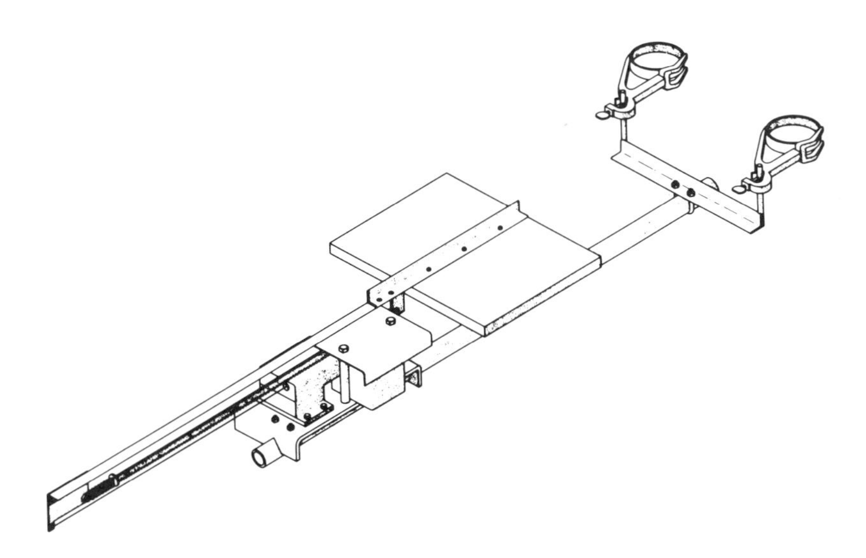

Figure 6. Petri Dish holder with automatic rainshield (32). Reproduced with permission from Ref. 32. Copyright 1985, American Chemical Society.

Table III: Dry Deposition Data for Surrogate Surfaces

Surface Type	Depositing Material (and size if known	Deposition Velocity (cm/sec)	Reference
Microscope slide	Pollen	0.23-91.	Raynor, 1974 (13)
Teflon plate	Cd,Pb,Zn	0.065-1.1	Davidson, 1977; Davidson and Friedlander, 1978 (14, 15)
Teflon plate	Ag,Al,As,Ba,Ca, Cd,Cu,Fe,Mg,Na, Pb,Ti,Zn	0.081-20.	Davidson et al., 1985a (16)
Teflon plate	Ba,Ca,Cs,K,Pb,Rb Sr	0.14-2.3	Elias and Davidson 1980; Davidson and Elias, 1982 (17, 18)
Teflon plate	SO_4^{2-}	0.17-0.42	Davidson et al., 1985b (20)
Teflon plate	SO_4^{2-}	0.075-0.36	Dolske and Gatz 1982 (19)
Teflon plate	SO_4^{2-}	0.038-0.29	Smith and Friedman, 1982; Dolske and Gatz, 1982 (21, 19)
Polyethylene plate	Na^+	0.8-8.1	McDonald et al., 1982 (22)
Filter paper (Whatman 541)	Ag,Al,As,Au,Br Ca,Cd,Ce,Cl,Co, Cr,Cs,Cu,Eu,Fe, Hg,I,In,K,La,Mg, Mn,Mo,Na,Ni,Pb, Rb,Sb,Sc,Se,Sm, Th,Ti,V,W,Zn	< 0.01-13.	Cawse et al., 1973, 1974, 1975, 1976 (23 -26)
Filter paper (Whatman 541)	Al,As,Au,Br,Ca, Cd,Ce,Cl,Co,Cr, Cs,Cu,Fe,Hg,I, In,La,Mn,Na,Ni, Pb,Rb,Sb,Sc,Se, Th,V,W,Zn	0.04-1.0	Peirson, 1973 (27)
Filter paper (Whatman 541)	$(NH_4)_2SO_4$, 0.7 um	0.005-0.007	Ibrahim et al., 1983 (28)

Table III: Dry Deposition Data
for Surrogate Surfaces (continued)

Surface Type	Depositing Material (and size if known	Deposition Velocity (cm/sec)	Reference
Filter paper (Whatman 541)	$(NH_4)_2\ SO_4$, 7.0 um	0.027-0.18	Ibrahim et al., 2983 (28)
Filter paper (Whatman 541)	Ag,Al,As,Cd,Co, Cr,Cu,Fe,In,Mn, Na,Ni,Pb,Sb,Sc, Se,V,Zn	1.0-20	Pattenden et al., 1982 (29)
Teflon filter (Zefluor)	SO_4^{2-}	0.053	Japar et al., 1985 (30)
Nylon filter (nylasorb)	SO_4^{2-}	0.36-0.75	Japar et al., 1985 (30)
Petri dish	SO_4^{2-}	0.14-0.57	Dolske and Gatz, 1982 (19)
Petri dish	Cd,Mn,Pb,Zn, SO_4^{2-}	0.05-6.4	Lindberg and Harriss, 1981 (31)
Petri dish	Ca,K,SO_4^{2-}	0.18-0.61	Lindberg and Lovett 1985 (32)
Pluviometer	Pb	0.13	Servant, 1976 (33)
Sangamo collector	$(NH_4)_2\ SO_4$, 0.7 um	0.012-0.031	Ibrahim et al., 1983 (28)
Sangamo collector	$(NH_4)_2\ SO_4$, 7.0 um	0.54-5.5	Ibrahim et al., 1983 (28)
O.M.E. collector	$(NH_4)_2\ SO_4$, 0.7 um	0.008-0.018	Ibrahim et al., 1983 (28)
O.M.E. collector	$(NH_4)_2\ SO_4$, 7.0 um	0.30-0.78	Ibrahim et al., 1983 (28)
Funnel	$(NH_4)_2\ SO_4$, 0.7 um	0.007-0.014	Ibrahim et al., 1983 (28)
Dustfall bucket	SO_4^{2-}	0.40-3.9	Dolske and Gatz, 1982 (19)
Dustfall bucket	SO_4^{2-}	0.10-1.2	Feeley et al., 1985 (37)

the literature. The values of $\bar{d}p$ for 38 trace elements and SO_4^{2-} are shown in Table 4. It is acknowledged that size distributions vary greatly temporally and spatially, hence the application of average dp to these surrogate surface measurements may have significant error.

Table IV: Elemental Average Mass Median Diameter, $\bar{d}p$ after Milford and Davidson, 1985 (39,40)

Element	Mass Median Diameter (um)	Element	Mass Median Diameter (um)	Element	Mass Median Diameter (um)
Al	4.5	Fe	3.4	Sc	4.4
As	1.1	Ga	6.0	Se	0.68
Ba	3.3	Hf	7.6	Si	3.9
Br	0.89	Hg	0.61	Sm	2.8
Ca	4.6	I	1.0	Sr	12.
Cd	0.68	In	1.8	Ta	1.7
Ce	5.1	K	3.8	Th	2.7
Cl	3.0	Mg	6.3	Ti	6.5
Co	2.6	Mn	2.1	U	1.6
Cr	1.1	Na	3.8	V	1.4
Cs	1.9	Ni	0.98	W	0.43
Cu	1.3	Pb	0.55	Zn	1.1
Eu	2.6	Sb	0.86	SO_4^{2-}	0.55

Results for the Whatman 541 filter paper (23-28), flat plates (14-22), and Petri dishes (31,32) are shown in figures 7, 8, and 9 respectively. The data for the filter paper and flat plates have been presented as separate points for each measurement reported in the literature. For the Petri dishes, only the mean values and standard errors were available. Least squares regression lines plotted through the points are also shown. Figure 10 presents all three regression lines plotted on the same axes with the predicted sedimentation velocity line. It must be cautioned that a comparison of the regression lines is tenuous due to the spread in the data represented by each line.

Previous work suggests that sedimentation is an important mechanism for relatively smooth, flat surfaces (15, 16, 20). One would therefore expect particle size to be a key parameter influencing the deposition rate. It is of interest that the slope of the flat plate curve is 1.7, close to the value expected for sedimentation, although it must be cautioned that the deposition velocities reported are also substantially greater than those predicted solely by sedimentation. This is most likely due to the use of MMD for the x-axis in the figure: in reality, the small fraction of large particles is probably responsible for most of the mass deposition. If impaction, interception, and other mechanisms are important, these mechanisms would increase the measured deposition velocities above that expected on the basis of sedimentation alone.

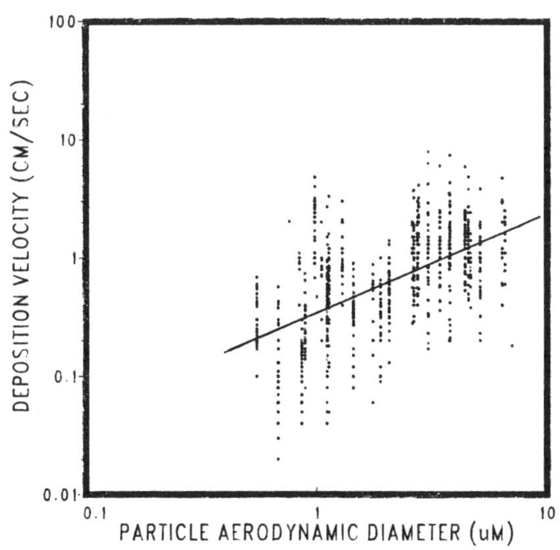

Figure 7. Deposition Velocity vs. Particle Aerodynamic Diameter, Type Surrogate Surface: Filter Paper (23-28).

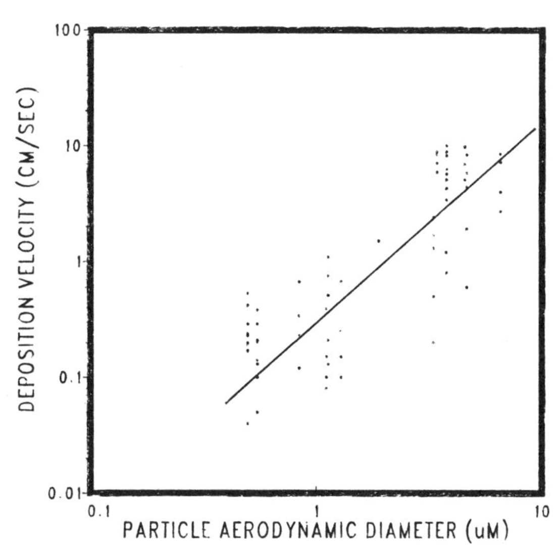

Figure 8. Deposition Velocity vs. Particle Aerodynamic Diameter, Type Surrogate Surface: Flat Plates (14-22).

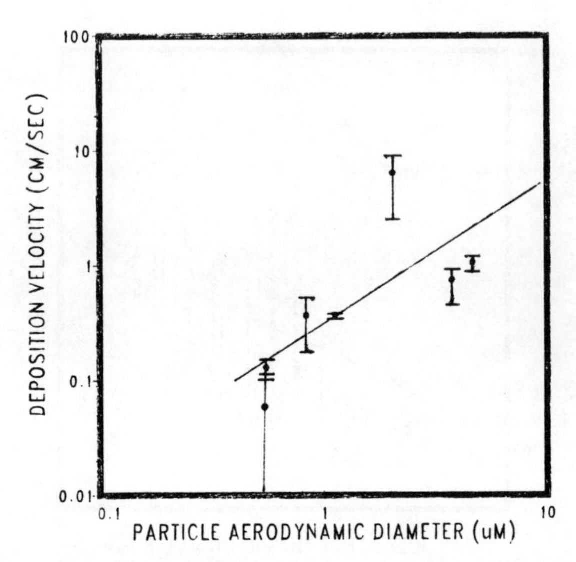

Figure 9. Deposition Velocity vs. Particle Aerodynamic Diameter, Type Surrogate Surface: Petri Dishes (31, 32).

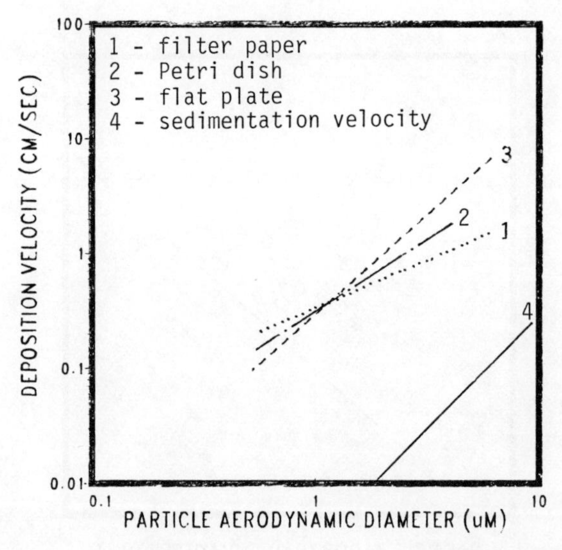

Figure 10. Deposition Velocity vs. Particle Aerodynamic Diameter, Composite Drawing.

It is of interest to compare the flat plate and the filter paper data. Wind tunnel experiments have shown that deposition velocities to filter paper are greater than those to smoother surfaces over a range of particle sizes from 0.08-30 um (42, 43). This was attributed to the role of the small fibers protruding from the surface of the paper in assisting the capture and preventing the reentrainment of the particles. However, Figure 10 shows that species with supermicron MMD have the deposition velocities for the flat plates which are greater than those for the filter paper. This may be partially due to the use of a fixed rainshield 12 cm. above the surface in the filter paper experiments. Note that all of the data points in Figure 7 involved the use of rainshields, while the data of Figure 8 involved very limited or no shielding. As an illustration of the possible effects of fixed rainshields, Figure 11 presents data for identical shielded and unshielded Teflon plates exposed simultaneously. These graphs show SO_4^{2-} deposition rates obtained as part of the Illinois Dry Deposition Intercomparison Study (20). The MMD of SO_4^{2-} during the time periods overlapping these experiments ranged from 0.63-1.7 um, and the rainshields were placed either 30 or 45 cm above the surfaces. The results suggest that the filter paper deposition of Figure 7 may have been greatly reduced by the presence of rainshields.

It is also of interest to compare the flat plate and Petri dish data. Figure 12 presents evidence that the geometry of the surrogate surfaces may significantly affect deposition. The curves show deposition velocity for SO_4^{2-} as a function of increasing rim height for four surfaces exposed simultaneously in the Illinois Dry Deposition Intercomparison Study. The effects of increased turbulence caused by the higher rims on the Petri dishes and dustfall buckets, as well as possible reduced bounceoff, may be responsible for the observed trend (20). Figure 10 shows that the regression line for Petri dishes is above that of flat plate for species with submicron MMD, consistent with the curves of Figure 12 since SO_4^{2-} is generally submicron. However, the regression line for Petri dishes falls below that of flat plates at larger MMD. This may be due to the location of the Petri dishes within a forest canopy, where shielding by foliage may have decreased concentrations of supermicron particles to a greater extent than submicron material. Another possible factor is the location of the Petri dishes 15-19 m above ground, compared with most flat plate data obtained at a height at heights of 1.0-1.5 m.

Finally, it is interesting to note that the intersection of the three regression lines in Figure 10 occurs at 1 um. This suggests that species with MMD near 1 um have similar deposition velocities for all of these surrogate surfaces, although species with smaller or larger MMD show significant differences.

Conclusion

This paper has presented a review of the various methods of monitoring dry deposition using surrogate surfaces, and has summarized data obtained using some of these surfaces. The lack of a standardized surface design and standardized field deployment procedures make comparison of the results obtained by the surfaces difficult. Nevertheless, the data show some consistencies: deposition velocity increases with increasing MMD of depositing species, and there is a definite pattern when data for filter paper, flat plates, and Petri dishes are compared.

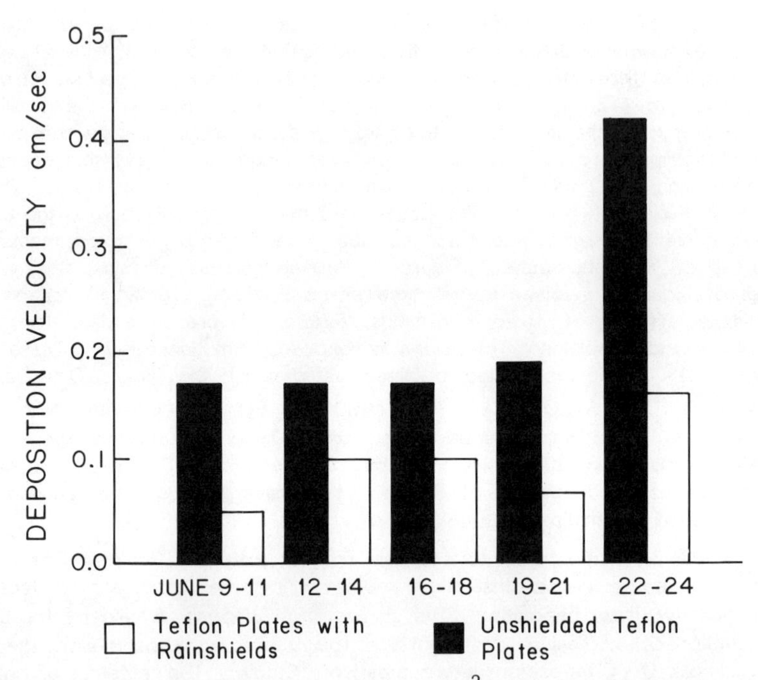

Figure 11. Dry Deposition Velocities of SO_4^{2-} for shielded and unshielded Teflon plates (20). Reproduced with permission from Ref. 20. Copyright 1985, American Geophysical Union.

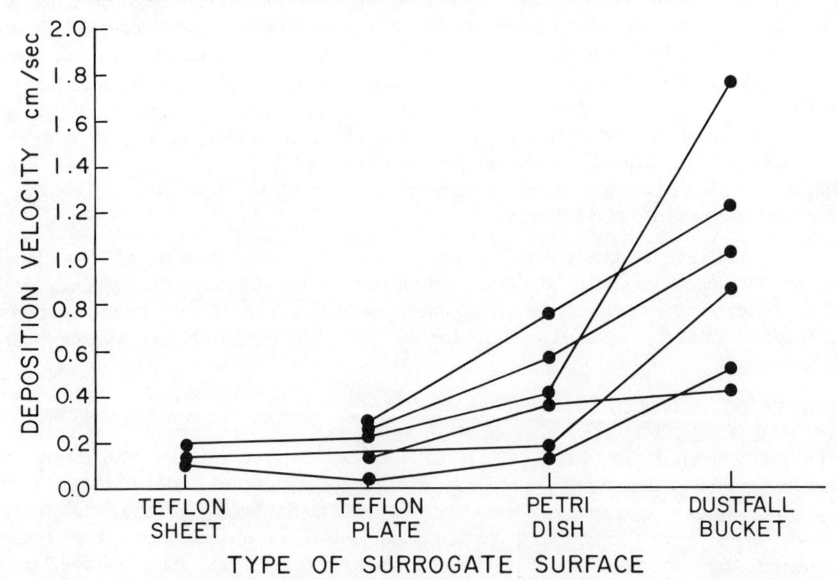

Figure 12. Dry Deposition Velocities of SO_4^{2-} as a function of increasing rim height (20). Reproduced with permission from Ref. 20. Copyright 1985, American Geophysical Union.

It would be desirable to develop a surrogate surface whose deposition characteristics resemble those of certain natural surfaces. Unfortunately, the complexities of airflow and deposition patterns over even the simplest natural surfaces make this a formidable task. As an alternative, it may be feasible to develop a simple surrogate surface whose deposition properties for each monitored species vary in a predictable way. Such a design would lessen the difficulties of examining complex airflow patterns, and could be accompanied by standardized methodologies for the use of rainshields, mounting height, and other parameters. A device of this type would be highly useful for examining the deposition of large particles and for certain reactive gasses (whose transport rates may not be sensitive to surface geometry), and may be valuable in assessing long-term trends in dry deposition to surfaces.

Acknowledgments

The authors gratefully acknowledge the efforts of S.E. Lindberg, J.M. Dasch, and V.C. Bowersox in providing additional references and information. Additionally, we would like to thank S. Knapp for assisting in the preparation of this manuscript. This work was funded by National Oceanic and Atmospheric Administration Grant NA84-WC-C-06140 and National Institutes of Health Training Grant GM0-7477.

Literature Cited

1 Slinn, W.G.N. In "Atmospheric Sciences and Powder Production"; Randerson D., Ed.; Technical Information Center: Oak Ridge TN, 1984.

2 Davidson, C.I.; Miller, J.M.; Pleskow, M.A. Water Air Soil Pollut., 1982, 18, 25-43.

3 Slinn, W.G.N. Atmos. Environ. 1982, 16, 1785-94.

4 Slinn, W.G.N. Water Air Soil Pollut. 1977, 7, 513-43.

5 Waldman, J.M.; Munger, J.W.; Jacob, D.J.; Flagan, R.C.; Morgan, J.J.; Hoffman, M.R. Science 1982, 218, 677-80.

6 Waldman, J.M.; Munger, J.W.; Jacob, D.J.; Hoffman, M.R. Proc. 4th Int. Conf. on Precipitation Scavenging, Dry Deposition, and Resuspension, 1982, p.137-46.

7 Jacob, D.J.; Hoffman, M.R. Proc. 4th Int. Conf. on Precipitation Scavenging, Dry Deposition, and Resuspension, 1982, p.149-57.

8 "The Design of the National Trends Network for Monitoring the Chemistry of Atmospheric Precipitation," U.S. Geological Survey Circular 964, 1985.

9 Jacob, D.J.; Flagan, R.C.; Waldman, J.M.; Hoffman, M.R. Proc. 4th Int. Conf. on Precipitation Scavenging, Dry Deposition, and Resuspension, 1982, p. 125-34.

10 Hicks, B.B.; Wesely, M.L.; Durham, J.L. "Critique of Methods to Measure Dry Deposition," 1980, EPA-600/9-80-050, Environ. Prot. Agency, Washington, D.C..

11 Davidson, C.I.; Lindberg, S.E. in "Critique of Methods to Measure Dry Deposition," 1980, EPA-600/9-80-050, Environ. Prot. Agency, Washington, D.C., pp. 66-70.

12 Semonin, R.G.; Stensland, G.J.; Bowersox, V.C.; Peden, M.E.; Lockard, J.M.; Doty, K.G.; Gatz, D.F.; Chu, L.; Backman, S.R.; Stahlhut, R.K. "Study of Atmospheric Pollution Scavenging," Illinois State Water Survey Contract Report 347, 20th Progress Report, Contract No. DE-AC02-76EV01199, Champaign IL, 1984.

13 Raynor, G.S. "Experimental Studies of Pollen Deposition to Vegetated Surfaces", Symposium on Atmosphere-Surface Exchange of Particulate and Gaseous Pollutants Symposium, September, 1974. Energy Research and Development Administration Symposium Series CONF-740921, National Technical Information Service, U.S. Department of Commerce, Springfield, VA, pp. 264-79.

14 Davidson, C.I. Powder Tech. 1977, 18, 117-26.

15 Davidson, C.I.; Friedlander, S.K. J. Geophys. Res. 1978, 83, 2343-52.

16 Davidson, C.I.; Gould, W.D.; Mathison, T.P.; Wiersma, G.B.; Brown K.W.; Reilly, M.T. Environ. Sci. Technol. 1985, 19, 27-34.

17 Elias, R.W.; Davidson, C.I. Atmos. Environ. 1980, 14, 1427-32.

18 Davidson, C.I.; Elias, R.W. Geophys. Res. Lett. 1982, 9, 91-93.

19 Dolske, D.A.; Gatz, D.F. ACS symposium on Acid Rain, 1982.

20 Davidson, C.I.; Lindberg, S.E.; Schmidt, J.A.; Cartwright, L.G.; Landis, L.R. J. Geophys. Res. 1985, 90, 2123-30.

21 Smith, B.E.; Friedman, E.J. "The Chemistry of Dew as Influenced by Dry Deposition: Results of Sterling, VA and Champaign IL Experiments," WP-82W00141, MITRE Corp., McLean, VA, 1982.

22 McDonald, R.L.; Unni, C.K.; Duce, R.A. J. Geophys. Res. 1982, 87, 1246-50.

23 Cawse, P.A. "A Survey of Atmospheric Trace Elements in the U.K. (1972-1973)," AERE-R 7669, AERE Harwell, Oxfordshire, England, 1974.

24 Cawse, P.A. "A Survey of Atmospheric Trace Elements in the U.K.: Results for 1974," AERE-R 8038, AERE Harwell, Oxfordshire, England, 1975.

25 Cawse, P.A. "A Survey of Atmospheric Trace Elements in the U.K.: Results for 1975," AERE-R 8398, AERE Harwell, Oxfordshire, England, 1976.

26 Cawse, P.A. "A Survey of Atmospheric Trace Elements in the U.K.: Results for 1976," AERE-R 8869, AERE Harwell, Oxfordshire, England, 1977.

27 Peirson, D.H.; Cawse, P.A.; Salmon, L.; Cambray, R.S. Nature 1973, 241, 252-56.

28 Ibrahim, M.; Barrie, L.A.; Fanaki, F. Atmos. Environ. 1983, 17,781-788.

29 Pattenden, N.J.; Branson, J.R.; Fisher, E.M.R. In "Deposition of Atmospheric Pollutants", Georgi, H.W.; Pankrath, J., Eds.; Reidel Journals: Boston, 1982; pp. 173-184.

30 Japar, S.M.; Brachaczek, W.W.; Gorse, R.A. 1985, J. Geophys. Res. (submitted for publication).

31 Lindberg, S.E.; Harriss, R.C. Water Air Soil Pollut. 1981, 16, 13-31.

32 Lindberg, S.E.; Lovett, G.M. Environ. Sci. Technol. 1985, 19, 238-244.

33 Servant, J. "Deposition of Atmospheric Lead Particles to Natural Surfaces in Field Experiments", Symposium on Atmosphere-Surface Exchange of Particulate and Gaseous Pollutants Symposium, 4-6 September, 1974. Energy Research and Development Administration Symposium Series CONF-740921, National Technical Information Service, U.S. Department of Commerce, Springfield, VA, 1976, pp. 87-95.

34 Cadle, S.M.; Dasch, J.M. "Wintertime Wet and Dry Deposition in Northern Michigan," GMR 5000, General Motors Corp., Warren MI, 1985.

35 Dasch, J.M., "Measurement of Dry Deposition to a Deciduous Canopy," GMR 5019, General Motors Corp., Warren MI, 1985.

36 Dasch, J.M. Environ. Sci. Technol. 1985, 19, 721-725.

37 Feely, H.W.; Bogen, D.C.; Nagourney, S.J.; Torquato, C.C. J. Geophys. Res. 1985, 90, 2161-65.

38 Garland, J.A. 4th Int. Conf. on Precipitation Scavenging, Dry Deposition, and Resuspension, 1982, p. 849-57.

39 Milford, J.B.; Davidson, C.I. J. Air Pollut. Contol Assoc. 1985, in press.

40 Milford, J.B.; Davidson, C.I. "The sizes of Airborne Sulfate- and Nitrate-Containing Particles: A Review" 1985, manuscript in preparation.

41 Wells, A.C.; Chamberlain, A.C. Brit. J. Appl. Phys. 1967, 18, 1973.

42 Clough, W.S. Aerosol Sci. 1973, 4, 227-34.

RECEIVED January 2, 1986

4

Fog and Cloud Water Deposition

Michael R. Hoffmann

Environmental Engineering Science, W. M. Keck Laboratories, California Institute of Technology, Pasadena, CA 91125

This paper is a summary of our findings from a four-year study of the chemical composition of fog and cloud water in California. Fog water was sampled at a number of sites with a rotating arm collector, which was developed in our laboratory and collects representative samples. Field investigations in the Los Angeles basin, the San Gabriel Mountains, and the San Joaquin Valley revealed very high ionic concentrations in polluted fogs, often coupled with very high acidities. Fogs and stratus clouds in the Los Angeles basin typically had pH values ranging from 2 to 4. Acidities were not as high in the San Joaquin Valley, mostly because of scavenging by the fogs of ammonia from agricultural sources. We showed that fogwater deposits efficiently on surfaces during fog events; this deposition was observed to be an important pollutant sink during stagnation episodes in the San Joaquin Valley, but at the same time it could be an important source of acid input to surfaces in some areas. Insight into the oxidation of S(IV) to S(VI), which is the major aqueous-phase source of acidity, was gained from field data, laboratory studies, and model development. Kinetic experiments showed that H_2O_2 was an important oxidant at low pH, and we predicted that metal-catalyzed autoxidation could also be an important source of sulfate. However, we found that the extreme acidities observed in fogs (below pH 3) require condensation on preexistent acidic nuclei and scavenging of gaseous nitric acid. Stabilization of S(IV) in the fog was observed, and this was attributed to the formation of S(IV)-aldehyde adduct.

0097-6156/86/0318-0064$07.75/0

Sulfur dioxide (SO_2) and nitrogen oxides (NO_x) are oxidized to sulfate and nitrate aerosols either homogeneously in the gas phase or heterogeneously in atmospheric microdroplets and hydrometeors (1-5). Gas-phase production of nitric acid appears to be the dominant source of aerosol nitrate because the aqueous phase reactions of NO_x(aq) are slow at the nitrogen oxide partial pressures typically encountered in the atmosphere (5,6). Conversely, field studies indicate that the relative importance of homogeneous and heterogeneous SO_2 oxidation processes depends on a variety of climatological factors such as relative humidity and the intensity of solar radiation (4, 7-10).

Cass (7) has shown that the worst sulfate pollution episodes in Los Angeles occur during periods of high relative humidity and when the day begins with low clouds or fog in coastal areas, while Cass and Shair (4) have reported that nighttime conversion rates (5.8%/hr) for SO_2 in the Los Angeles sea breeze/land breeze circulation system are statistically indistinguishable from typical daytime conversion rates (5.7%/hr) for the month of July. Zeldin et al. (11) have reported a high correlation between relative humidity and sulfate when a marine layer covers the Los Angeles Basin. They surmised that marine layers provide an ideal climate for the conversion of sulfur dioxide to sulfate. On certain days Zeldin et al. found high concentrations of both ozone and sulfate. A typical high sulfate-ozone day was described as a day with nighttime and early morning stratus clouds that enhance sulfate formation, followed by a sunny afternoon with sufficient radiation for the production of ozone by photochemical reactions. On these days the inversion strength was $\geq 7\ ^{o}C$ with inversion bases between 244 and 732m MSL. These results suggest that fog and low lying clouds may play an important role in the diurnal production of sulfate in the Los Angeles basin during certain times of the year when the meteorological conditions are propitious for fog and cloud formation. Results of these investigations combined with the data of Gartrell et al. (8), Smith and Jeffrey (9), Cox (10), Diffenhoeffer and dePena (12), Enger and Hogstrom (13), Wilson and McMurry (14), McMurry et al. (15), Hegg and Hobbs (16,17), Daum et al. (18), and Jacob et al. (19) indicate that aqueous-phase oxidation of SO_2 is a significant pathway for the formation of acidic sulfate in the atmosphere.

Our research group has characterized the chemistry of fogs, clouds, dew and rain at selected locations in California (19, 21-25). We have reported in the open literature that fog and cloud water often has extremely low pH values (e.g. $1.7 \leq pH \leq 4$) and extremely high concentrations of sulfate, nitrate, ammonium ion and trace metals. A representative set of values reported by Waldman et al. (21), Munger et al. (22) are illustrated in Figures 1-4. Of special interest are the high values observed for SO_4^{2-}, NO_3^-, S(IV), CH_2O, Fe, Mn, Pb and Cu in fog water. These values and their time-dependent changes (19) as shown in Figures 5 and 6 indicate that fogs provide a very reactive environment for the accumulation of HNO_3 and H_2SO_4. Concomitant incorporation of NH_3 gas and calcareous dust into the droplet phase neutralizes some of the acidity. In the pH domain typically encountered in fogs and clouds (pH 2-7), absorption of SO_2(g), HNO_3(g), H_2O_2(g), and NH_3(g) is thermodynamically favorable because of their relatively high Henry's Law coefficients.

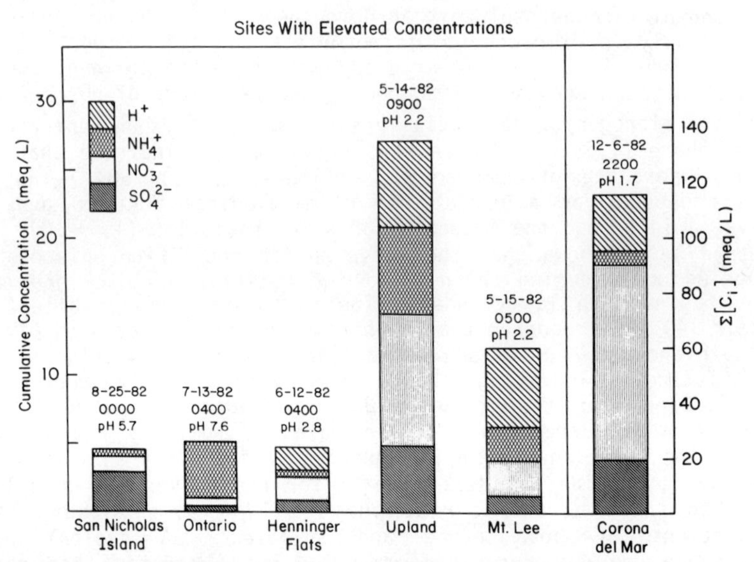

Figure 1. Cumulative bar diagram for the major chemical constituents in fog episodes with elevated concentrations.

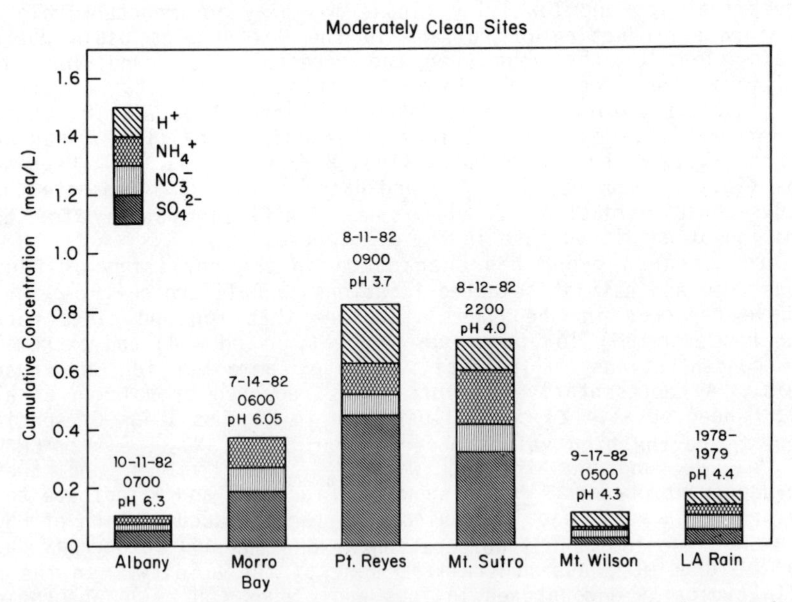

Figure 2. Cumulative bar diagram for the major chemical constituents in fog episodes with moderate concentrations.

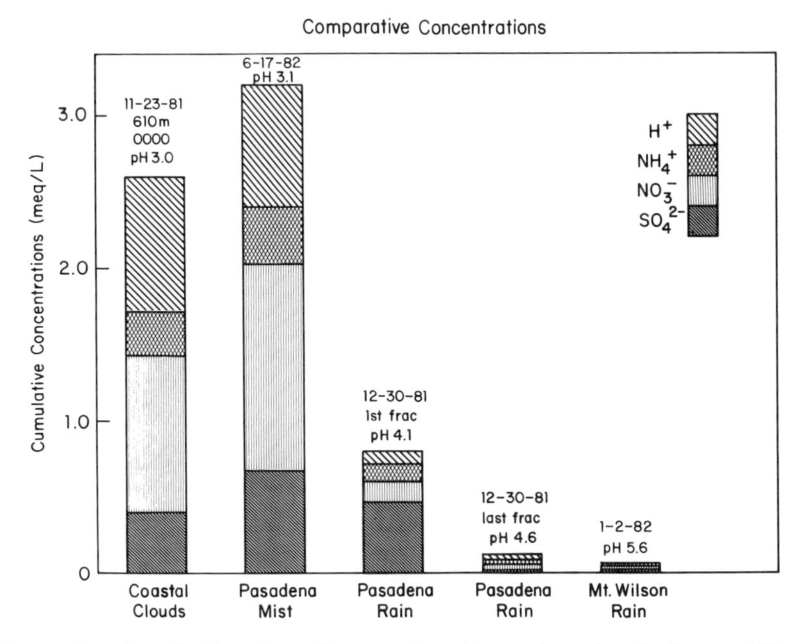

Figure 3. Cumulative bar diagram for the major chemical constituents in clouds, mist, and rain.

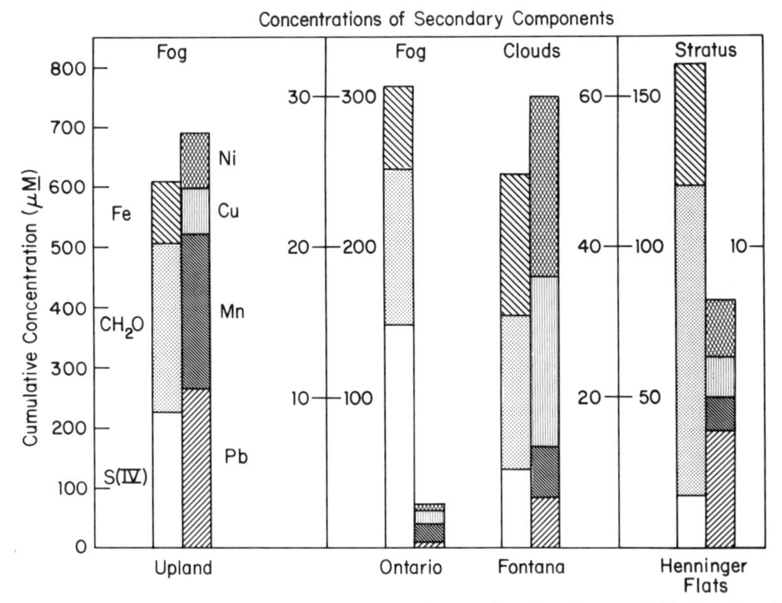

Figure 4. Concentrations of secondary chemical constitutents in fogs and clouds in southern California.

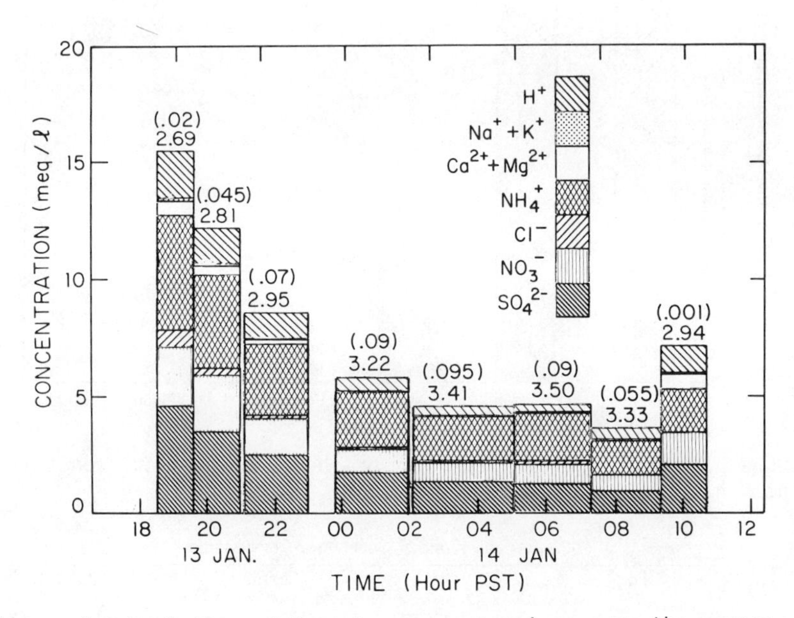

Figure 5. Evolution of fogwater concentrations over the course of a fog event at Bakersfield during the winter of 1983. Fogwater pH and the average liquid water content (LWC) in g m^{-3} is given on the top of each data bar.

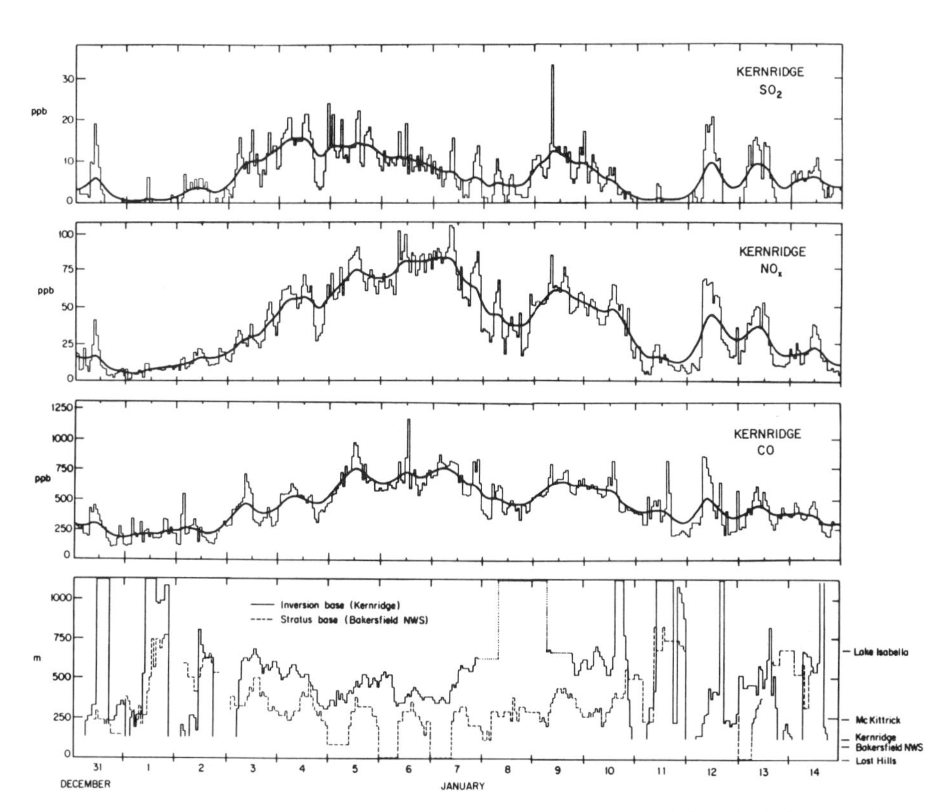

Figure 6. Variation of gas-phase components, mixing heights, and stratus base during a prolonged stagnation episode in the southern San Joaquin Valley at Kernridge.

Because of their similarity to clouds with respect to physical characteristics, fogs are likely to reflect the same chemical processes occurring in clouds and to some degree in aqueous microdroplets. Cloud and fog water droplets are in the size range of 2 to 50 μm whereas deliquescent haze aerosol will be in the range of 0.01 to 1 μm. On the other hand, raindrops are approximately 100 times larger than cloud and fog water droplets (e.g., 0.1 to 3 mm). In Los Angeles, we have found that fog water was more concentrated in the primary constituents than the overlying cloud water which was in turn more concentrated than rain water during overlapping periods of time (see Figure 1 a-d). Furthermore, Hegg and Hobbs (16) have observed sulfate production rates in cloud water over western Washington that ranged from 4.0 to 300%/hr and pH values from 4.3 to 5.9. In a later study (17), they reported even higher sulfate production rates in wave clouds, which were found to be in the range of 2 to 1,900 %/hr over a pH range of 4.3 to 7.0. The sulfate production rate measured with respect to SO_2 in ambient air appeared to increase with an increase in pH. Martin (26) has shown that the results of Hegg and Hobbs can be interpreted kinetically in terms of the open phase oxidation of S(IV) by O_3 in aqueous solution. Schwartz and Newman (27) have pointed out that the large inherent uncertainties in the aircraft measurement of Hegg and Hobbs may have resulted in conversion rates were statistically indistinguishable from zero (i.e. the reported data included negative conversion rates due to apparent sulfate loss with respect to time.) Similar observations were made by our research group (19,28) during "Tule" fog episodes in the San Joaquin Valley in the winters of 1982-1983 and 1983-1984. In Bakersfield, we found S(IV) conversion rates to be as high as 8%/hr at $0^{o}C$; however, a significant number of negative conversion rates were also observed such that the mean conversion rate was indistinguishable from zero on a statistical basis as illustrated in Figure 7. Waldman (20) has reported S(IV) conversion rates of 7-10 %/hr during the "Tule" fogs of 1984-1985. Ancillary evidence such as the increase in the sulfate equivalent fraction of the aerosol during the course of a pronounced stagnation episode supports the general notion of in situ sulfate production during fog and haze episodes in the San Joaquin Valley as shown in Figure 8. The observation of negative sulfate production rates indicates the relative importance of advection and mass transport in determining the observed fog and cloud water chemistry at a particular time and location. Current research efforts have been designed to address these problems.

Calvert (29) has pointed out that gas-phase reactions of SO_2 with ozone (O_3), hydroxyl radical (OH•), and hydroperoxyl radical (HO_2•) are too slow to account for the aforementioned rates of sulfate production. Consequently, the catalytic autoxidation of SO_2 in deliquescent haze aerosol and hydrometeors has been proposed as a viable non-photolytic pathway for the rapid formation of sulfuric acid in humid atmospheres (30-35). In addition, hydrogen peroxide and ozone have been given serious consideration as important aqueous-phase oxidants of dissolved SO_2 as discussed by Martin (35). Oxidation by H_2O_2 seems to be most favorable under low pH conditions (pH $\leq$ 4) because of a rapid rate of reaction and a negative pH-dependence that favors the facile conversion of HSO_3^- to sulfate.

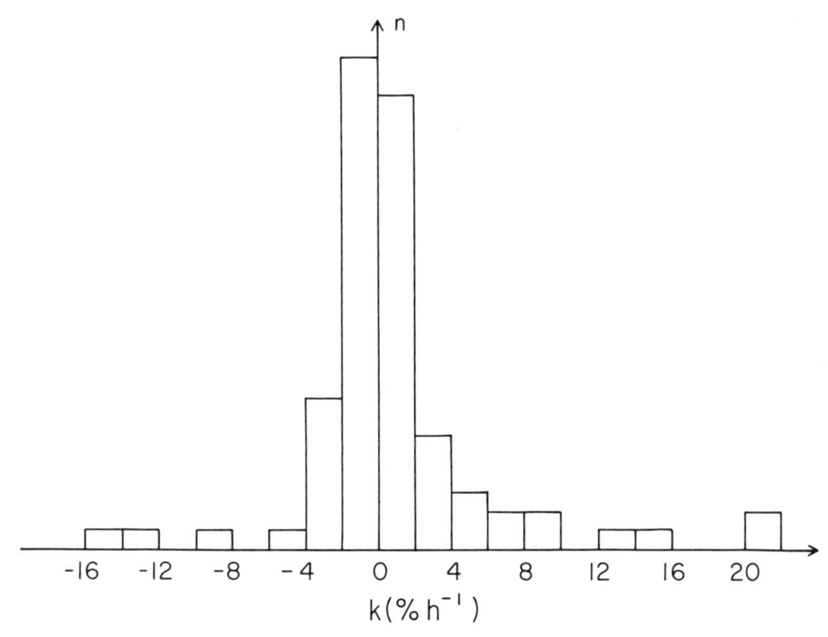

Figure 7. Frequency distribution of the pseudo first-order rate constant for in situ S(VI) production at Bakersfield. Data were obtained from 80 sequential fogwater samples.

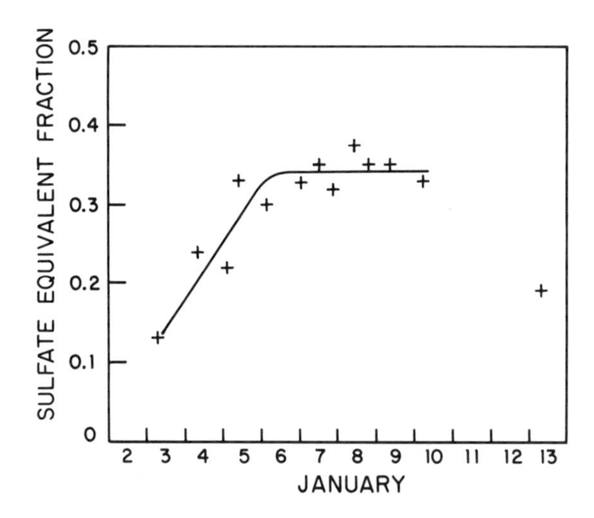

Figure 8. Evolution of the sulfate equivalent fraction in fog-water at Bakersfield during the winter of 1983.

In comparison, metal-catalyzed autoxidation and oxidation of S(IV) with O_3 tend to proceed more slowly with decreasing pH (36).

Limiting factors in the autoxidation pathways are the total concentration of the active metal catalyst and its equilibrium speciation as a function of pH. Los Angeles fog water contains high concentrations of iron, manganese, copper, nickel and lead, as shown in Figure 1d. Of these metals, Fe, Mn and Cu are expected to be the most effective catalysts for the reaction of S(IV) with molecular oxygen (23,35-37). Observed concentrations of Fe and Mn in fog of 400 μM and 15 μM, respectively, were not unusual (20,22,24,38). Model calculations indicate that metal-catalyzed autoxidations may contribute significantly to the overall sulfate formation rate in atmospheric droplets, particularly in the range of Fe and Mn concentrations observed in urban fog (20,23,24,37-39)

Carbonyl compounds, aldehydes and ketones, influence liquid-phase sulfur dioxide chemistry through their reactions with SO_2 to form stable α-hydroxyalkanesulfonates. Aldehydes are ubiquitous contaminants in the atmosphere. They exist at especially high concentrations (24-58 $\mu g/m^3$) in urban environments where vehicular emissions are a significant or even dominant source (40,41). In addition, aldehydes are generated via numerous pathways from a variety of precursors present in both clean and polluted atmospheres. These include the oxidation of alkanes and alkenes by OH• and O_3. Aldehydes are highly reactive species that decompose rapidly through photolytic and free-radical reactions. For example, the half-life of gaseous formaldehyde, $CH_2O(g)$, in the atmosphere is fairly short (2-3 hours). However, dissolution of CH_2O and hydration to form methylene glycol, $CH_2(OH)_2$, protects it from photochemical decomposition. Consequently, atmospheric droplets offer an ideal environment for sulfonic acid production.

Field measurements have detected formaldehyde at concentrations of greater than 100 μM in fog- and cloud-water samples collected in Southern California (22,42-44) as shown in Figures 4a and b). The concentrations of acetaldehyde, propionaldehyde, propenal (acrolein), n-butanal, n-pentanal, n-hexanal, and benzaldehyde occasionally approached or exceeded that of CH_2O (42). We have also shown that formaldehyde, acetaldehyde, and propanal are present in urban fog water samples at substantial concentrations (44). In addition we have shown that for each one of the aldehydes present the corresponding carboxylic acid is also present and that the aldehydes may be slowly oxidized to their respective carboxylic acids upon storage. Furthermore, the presence of CH_2O and H_2O_2 in conjunction with S(IV) at levels higher than those predicted by gas/liquid solubility equilibria suggests that hydroxymethanesulfonate (HMSA) production stabilizes a fraction of S(IV) with respect to oxidation (43). Our equilibrium calculations using available thermodynamic and kinetic data for the reaction of SO_2 and CH_2O demonstrate that elevated concentrations of S(IV) in fog water cannot be achieved without consideration of sulfonic acid production, $HORHSO_3^-$ (44) (see Figures 9-12). Recently, Munger et al. (45) have identified and quantified HMSA using ion-pairing chromatography.

Carbonyl-bisulfite addition products are highly stable in aqueous solution at low pH, but they undergo facile dissociation under alkaline conditions. The kinetics of hydroxymethanesulfonate

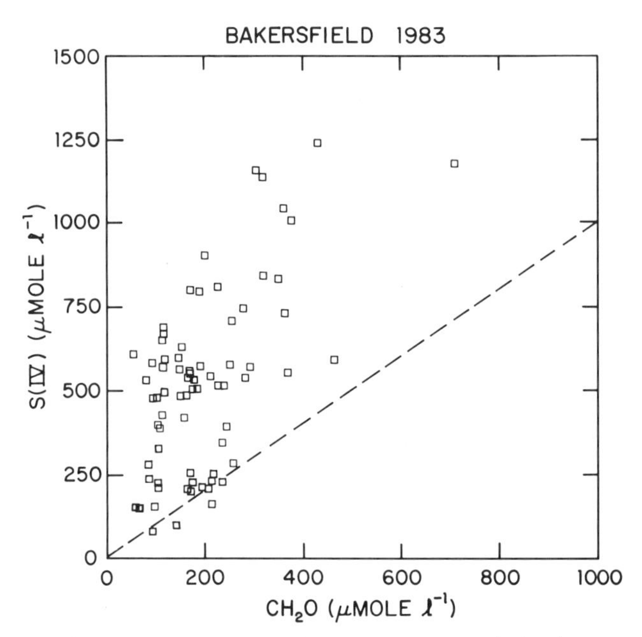

Figure 9. Plot of the observed aqueous-phase (S(IV) concentrations versus observed HCHO concentrations in fogwater collected in Bakersfield, CA.

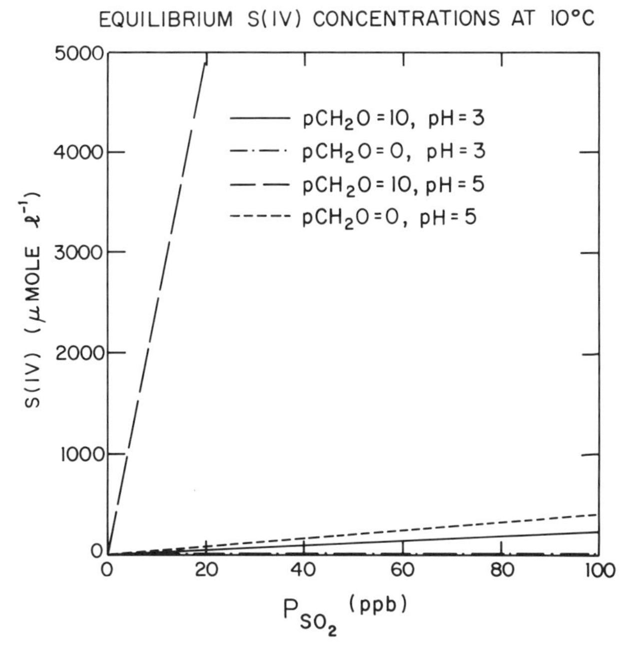

Figure 10. Calculated equilibrium relationship for the total aqueous-phase S(IV) as a function of P_{SO_2}.

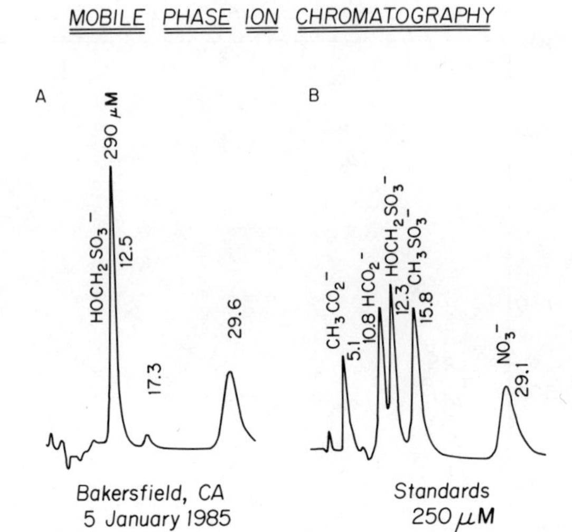

Figure 11. A representative chromatogram for the separation and identification of hydroxymethane sulfonate in fogwater collected in Bakersfield, CA.

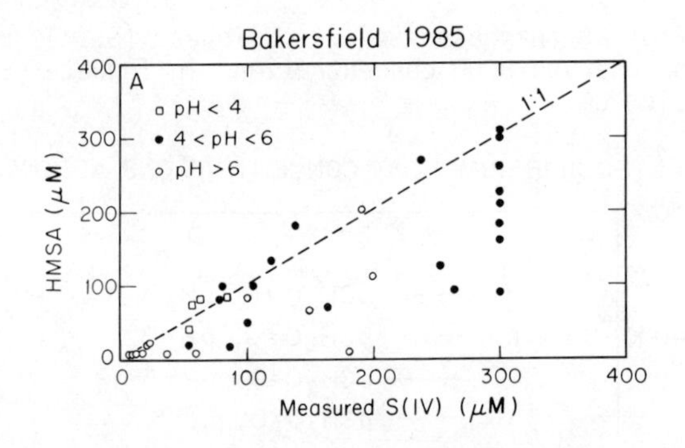

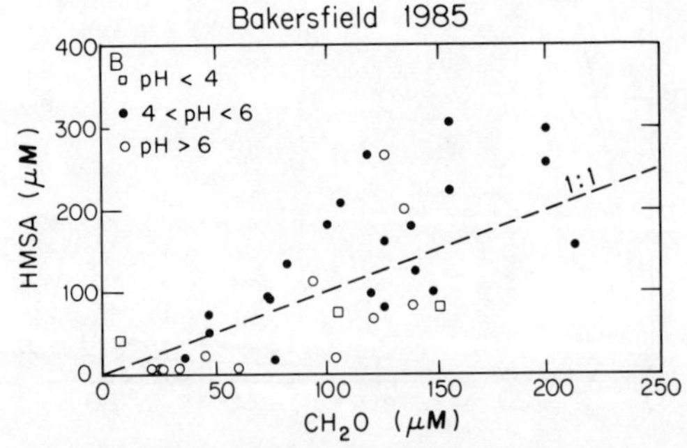

Figure 12. Concentration of hydroxymethanesulfonate (HMSA) vs. measured S(IV) and vs. measured (HCHO).

production under pH conditions characteristic of fog and cloud water has been studied in our laboratory. Boyce and Hoffmann (46) found that the formation of HMSA over the pH range 0.0 to 3.5 occurs by parallel reaction pathways involving nucleophilic addition of HSO_3^- and SO_3^{2-} to the carbonyl C-atom of formaldehyde as follows:

$$CH_2(OH)_2 \Longleftrightarrow CH_2O + H_2O \quad (K_d)$$

$$SO_2{\cdot}H_2O \Longleftrightarrow H^+ + HSO_3^- \quad (K_{a1})$$

$$HSO_3^- \Longleftrightarrow H^+ + SO_3^{2-} \quad (K_{a2})$$

$$HSO_3^- + CH_2O \Longleftrightarrow CH_2(OH)SO_3^- \quad (k_1), (k_{-1}), (K_1)$$

$$SO_3^{2-} + CH_2O \longrightarrow CH_2(O^-)SO_3^- \quad (k_2)$$

$$CH_2(O^-)(SO_3^-) + H^+ \Longleftrightarrow CH_2(OH)SO_3^-$$

where $k_1 = 7.90 \times 10^2 M^{-1}s^{-1}$ and $k_2 = 2.48 \times 10^7 M^{-1}s^{-1}$ at 25°C. The formation constant, K_1, has been determined recently by Deister et al.(47) and Kok et al. (48) to be 10^7 M^{-1}. This number is in agreement with those reported previously by Kerp (49) and Donally (50). Under more weakly acidic conditions (pH > 4), the dehydration of methylene glycol (equation K_d) may become rate-determining (51). Application of the rate constants and activation energy parameters obtained in the laboratory to the analysis of the field measurements discussed above indicates that HMSA formation may account for the occurrence of S(IV) at elevated concentrations (44). Kinetic data obtained for other aldehyde/sulfur(IV) reaction systems suggests that the mechanism outlined above can be generalized to describe the formation of a wide variety of α-hydroxyalkanesulfonates (52).

In order to develop a comprehensive physiochemical description of the complex SO_2 reaction network in atmospheric droplets, rate laws, mechanisms and activation energies are being determined for the various pathways of sulfur dioxide transformation in aqueous solution. However at this time complete information has been assembled in only a very few cases. The mechanism of S(IV) oxidation by hydrogen peroxide is fairly well understood (33,53-57). The reaction proceeds by nucleophilic displacement of water by H_2O_2 on bisulfite ion to form peroxymonosulfite anion ($HOOSO_2^-$), which rearranges to sulfate under the influence of specific and general acid catalysis (57-58). The significance of this latter feature for open atmospheric systems has been discussed by Schwartz (5), Martin (35), Jacob and Hoffmann (23), Hoffmann and Jacob (37), and McArdle and Hoffmann (57), and Hoffmann and Calvert (39). Hoffmann (59) has proposed a mechanism for the oxidation of S(IV) by ozone that proceeds via the simultaneous nucleophilic attack on ozone by SO_3^{2-}, HSO_3^-, and $SO_2{\cdot}H_2O$. Hoffmann (59) has analyzed the kinetic data of a number of different investigators (35,60-63) in terms of a single self-consistent rate expression.

For the metal-catalyzed autoxidation of S(IV), there is considerable ambiguity about the mechanism(s) of reaction. First-row transition-metal species can catalyze the reaction of aquated sulfur dioxide and O_2 through four distinctly different pathways as described by Hoffmann and Boyce (36) and Hoffmann and Jacob (37). These mechanisms include: a thermally-initiated free radical chain processes involving a sequence of one-electron transfer steps, heterolytic pathways in which formation of an inner-sphere metal-sulfite-dioxygen complex occurs as a prelude to two-electron transfer, heterogeneous catalysis through complexation of HSO_3^- and/or SO_3^{2-} at the surface of metal oxides and oxyhydroxides in suspension, and photochemical oxidation initiated by the absorption of light by S(IV), metal cations, metal-oxide semiconductors, and/or specific metal-sulfite complexes. Hoffmann and Jacob (37) have compared the theoretical kinetic expressions for each type of mechanism with the empirical rate data obtained in experimental studies of the catalytic autoxidation of SO_2.

Quantitative analysis of different reaction pathways for the transformation of aquated sulfur dioxide in atmospheric droplet systems has been a major objective of the research conducted in the principal investigator's laboratory for the last four years. Available thermodynamic and kinetic data for the aqueous-phase reactions of SO_2 have been incorporated into a dynamic model of the chemistry of urban fog that has been developed by Jacob and Hoffmann (23) and Hoffmann and Calvert (39). The fog and cloud water models developed by them are hybrid kinetic and equilibrium models that consider the major chemical reactions likely to take place in atmospheric water droplets. Model results have verified that extremely high acidity may be imparted to fog water droplets by condensation and growth on acidic nuclei or by in situ S(IV) oxidation. Based on both kinetic and equilibrium considerations the important oxidants in the aqueous phase were found to be O_2 as catalyzed by Fe(III) and Mn(II), H_2O_2, and O_3 (See Figures 13 and 14). The results of the model calculations show that metal-sulfite complexation (both with and without electron transfer) and hydroxyalkanesulfonate formation enhance water droplet capacity for SO_2, but did not slow down the net S(IV) oxidation rate leading to fog acidification. Nitrate production in the aqueous phase was found to be dominated by HNO_3 gas phase scavenging. Highly acidic fog water appears to form predominantly from condensation on highly acidic haze aerosol (i.e. fog condensation nuclei); in these cases (i.e. pH 1.7 fog at Corona del Mar), in situ S(IV) oxidation leads to little further acidification of the fog water. In field situations, the ultimate acidity level in fog depends upon the degree of neutralization of free acidity by ammonia or by scavenged alkaline aerosol. In Los Angeles more ammonia is available further inland from the coast. This may explain in part why higher fog water acidities are found along the coast than at inland sites. Similar observations have been made in the Southern San Joaquin Valley.

Waldman et al.(25) have studied the chemistry and microphysics of intercepted cloudwater on Los Angeles area mountain slopes. From 1982 to 1985, the observed pH values of the cloudwater ranged from 2.06 to 3.87 with the median value below pH 3 (See Figure 15). The

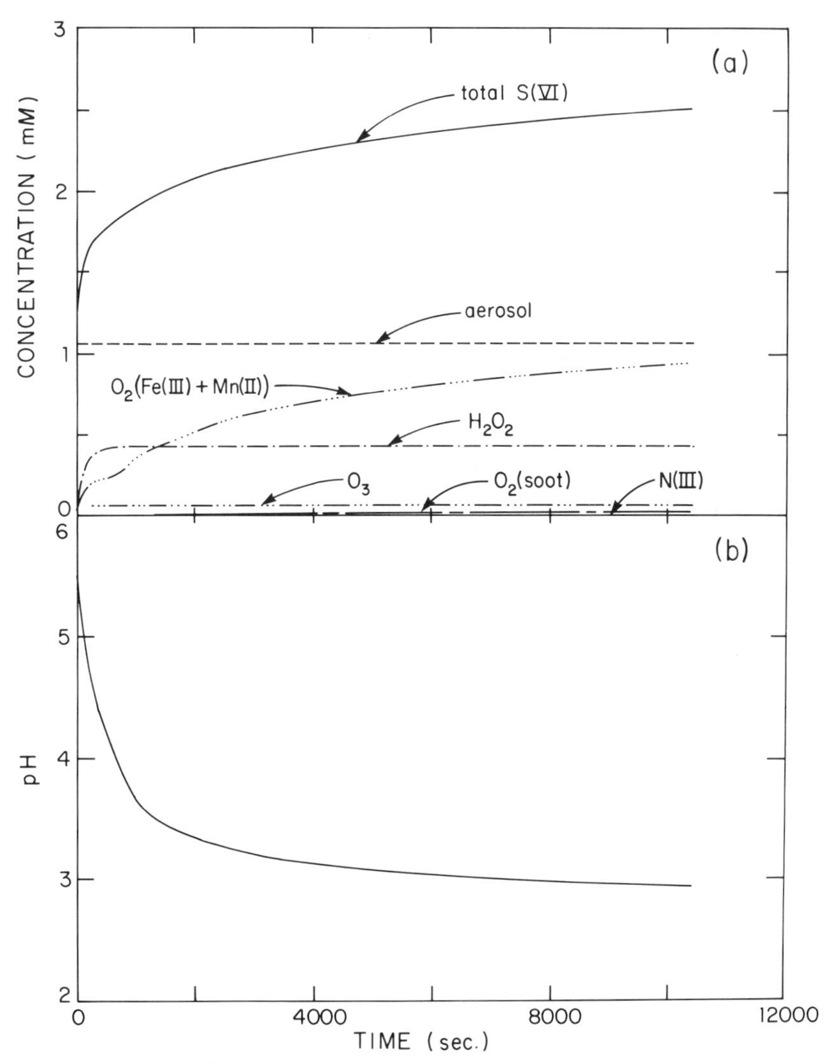

Figure 13. (a) Profile of concentration versus time of total S(VI) in fogwater and of the individual contributions to the total S(VI) due to different pathways. Model calculations are shown for different aqueous-phase oxidants. (b) Profile of pH vs. time for LWC = 0.1 g m^{-3} at T = 10 ^{o}C.

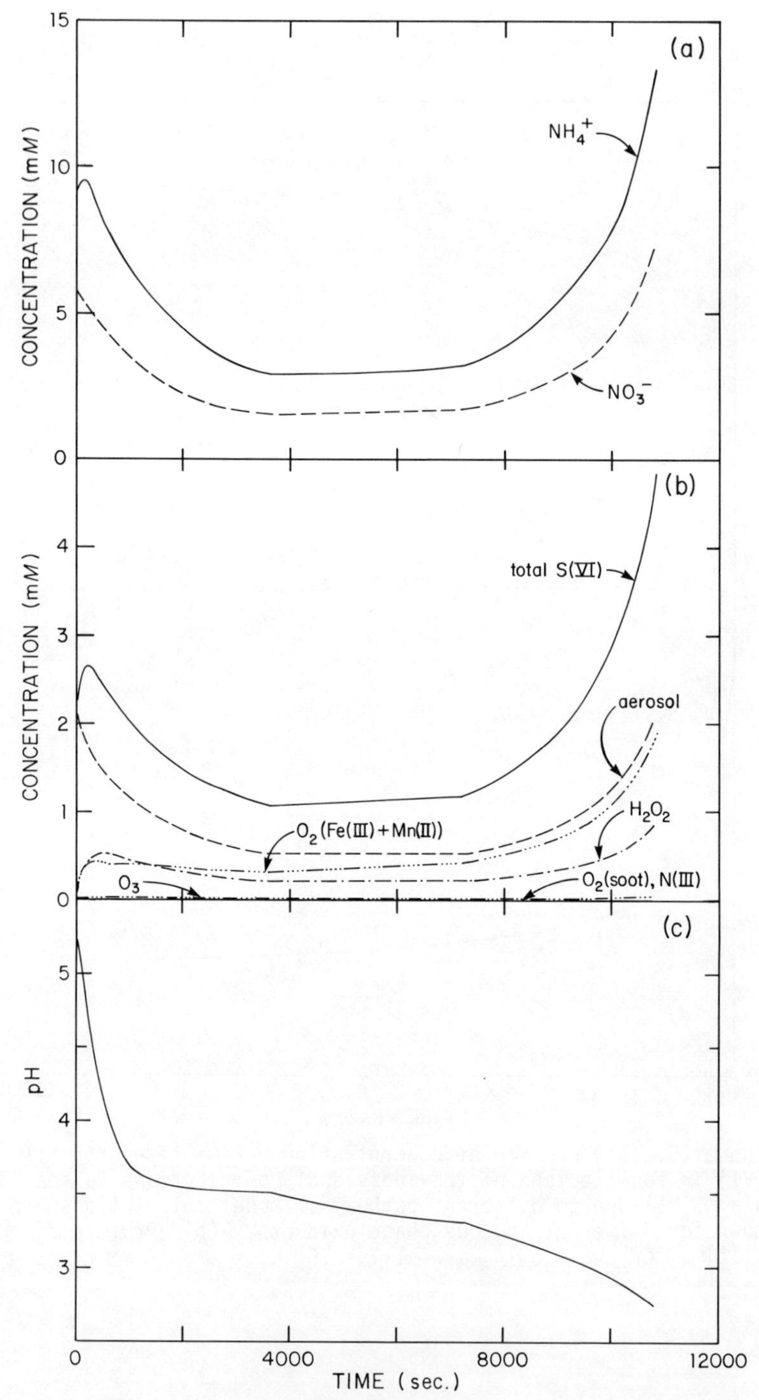

Figure 14. Profiles of concentration vs. time for pH, S(VI), N(-III), and N(V) as predicted by model calculations.

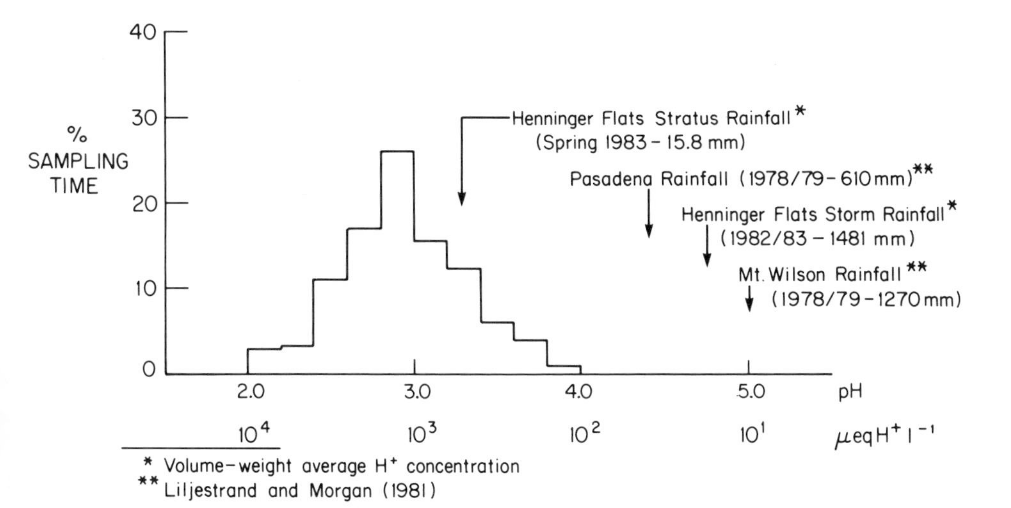

Figure 15. Histograph of pH frequency of cloudwater samples collected at Henninger Flats near Altadena, CA. Volume-weighted averages are indicated with arrows.

equivalent ratio of nitrate to sulfate in cloudwater at Henninger Flats (MSL = 2,520 ft.) was close to two, while at the same site the $[NO_3^-]/[SO_4^=]$ ratio in rainwater was ~ 1. However, the nitrate/sulfate ratio observed in dry aerosol was significantly lower than that observed in cloudwater; the additional nitrate found in cloudwater appears to be derived from the scavenging of gaseous nitric acid (See Figure 16). In addition, a higher fraction of nitrate aerosol appears to be scavenged by cloud droplets. This observation is consistent with current theories of homogeneous versus heterogeneous gas-to-particle pathways open to sulfur dioxide versus nitrogen oxides.

Cloud droplet capture in the form of intercepted fog appears to be a seasonably important sink for pollutant emissions in the LA Basin. At Henninger Flats up to 50% of the total wet deposition of H^+, NO_3^-, and $SO_4^=$ may be due to cloud interception; low intensity springtime drizzle accounted for 20% of the deposition measured in precipitation. The intercepted cloudwater that deposited on pine needles was collected and analyzed. The acidity of the water dripping from trees was very similar to that of the suspended cloudwater. The concentrations of major chemical components were found to be significantly greater than in the overlying cloudwater. The additional solute in the drippings is thought to be derived from previously deposited material and the evaporated residue of intercepted cloudwater. Even after sufficient rainfall had removed most of the accumulated residue, the concentrations of major cations such as Ca^{2+}, Mg^{2+}, and K^+ showed relative increases compared to suspended cloudwater samples. These increases may be attributed in part to ion-exchange of H^+ for K^+, Mg^{2+}, and Ca^{2+} from the pine needles (64). The potential for harm to sensitive plant tissue appears to be high given prolonged exposure to the severe microenvironments observed on the slopes of the San Gabriel Mountains and in the Angeles National Forest.

The Caltech rotating arm collector (RAC) was calibrated precisely using a scale model rotating arm device and a chemically tagged monodisperse aerosol. Jacob et al. (65) have characterized the performance of the rotating arm collector in great detail. The rotating arm collector was designed to meet the following criteria:

1. The aerodynamic heating associated with the flow of air towards the impactor surface must be small enough not to cause droplet evaporation.
2. The collected droplets must be rapidly sheltered from the changing air masses to prevent evaporation and chemical contamination.
3. The lower size cut must be sharp and in the range of 1-10 um, and no sampling biases must be introduced for the droplets up to at least 50 µm.
4. The sampling rate must be high enough to collect sufficient amounts of sample for chemical analysis while allowing a reasonable time resolution.

These criteria were met with the exception of criteria 3. Jacob et al.(65) found that the scale model version of the RAC had a particle size cut of 20 µm rather than the desired 1-10 µm cut.(See Figure 17) The RAC performs well in preserving the chemical integrity of the collected droplets and provides a high sampling rate; however, it has the drawback that it does not collect efficiently the smaller

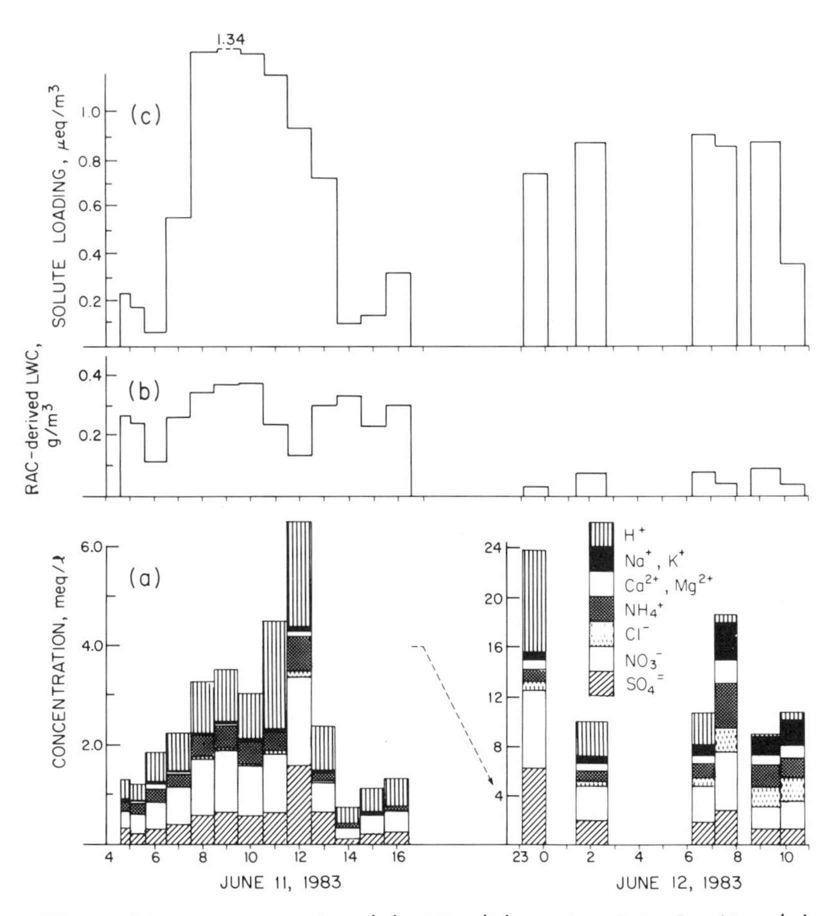

Figure 16. Concentration (a), LWC (b), and solute loading (c) for sequential cloudwater samples.

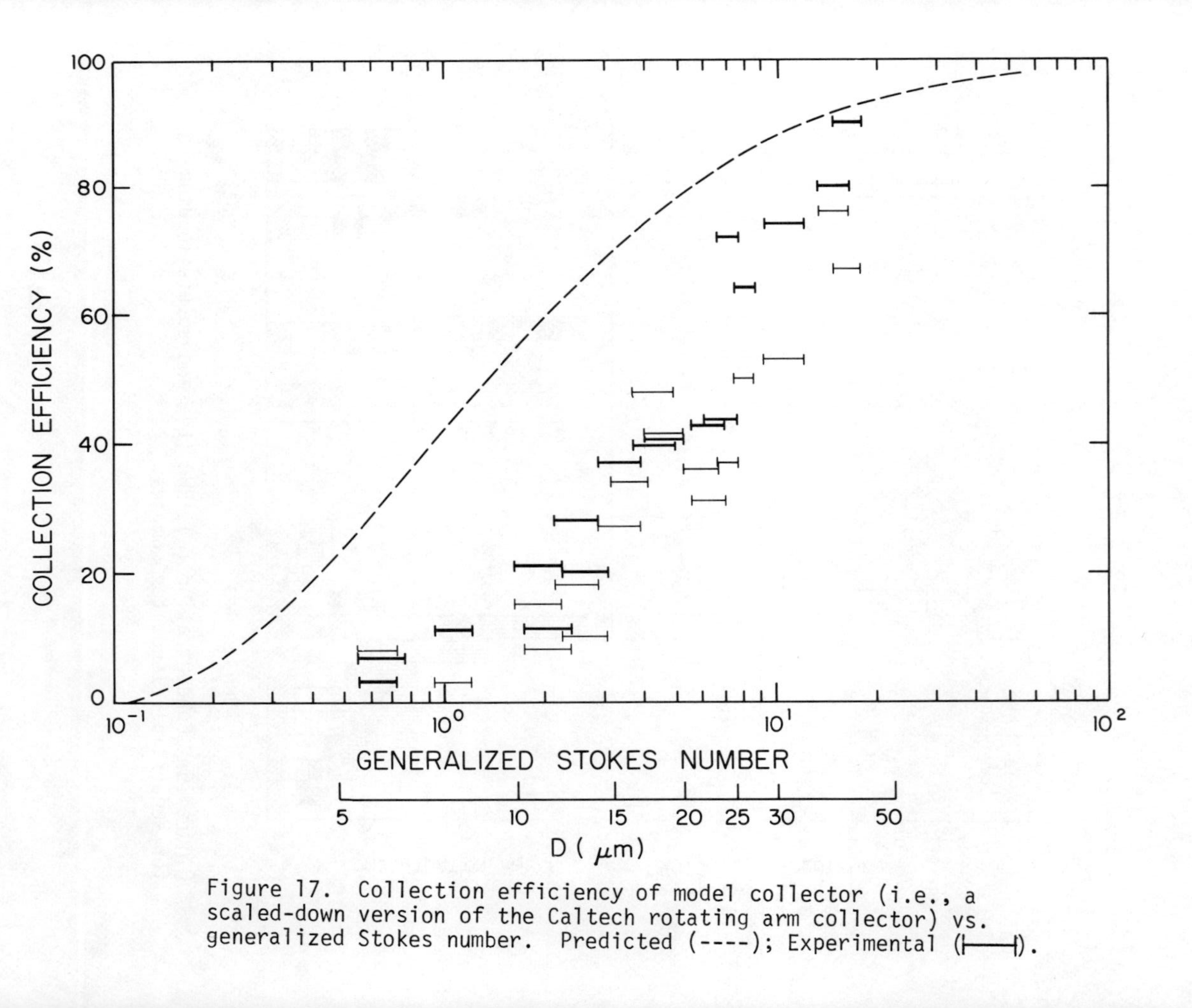

Figure 17. Collection efficiency of model collector (i.e., a scaled-down version of the Caltech rotating arm collector) vs. generalized Stokes number. Predicted (----); Experimental (|—|).

droplets in the fog. This has not proved to be a significant problem in that preliminary data suggest that there is very little variation of chemical composition of fog water droplets with size. Changes in the collector geometry, such as streamline shaping to reduce drag, would produce a smaller size cut. Jet impactors and screen collectors are being evaluated as an alternative to rotating arm collectors.

In June of 1983, five different types of fog/cloud water collectors were compared under field conditions at Henninger Flats. Collectors designed by AeroVironment(AV), Caltech(CIT), the Desert Research Institute(DRI), Global Geochemistry(GGC), and the State University of New York(SUNY)-Albany were tested against one another. The mass and pH of the samples collected were measured on site while the detailed chemical analyses were performed by an independent laboratory, Rockwell International. Results of the intercomparison study, sponsored the the Coordinating Research Council, showed that the Caltech and DRI collectors performed the best over the broadest range of conditions (66,67). At low liquid water content (i.e. LWC < 0.03 g/m^3), the SUNY-Albany rotating string collector, the DRI jet impactor, and the Caltech rotating arm collector had reasonable collection rates. The Global Geochemistry mesh collector and the AeroVironment rotating rod collector were not effective under low LWC conditions. With respect to sulfate, nitrate, and pH values obtained under a wide variety of conditions over 38 hours of fog, the Caltech and DRI collectors showed the most consistency and closest agreement. The Suny-Albany and the AerVironment collectors showed the greatest deviations from the median values. For example the slope of the observed [$SO_4^{=}$] vs. median [$SO_4^{=}$] over the range of concentration from 100 to 2000 μM for the SUNY, Caltech, DRI, AeroVironment, and Global collectors, respectively, were 0.69(SUNY), 0.95(CIT), 0.91(DRI), 1.66(AV), and 1.21(GGC). The high values for the AV and GGC collectors indicate that concentration effects due to evaporation are significant problems in these collectors. In the case of nitrate correlations, the slopes of the observed concentrations vs. the median values for the same sequence of collectors were found to be 0.83(SUNY), 0.94(CIT), 0.93(DRI), 2.43(AV), and 1.17(GGC). When identical RAC's were placed at different locations at Henninger Flats the results correlated quite well. There appeared to be no statistical difference between the separated collectors (66,67).

Summary of Major Observations and Conclusions

1. Fogwater collected at sites in the South Coast Air Basin of Los Angeles was consistently acidic, with pH values typically ranging from 2 to 4. The highest acidities were observed during smog episodes. The main contributors to the acidity were nitric and sulfuric acids, with a typical equivalent ratio of 3:1. Secondary sulfate and nitrate aerosol accounted for over 80% of the fogwater loading.
2. Fogwater collected at non-urban coastal sites was usually acidic (pH range 3 to 7). Impact of emission centers on distant coastal locations was documented. The low alkalinity of marine atmospheres make them particularly

susceptible to acidification. Oxidation of oceanic dimethylsulfide could be a natural source of sulfuric acid.

3. Stratus clouds collected at 2500' MSL over the Los Angeles Basin were consistently acidic, pH range 2 to 4. Cloudwater concentrations were in the same range as those observed in the basin itself.
4. Fogwater collected in the Southern San Joaquin Valley was not consistently acidic; pH values ranged from 2.5 to 7.5. Millimolar concentrations of sulfate were typically observed, but high ammonia emissions from livestock and cropland neutralized the acid input. Visalia, which is some distance from the major emission sources, had alkaline fogwater (pH 6 - 7.5). McKittrick, located in an oil field with little surrounding agricultural activity, had acidic fogwater (pH 2.5 - 4.5).
5. Liquid water content (LWC) was the major factor affecting ionic concentrations in fogs. As the fog formed, droplet growth diluted the droplets; as the fog dissipated, the droplets became more concentrated.
6. Evidence was found for the major processes responsible for the acidification of fogwater: (i) the scavenging of acidic precursor aerosol, (ii) the scavenging of gaseous nitric acid, and (iii) oxidation of reduced sulfur components to sulfate. Conversion of $SO_2(g)$ to sulfate in fogwater does not appear to proceed faster than 10% $hour^{-1}$ and therefore cannot account for the high acidities observed at the beginning of fog events; however, sulfate production in the precursor air parcel can lead to sulfuric acid fog condensation nuclei.
7. Modeling of fogwater chemistry indicated that the high acidities observed can be explained by either of the three processes listed above. The main aqueous-phase S(IV) oxidants were found to be hydrogen peroxide, ozone, and oxygen (catalyzed by trace metals). Aqueous-phase production of nitrate was found to be unimportant.
8. The rotating arm collector designed in our laboratory and used to sample fogwater was fully characterized. Evaporation of droplets during all stages of collection was shown to be negligible. Experimental calibration indicated a lower size cut of 15-20 microns. Field data show an overall liquid water collection rate of about 60%. A field intercomparison of fogwater collectors used by various investigators confirmed that our sampler collects representative samples.
9. A screen collector (lower size cut 2 microns, sampling rate 20 m^3 min^{-1}) was designed (68) and has been used in the field. Side-by-side comparison indicates that samples collected with this collector and with the rotating arm have similar concentrations.
10. Concentrations of S(IV) in fogwater were far in excess of those expected to be in equilibrium with ambient $SO_2(g)$. Elevated formaldehyde concentrations suggest the formation of a formaldehyde-S(IV) complex; kinetic and model studies have shown that this complex is very stable and that its formation leads to high aqueous-phase S(IV) concentrations.

11. Extensive Bakersfield fogwater data indicated an important removal to the ground of pollutants scavenged by the fog droplets. This was ascribed to the slow residence time of the supermicron fog droplets in the atmosphere. In a stagnant atmosphere this deposition was suspected to alleviate build-up of suspended particles. On the mountain slopes surrounding the Los Angeles Basin, such non-recipitating wet deposition was shown to be a significant source of overall pollutant deposition.
12. Concentrations of NH_4^+, NO_3^- and SO_4^{2-} in urban fogwater samples are routinely on the order of 10^{-3} M.
13. The relative importance of NO_3^- and SO_4^{2-} reflects their emission pattern in the vicinity. Nitrate exceeds SO_4^{2-} by a facto.

13. The relative importance of NO_3^- and SO_4^{2-} reflects their emission pattern in the vicinity. Nitrate exceeds SO_4^{2-} by a factor of 2-3 in Los Angeles where vehicle emissions of NO_x are significant. Sulfate equals or exceeds nitrate in the Southern San Joaquin Valley where emissions from oil-production facilities are important.
14. Fog- and cloud-water in the Los Angeles Basin routinely has a pH < 4 with the lowest value being below 2.
15. Ammonia emissions in the Southern San Joaquin Valley are sufficient to neutralize most of the acidity present. Acid anion concentrations in Bakersfield are comparable to those in LA, but very few fogwater samples had pH < 4; ammonium was about equal to the sum of NO_3^- and SO_4^{2-}.
16. Droplet growth and evaporation is a major factor determining fog water concentrations -- the highest concentrations are observed as fog dissipates.
17. Deposition of fog droplets appears to be significant. The mass of solute per volume of air decreases over the course of a fog event. Repeated fogs may diminish the buildup of pollutants during stagnation episodes.
18. No statistical evidence for aqueous-phase sulfur oxidation can be found for events. However over the course of stagnation episodes in the Southern San Joaquin Valley the sulfate fraction in the aerosol increases.
19. Concentrations of S(IV) and CH_2O in fog and cloudwater on the order of 10^{-4} M are routinely found in urban areas. Peak values are about 10^{-3} M.
20. The partial pressure of SO_2 during fog is much too low to support all of the S(IV) as free S(IV) ($SO_2 \cdot H_2O + HSO_3^- + SO_3^{2-}$). These observations of high S(IV) can best be explained by the formation of S(IV)-aldehyde adducts.
21. The solute loading (mass/m^3 air) in fog is comparable to that in the aerosol.
22. The NO_3^-/SO_4^{2-} ratio in fog is higher than in the dry aerosol proceeding the fog, which suggests that gaseous HNO_3 is incorporated into the fog.
23. Deposition from fog by sedimentation or impaction may be comparable to rainfall deposition at some mountain sites. Trees are very efficient collectors and are often bathed with impacted fog. Fogwater impacting on vegetation

routinely had a pH of 3, which may be injurious to sensitive species.

24. In addition to formaldehyde, fog- and cloudwater contain a variety of higher aldehydes. Acetaldehyde and proponal (or acrolein) often have concentrations comparable to formaldehyde.
25. Low molecular weight carboxylic acids are present in fog and cloudwater at about 10^{-4} M. Formic and acetic acid dominate.

Acknowledgments

We are grateful to the California Air Resources Board and the U.S. Environmental Protection Agency for their support of the research described above. In particular we would like to thank Drs. J. Holmes, D. Lawson, R. Papetti, and Mr. E. Fujita for their assistance in making this research possible.

Literature Cited

1. Calvert, J. G., Su, F., Bottenheim, J. W., and Strausz, O. P. "Mechanisms of Homogeneous Oxidation of Sulfur Dioxide in the Troposphere," Atmos. Environ. 1978, 12, 197-226.

2. Middleton, P., Kiang, C. S., and Mohnen, V. A. "Theoretical Estimates of the Relative Importance of Various Urban Sulfate Aerosol Production Mechanisms," Atmos. Environ. 1980, 14, 463-472.

3. Moller, D. "Kinetic model of atmospheric oxidation based on published data," Atmos. Environ. 1980, 14 1067-1076.

4. Cass, G. R. and Shair, F. H. "Sulfate accumulation in a sea breeze/land breeze circlation system," J. Geophys. Res. 1984, 89D, 1429-1438.

5. Schwartz, S. E. (1984) "Gas-Aqueous Reactions of Sulfur and Nitrogen Oxides in Liquid-Water Clouds," Acid Precipitation edited by J. G. Calvert, Butterworth Publishers, Boston, MA.

6. Schwartz, S. E. and White, W. H. "Solubility equilibria of the nitrogen oxides and oxyacids in dilute aqueous solution," Adv. Env. Sci. Eng. 1981, 4, 1-45.

7. Cass, G. R. "Methods for Sulfate Air Quality Management with Applications to Los Angeles," Ph.D. Thesis, California Institute of Technology, Pasadena, CA, 1977.

8. Gartrell, J. E., Thomas, J. W., and Carpenter, S. B. "Atmospheric Oxidation of SO_2 in Coal-Burning Power Plant Plumes," Amer. Ind. Hyg. Assoc. Quart. 1963, 24, 254-262.

9. Smith, F. B. and Jeffrey, G. H. "Airborne Transport of Sulphur Dioxide from the U.K.," Atmos. Environ. 1975, 9, 643-659.

10. Cox, R. A. "Particle Formation from Homogeneous Reactions of Sulfur Dioxide and Nitrogen Dioxide," Tellus 1974, 26, 235-240.

11. Zeldin, M. D., Davidson, A., Brunelle, M. F., and Dickinson, J. E. "A Meteorological Assessment of Ozone and Sulfate Concentrations" in Southern California, Evaluation and Planning and Division Report 76-1 Southern California Air Pollution Control District, El Monte, CA, 1979.

12. Diffenhoefer, A.C. and dePena, R. G. "A Study of Production and Growth of Sulfate Particles in Plumes from a Coal-fired Power Plant," Atmos. Environ. 1978, 12, 297-306.

13. Enger, L. and Hogstrom, U. "Dispersion and Wet Deposition of Sulfur from a Power Plant Plume," Atmos. Environ. 1981, 15, 297-306.

14. Wilson, J. C. and McMurry, P. H. "Studies of Aerosol Formation in Power Plant Plumes. I. Secondary Aerosol Formation in the Navajo Generating Station Plume," Atmos. Environ. 1981, 15, 2329-2339.

15. McMurry, P. H., Rader, D. J., and Smith, J. L. "Studies of Aerosol Formation in Power Plant Plumes. I. Parameterization of Conversion Rate for Dry, Moderately Polluted Ambient Conditions," Atmos. Environ. 1981, 15, 2315-2329.

16. Hegg, D. A. and Hobbs, P. V. "Cloudwater Chemistry and the Production of Sulfates in Clouds," Atmos. Environ. 1981, 15, 1597-1604.

17. Hegg, D. A. and Hobbs, P. V. "Measurements of Sulfate Production in Natural Clouds," Atmos. Environ. 1982, 16, 2663-2668.

18. Daum, P. H., Schwartz, S. E., and Newman, L. "Acidic and Related Constituents in Liquid Water Stratiform Clouds," J. Geophys. Res. 1984, 89D, 1447-1458.

19. Jacob, D. J., Waldman, J. M., Munger, J. W., and Hoffmann, M. R. "A Field Investigation of Physical and Chemical Mechanisms Affecting Pollutant Concentrations in Fog Droplets," Tellus, 1984, 36B, 272-285.

20. Waldman, J. M. "Depositional Aspects of Fogs and Clouds," Ph.D. Thesis, California Institute of Technology, Pasadena, CA, 1985.

21. Waldman, J. M., Munger, J. W., Jacob, D. J., Flagan, R. C., Morgan, J. J., and Hoffmann, M. R. "Chemical Composition of Acid Fog," Science 1982, 218, 677-680.

22. Munger, J. W., Jacob, D. J., Waldman, J. M., and Hoffmann, M. R. "Fogwater Chemistry in an Urban Atmosphere," J. Geophys. Res. 1983, 88C, 5109-5121.

23. Jacob, D. J. and Hoffmann, M. R. "A dynamic model for the production of H^+, NO_3^-, and SO_4^{2-} in urban fog," J. Geophys. Res. 1983, 88C, 6611-6621.

24. Jacob, D. J., Waldman, J. M., Munger, J. W., and Hoffmann, M. R. "Chemical Composition of Fogwater Collected Along the California Coast," Environ. Sci. Tech. 1985, 19, 730-735.

25. Waldman, J. M., Munger, J. W., Jacob, D. J., and Hoffmann, M. R. "Chemical Characterization of Stratus Cloudwater and Its Role as a Vector for Pollutant Deposition in a Los Angeles Pine Forest," Tellus, 1985, 37B, 91-108.

26. Martin, L. R. "Comment on Measurements of Sulfate Production in Natural Clouds," Atmos. Environ., 1983, 17, 1603-1604.

27. Schwartz, S. E. and Newman, L. "Comment on Measurements of Sulfate Production in Natural Clouds," Atmos. Environ., 1983, 17, 2629-2632.

28. Jacob, D. J., Munger, J. W., Waldman, J. M., and Hoffmann, M. R. "Aerosol composition in a stagnant air mass impacted by dense fogs: Preliminary results," in Proc. 77th Annual Meeting of the Air Pollution Control Assoc., San Francisco, CA, June 24-29, Paper 24.5, 1984.

29. Calvert, J. J. in Acid Precipitation, SO_2, NO, and NO_2 Oxidation Mechanisms: Atmospheric Considerations, Butterworth Publishers, Stoneham, MA, 1984.

30. Hegg, D. A. and Hobbs, P. V. "Oxidation of Sulfur Dioxide in Aqueous Systems with Particular Reference to the Atmosphere," Atmos. Environ. 1978, 12, 241-253.

31. Kaplan, D. J., Himmelblau, D. M., and Kanaoka, C. "Oxidation of Sulfur Dioxide in Aqueous Ammonium Sulfate Aerosols Containing Manganese as a Catalyst," Atmos. Environ. 1981, 15, 763-773.

32. Penkett, S. A., Jones, B. M. R., and Eggleton, A. E. J. "A Study of SO_2 Oxidation in Stored Rainwater Samples," Atmos. Environ., 1979, 13, 139-147.

33. Penkett, S. A., Jones, B. M. R., Brice, K. A., and Eggleton, A. E. J. "The Importance of Atmospheric Ozone and Hydrogen Peroxide in Oxidizing Sulfur Dioxide in Cloud and Rainwater," Atmos. Environ 1979, 13, 123-137.

34. Beilke, S. and Gravenhorst, G. (1978) "Heterogeneous SO2-Oxidation in the Droplet Phase, Atmos. Environ. 1978, 12, 231-239.

35. Martin, L. R. "Kinetic Studies of Sulfite Oxidation in Aqueous Solution," in Acid Precipitation, edited by J. G. Calvert, Butterworth Publishers, Stoneham, MA, 1984, 63-100.

36. Hoffmann, M. R., and Boyce, S. D. "Catalytic Autooxidation of Aqueous Sulfur Dioxide in Relationship to Atmospheric Systems," in Trace Atmospheric Constituents: Properties, Transformations, and Fates, S. E. Schwartz, ed. Adv. Environ. Sci. Technol. 1983, 12, 147-189.

37. Hoffmann, M. R. and Jacob, D. J. "Kinetics and Mechanisms of the Catalytic Oxidation of Dissolved Sulfur Dioxide in Aqueous Solution: An Application to Nighttime Fog-water Chemistry," in Acid Precipitation, edited by J. G. Calvert, Butterworth Publishers, Boston, MA., 1984.

38. Jacob, D. J. "The Origins of Inorganic Acidity in Fogs," Ph.D. Thesis, California Institute of Technology, Pasadena, CA, 1985.

39. Hoffmann, M. R. and Calvert, J. G. "Chemical Transformation Modules for Eulerian Acid Deposition Models Vol. II: The Aqueous-Phase Chemistry," EPA/NCAR Report DW 930237, March 1985.

40. Grosjean, D. "Formaldehyde and Other Carbonyls in Los Angeles Ambient Air," Environ. Sci. & Tech., 1982, 16, 254-262.

41. National Research Council, Formaldehyde and Other Aldehydes, National Academy Press, Washington, DC, 1981.

42. Grosjean, D. and Wright, B. "Carbonyls in Urban Fog, Ice Fog, Cloudwater, and Rainwaater, Atmos. Environ. 1983, 17, 2093-2096.

43. Richards, L. W., Anderson, J. A., Blumenthal, D. L., McDonald, J. A., Kok, G. L., and Lazrus, A. L. "Hydrogen Peroxide and Sulfur(IV) in Los Angeles Cloudwater," Atmos. Environ. 1983, 17, 2093-2096.

44. Munger, J. W., Jacob, D. J., and Hoffmann, M. R. "The Occurrence of Bisulfite-aldehyde Addition Products in Fog- and Cloudwater," J. Atmos. Chem. 1984, 1, 335-350.

45. Munger, J. W., Tiller, C., and Hoffmann, M. R. "Identification of Hydroxymethanesulfonate in Fog Water," in press, Science 1986.

46. Boyce, S. D. and Hoffmann, M. R. "Kinetics and mechanism of the formation of hydroxymethanesulfonic acid at low pH," J. Phys. Chem., 1984, 88, 4740-4746.

47. Deister, U., Neeb, R., Helas, G., Warneck, P. "The Equilibrium $CH_2)OH)SO_3^- + H_2O$ in Aqueous Solution: Temperature Dependence and Importance in Cloud Chemistry," J. Phys. Chem., 1986, in press.

48. Kok, G. L., Gitlin, S. N., and Lazrus, A. L. "Kinetics of the Formation and Decomposition of Hydroxymethanesulfonate," J. Geophys Res. 1986, in press.

49. Kerp, W. Arbb Kaisel. Gesundh., 421, 1984, 180; Chem. Zentralblatt 1904, 75/II, 56-59.

50. Donally, L. H. Ind. Eng. Chem. Anal. Ed. 1933, 91-

51. Olson, T. M. and Hoffmann, M. R. "On the Kinetics of Formaldehyde-S(IV) Adduct Formation in Slightly Acidic Solution," Atmos. Environ. 1986, in review.

52. Olson, T. M., Boyce, S. D., and Hoffmann, M. R. "Kinetics, Thermodynamics, and Mechanism of the Formation of Benzaldehyde-S(IV) Adducts," J. Phys. Chem. 1986, in press.

53. Kunen, S. M., Lazrus, A. L., Kok, G. L., and Heikes, B. G. "Aqueous Oxidation of SO_2 by Hydrogen Peroxide," J. Geophys. Res. 1983, 88, 3671-3674.

54. Martin, L. R. and Damschen, D. E. "Aqueous Oxidation of Sulfur Dioxide by Hydrogen Peroxide at Low pH," Atmos. Environ. 1981, 15, 1615-1622.

55. Halperin, J. and Taube, H. "The Transfer of Oxygen Atoms in Oxidation-Reduction Reactions. IV. The Reaction of Hydrogen Peroxide with Sulfite, Thiosulfate and of Oxygen, Manganese Dioxide and Permanganate with Sulfite," J. Am. Chem. 1975, 380-382.

56. Hoffmann, M. R. and Edwards, J. O. "Kinetics and Mechanism of the Oxidation of Sulfur Dioxide by Hydrogen Peroxide in Acidic Solution," J. Phys. Chem. 1975, 79, 2096-2098.

57. McArdle, J. V. and Hoffmann, M. R. "Kinetics and Mechanism of the Oxidation of Aquated Sulfur Dioxide by Hydrogen Peroxide at Low pH," J. Phys. Chem. 1983, 87, 5425-5429.

58. Mader, P. M. "Kinetics of the Hydrogen Peroxide-sulfite Reaction in Alkaline Solution," J. Am. Chem. Soc. 1958, 80, 2634-2639.

59. Hoffmann, M. R. "On the Kinetics and Mechanism of Oxidation of Aquated Sulfur Dioxide by Ozone," Atmos. Environ. 1986, in press.

60. Larson, T. V., Horike, N. R., and Halstead, H. "Oxidation of Sulfur Dioxide by Oxygen and Ozone in Aqueous Solution: A Kinetic Study with Significance to Atmospheric Processes," Atmos. Environ. 1978, 12, 1597-1611.

61. Maahs, H. G. "Measurements of the Oxidation Rate of Sulfur(IV) by Ozone in Aqueous Solution and Their Relevance to SO2 Conver-

sion in Nonurban Tropospheric Clouds," Atmos. Environ. 1983, 17, 341-345.

62. Erickson, R. E., Yates, L. M., Clark, R. L., and McEwen "The Reaction of Sulfur Dioxide with Ozone in Water and Its Possible Atmospheric Significance," Atmos. Environ., 1977, 11, 813-817.

63. Hoigne, J., Bader, H., Haag, W. R., and Staehelin, J. "Rate Constants of Reactions with Ozone with organics and Inorganic Compounds in Water III. Inorganic Compounds and Radicals," Water Res. 1985, 19, 993-1004.

64. Tukey, Jr., H. B. "The Leaching of Substances from Plants," Ann. Rev. Plant Physiol. 1970, 71, 305-324.

65. Jacob, D. J., Wang, R-R. T., and Flagan, R. C. "Fogwater Collector Design and Characterization, Environ Sci. & Technol. 1985, 18, 827-833.

66. Hering, S. V. and Blumenthal, D. L. "Sampler Intercomparison Study," Final Report to Coordinating Research Council, Atlanta, GA, 1985.

67. Hering, S. V., Pettus, K., Gertler, A., Brewer, R. L., Hoffmann, M. R., and Kadlecek, J. A. "Field Intercomparison of Five Types of Fog Water Collectors," Environ. Sci. Technol. 1986, in review.

68. Jacob, D. J., Waldman, J. M., Haghi, M., Hoffmann, M. R., and Flagan, R. C. "Instrument to Collect Fogwater for Chemical Analysis," Rev. Sci. Instrum. 1985, 56, 1291-1293.

RECEIVED February 3, 1986

5

Urban Dew: Composition and Influence on Dry Deposition Rates

Patricia A. Mulawa, Steven H. Cadle, Frank Lipari, Carolina C. Ang, and René T. Vandervennet

Environmental Science Department, General Motors Research Laboratories, Warren, MI 48090-9055

The composition of dew collected from a Teflon surface was compared to summer rainwater concentrations at a site in Warren, Michigan. This comparison showed that natural dew is similar to rainwater with the exception that dew has much higher concentrations of Ca^{+2} and Cl^- and much lower acidity. Dry deposition rates of several species were measured to artificially-generated dew and a dry surface. It was found that deposition rates were 2 to 20 times greater to the artificial dew than to the dry surface indicating that the presence of dew enhances both the retention of dry deposited particles and the absorption of water soluble gases. Measurement of the atmospheric concentrations of the depositing species permitted the calculation of deposition velocities for particulate Cl^-, NO_3^-, SO_4^{-2}, Ca^{+2}, Mg^{+2}, Na^+, and NH_4^+. Deposition velocities for gaseous HNO_3, HCl, SO_2 and NH_3 were also determined after correction for particle deposition. These results indicate that acid dew is not a problem at this site. However, the ability of dew to increase the deposition rate of acids and acid precursors to some surfaces suggests that dew may be more acidic at sites with lower deposition rates of basic particles.

Recently, interest in acid deposition has broadened to include special acidic events such as dew, frost, and fog. Little is known about the frequency with which acidic dew occurs, its composition, or its effect on dry deposition rates. However researchers have long recognized that surface wetness contributes to the corrosion of metal surfaces (1) and to the deterioration of stonework (2).

0097-6156/86/0318-0092$06.00/0

Additionally, acid dew may also be involved in plant effects since it has been reported that acid rain can damage protective surfaces on leaves, interfere with guard cells, and poison plant cells (3).

A few studies on the composition of dew have been reported. Yaalon and Ganor (4), Brimblecombe and Todd (5), Anderson and Landsberg (6) and Smith and Friedman (7) collected dew from a variety of surfaces and report median pH values in the range of 5.7 to 7.7. Wisniewski (8) reviewed the sparse acid dew literature and calculated that dew could have a pH as low as 2 based solely on the oxidation of all deposited SO_2 to H_2SO_4 and no subsequent neutralization. Recently Pierson et al. (9) found the pH of dew samples from Alleghany Mountain in Pennsylvania ranged from 3.5 to 5.3 with a volume weighted average of 4.0.

Cadle and Groblicki (10) determined the composition of dew deposited naturally on glass, Teflon, and plastic surfaces in Warren, MI. Dew composition was compared to wet and dry deposition obtained the previous year at the same site. In this paper, the comparison of dew and rain composition is updated and the results of a new study of the composition of artificially-generated dew are reported. Deposition velocities to the dew of SO_2, HNO_3, HCl, NH_3, Ca^{+2}, Mg^{+2}, Na^+, and K^+ are also presented.

Experimental

Site. Samples were collected from a site located on a 330 acre parcel of undeveloped land in Warren, MI, a suburb north of Detroit. Most of the surrounding area is highly developed. A major surface street 300 m south of the site has a traffic flow of 20 000 vehicles/day. Another street, 800 m east of the site has a traffic flow of 38 000 vehicles/day. Annual emissions of NO_x, SO_2, and TSP for the surrounding area have been presented elsewhere (11).

Dew Collection. In our previous work, natural dew was collected from a Teflon surface. The Teflon collector consisted of a sheet of aluminum backed FEP Teflon bonded to a 1 m^2 copper plate mounted on a plywood base. The collector was tilted 30° from horizontal with the centerpoint 1 m above the ground.

In this work the intent was to perform a more comprehensive analysis of the dew and to compare deposition velocities to a wet and a dry surface. In order to bring more control to the experiment, dew was generated artificially by attaching cooling coils to the Teflon covered copper plate. The dry plate counterpart consisted of a 1 m^2 glass collector covered with Teflon and mounted in the same manner as the copper plate. All deposition rates are based on the actual area of the plate rather than the projected horizontal area, which was 0.87 m^2.

The natural dew collection procedure used previously consisted of washing the collector with deionized water late in the afternoon. Dew was collected the following morning at approximately 7:00 a.m.

The average elapsed time between cleaning and dew collection was 17 h. Since dew was not present the entire time, the dew concentrations reported below include some material which was deposited on the dry collection plate and subsequently dissolved in the dew. The samples were filtered through 0.2 μm Gelman Acrodisc filters prior to determining pH. The samples were then refrigerated until the remaining analyses could be performed.

Generation of artificial dew started at approximately 7:30 a.m. and lasted for periods of 1.5 to 4.5 h. The average generation time was 3.1 h. Artificial dew generation was done only during periods with wind speeds less than 2 mph. Higher wind speeds limited our ability to generate dew. Also, this procedure minimized turbulent mixing and thus approached more realistic nighttime deposition rates. The collector was washed with deionized water immediately before use. Samples were processed in the same manner as the natural dew samples. The dry collection plate was likewise cleaned with deionized water immediately prior to the onset of artificial dew formation. Thus, the two collection plates were exposed for essentially the same time periods. At the end of an experiment the artificially-generated dew was collected and the dry plate was misted twice with deionized water. This wash water was collected in the same manner as dew samples. Contact time between the water and the plate was approximately 5 minutes.

Atmospheric Concentrations. Atmospheric concentrations of the major depositing species were determined during the artificial dew formation period. The species measured were NO, NO_2, O_3, HNO_3, SO_2, HCl, NH_3, and particulate NO_3^-, SO_4^{-2}, Cl^-, NH_4^+, Ca^{+2}, Mg^{+2}, Na^+, and K^+. NO, NO_2, and O_3 were monitored continuously with a Monitor Labs dual channel chemiluminescence NO_x analyzer and an AID portable ozone analyzer. SO_2 was determined with the carbonate-glycerol impregnated filter technique. HNO_3 was determined by the denuder difference method (12). HCl was determined with the Na_2CO_3 impregnated filter technique (13). The oxalic acid impregnated filter method (14) was used to determine NH_3. Total particulate was collected on 1-μm pore size Ghia Tefweb filters. Aqueous extracts of the filters were analyzed to determine particulate NO_3^-, SO_4^{-2}, Cl^-, NH_4^+, Ca^{+2}, Mg^{+2}, Na^+, and K^+ concentrations.

Analysis. SO_3^{-2}, SO_4^{-2}, NO_3^-, and Cl^- were measured on a Dionex 2110i ion chromatograph. SO_3^{-2} responses were attributed to the collective presence of S(IV) species, without any attempt to identify the specific form, and hereafter will be referred to as S(IV). Ca^{+2}, Mg^{+2}, Na^+, and K^+ were determined on a Perkin-Elmer atomic absorption spectrometer. H^+ concentrations were calculated from pH measurements obtained using an Orion combination electrode and pH meter. NH_4^+ was determined using a modified indophenol blue method on either a Technicon Autoanalyzer or a Lachat flow injection analyzer.

Results and Discussion

Natural dew samples were collected between June 14, 1982 and October 14, 1982. Dew and rain event frequencies were recorded for 74 days. Dew occurred on 61% of the days and rain occurred on 15% of the days. Artificial dew samples were generated between June 21, 1984 and August 17, 1984. Dew and rain event frequencies were not recorded. However, this period was typical of a normal summer for the area.

Dew, Rainwater and Dry Plate Wash Composition. Table I contains the average concentrations of measured ions in natural dew, artificially-generated dew and dry plate washes. Also included are the volume weighted average summer rainwater concentrations of these same species for the period June 1981 through July 1983 (15).

Table I. Average Dew, Rainwater and Dry Plate Concentrations

Species	Average Natural Dew Concentration µeq/L	Average Rainwater Concentration µeq/L	Average Artificial Dew Concentration µeq/L	Average Dry Plate Conc. µeq/L
Ca^{+2}	690 ± 935	24	155 ± 121	62 ± 43
Mg^{+2}	31 ± 11	6.7	26 ± 16	6.9 ± 4.4
K^{+}	4.1 ± 3	1.5	19 ± 29	9.5 ± 6.5
Na^{+}	20 ± 6	5.1	37 ± 75	18 ± 10
NH_4^{+}	65 ± 26	31	140 ± 108	6.9 ± 6.6
Cl^{-}	106 ± 252	6.7	52 ± 39	24 ± 17
NO_3^{-}	166 ± 282	40	52 ± 38	17 ± 13
S(IV)	ND*	ND	50 ± 57	8.2 ± 5
SO_4^{-2}	242 ± 312	96	112 ± 69	16 ± 12

* not determined

The pH of natural dew samples ranged from 3.62 to 8.20 with a median of 6.5 (0.3 µeq H^{+}/L). The average natural dew volume was 82.4 mL. The average pH of the rainwater samples was 4.14 (73 µeq H^{+}/L). Natural dew concentrations of NH_4^{+}, Mg^{+2}, Na^{+}, K^{+}, NO_3^{-}, and SO_4^{-2} were 2 to 4.6 times higher than their concentration in rainwater, while Ca^{+2} and Cl^{-} were 29 and 16 times more concentrated, respectively, in the dew. Ca^{+2} concentrations are elevated because Ca^{+2} is present in predominately large particles at this site (16) and large amounts of dry deposited Ca^{+2} are incorporated into the dew. Some large particle chloride is also present at this site during the summer which may account for the higher dew chloride concentrations. S(IV) and NO_2^{-} were frequently present in natural dew samples. Semi-quantitative analysis by ion chromatography indicated that S(IV) species can be a major component in the dew at this site, but that NO_2^- was a minor component.

Artificial dew and dry plate washes were collected on 24 mornings. The average dew volume was 82 mL and the average dry plate wash volume was 76 mL. The pH of artificial dew ranged from 4.42 to 8.16 with a median of 5.30 (5.0 μeq H^+/L). The pH of dry plate washes ranged from 5.28 to 9.04 with a median of 6.3 (0.5 μeq H^+/L). Thus the artificial dew samples were more acidic than either the natural dew or the dry plate washes. Inspection of Table I shows that the higher acidity of the artificial dew samples can be explained by the relatively lower Ca^{+2} concentrations in the artificial dew. The average ratios of equivalents of Ca^{+2} to the equivalents of SO_4^{-2} in artificial dew, natural dew, and dry plate samples were 1.59, 3.55, and 4.66, respectively.

Average artificial dew concentrations were 2 to 20 times higher than average dry plate concentrations for the measured ions. The largest differences appear for NH_4^+, SO_4^{-2}, and S(IV). There are two explanations for these differences. The most likely is that the presence of moisture on the surface of the collector increases the retention of dry deposited particles and enhances the capture of gases. This possibility will be explored in greater detail below. The second possibility is that the dry deposited material on the dry plate was not completely dissolved by the wash procedure which consisted of misting twice with a spray of fine droplets and physically sweeping the water around the plate during collection. We do not believe that incomplete dissolution was a problem since collection of second washes were very clean.

Deposition Rates. Deposition rates which were calculated from the sample concentrations, volumes, and the sample times are tabulated in Table II. Deposition rates to the artificial dew samples were 1.4 to 7.2 times greater than the rates to the natural dew samples for all species except K^+ and Na^+. The K^+ and Na^+ were deposited at rates 19 and 16 times greater to the artificial dew than to the natural dew, respectively. There is considerable uncertainty in these values because there were relatively few measurements of these species made in the natural dew samples and because the values were frequently close to the blank concentrations. Three factors contribute to the higher deposition rates to the artificial dew samples. First, the artificial dew samples were wet the entire sampling time whereas the natural dew samples were wet only part of the average 17 hour collection period. For many species a wet surface is expected to be a better collector than the dry surface. For example, Dasch (17) found that all these species had higher deposition rates to water than to a Teflon filter. Therefore, the deposition rates to natural dew reported in Table II should be taken as lower limits. Second, meteorological conditions influence the deposition rates. Greater atmospheric turbulence in the morning compared to the night would

Table II. Average Deposition Rates

Species	Number of Samples	Nat. Dew Deposition Rate μeq/m²/h	Number of Samples	Art. Dew Deposition Rate μeq/m²/h	Number of Samples	Dry Plate Deposition Rate μeq/m²/h
Ca^{+2}	31	2.06 ± 1.09	24	2.87 ± 1.65	24	1.64 ± 1.34
Mg^{+2}	9	0.14 ± 0.12	24	0.54 ± 0.31	24	0.17 ± 0.13
K^+	9	0.02 ± 0.02	24	0.32 ± 0.26	24	0.25 ± 0.18
Na^+	9	0.04 ± 0.05	24	0.62 ± 0.66	24	0.47 ± 0.34
NH_4^+	14	0.34 ± 0.21	24	2.45 ± 1.31	24	0.20 ± 0.24
Cl^-	39	0.23 ± 0.23	24	0.86 ± 0.44	24	0.69 ± 0.62
NO_3^-	39	0.45 ± 0.33	24	1.06 ± 0.64	24	0.46 ± 0.50
S(IV)	0	ND*	23	0.92 ± 0.83	5	0.19 ± 0.13
SO_4^{-2}	39	0.82 ± 0.88	24	2.26 ± 1.28	24	0.45 ± 0.47

* not determined

increase the deposition rates to the artificial dew samples. Third, diurnal variations in atmospheric concentrations have not been taken into account. For example, since the artificial dew samples were generated during the morning rush hour they were exposed to higher average NO_x levels than the natural dew samples.

The greatest differences between the deposition rates to the artificial dew and to the dry plate were for NH_4^+, S(IV), and SO_4^{-2}. The ratios of the artificial dew and the dry plate rates for these species were 12.3, 4.8, and 5.0, respectively. This difference in rates is most likely due to the enhanced retention of water soluble gases. Therefore, these results suggest that dry deposited NH_3 and SO_2 are important contributors to the dew composition. As noted above, soluble oxidant gases such as O_3 and H_2O_2 may also play an important role in determining the concentration of SO_4^{-2}.

Deposition Velocities. Deposition velocities, v_d's, were calculated from the artificial dew deposition rates discussed above and the atmospheric concentrations of the depositing species measured during dew generation. The average atmospheric concentrations and their ranges appear in Table III. Average v_d's are shown in Table IV. The calculation of V_d's for Ca^{+2}, Mg^{+2}, Na^+, and K^+ is straightforward since there is only one dry deposition source for these species-atmospheric particles. The average v_d's to the Teflon plate were 0.69, 0.33, 0.23 and 0.15 cm/s for K^+, Na^+, Ca^{+2}, and Mg^{+2}, respectively. V_d's to the dew were higher with average values of 0.88, 0.42, 0.46 and 0.41 cm/s for K^+, Na^+, Ca^{+2} and Mg^{+2}, respectively. Since particle size distributions were not measured during this study, it is not possible to determine if the differences in v_d's are due to the differences in the particle sizes.

The v_d calculations for HNO_3, SO_2, HCl and NH_3 become more complex due to the likelihood that more than one species contributed to the observed artificial dew deposition rates. If we assume that

Table III. Average Ambient Concentrations

Species	No. of Samples	Average Concentration $\mu g/m^3$	Range
SO_4^{-2}	24	6.51 ± 6.74	0.17 - 28.5
NO_3^-	24	2.97 ± 3.24	0.09 - 14.4
Cl^-	24	0.28 ± 0.25	0.01 - 0.74
NH_4^+	24	4.09 ± 3.54	0.06 - 15.5
Ca^{+2}	24	8.02 ± 9.70	1.13 - 37.9
Mg^{+2}	24	0.50 ± 0.29	0.07 - 1.63
K^+	20	0.82 ± 1.18	0.16 - 5.97
Na^+	20	1.59 ± 1.09	0.33 - 4.15
HCl	24	0.57 ± 0.36	0.07 - 1.37
HNO_3	20	2.55 ± 2.36	0.0 - 7.08
NH_3	23	0.40 ± 0.47	0.02 - 2.22
NO_2*	20	11.1 ± 4.08	4.8 - 22.9
NO*	24	13.4 ± 12.1	1.2 - 43.7
O_3*	21	28.4 ± 18.5	0.0 - 70
SO_2*	23	9.19 ± 9.12	0.0 - 32.5

* concentration in ppb

adsorption of these gases onto Teflon is minimal then the observed dry plate deposition of NO_3^-, SO_4^{-2}, Cl^- and NH_4^+ can be attributed to particulate deposition. Using this approach the average v_d's for particulate NO_3^-, SO_4^{-2}, Cl^- and NH_4^+ were estimated to be 0.33, 0.10, 2.36 and 0.06 cm/s, respectively. Since most of the SO_4^{-2} and NH_4^+ has been shown to be in the submicron size range at this site (16,18), their v_d's appear to be reasonable. Similarly, the higher v_d's for NO_3^- and Cl^- are consistent with the fact that they are present as larger particles. These v_d's are then used to correct the observed artificial dew HNO_3, SO_2, HCl and NH_3 deposition rates for the corresponding particulate dry deposition contribution. The corrected v_d's for HNO_3 and SO_2 averaged 0.39 and 0.15 cm/s, respectively, reflecting a 47% and a 17% decrease in the v_d after correction for particulate NO_3^- and SO_4^{-2} input. Likewise, NH_3 v_d and HCl v_d decreased 36 and 54% after correction for particulate NH_4^+ and Cl^- to 1.90 and 0.73 cm/s, respectively.

Conclusions

Concentrations of all measured species were higher in natural dew than in rain. The biggest differences were much higher Ca^{+2} and Cl^-

Table IV. Estimated Deposition Velocities to Artificial Dew

Species	Surface	V_d, cm/s
HNO_3	dew	0.39 ± 0.31
SO_2	dew	0.15 ± 0.11
HCl	dew	0.73 ± 0.11
NH_3	dew	1.9 ± 1.55
NO_3^-	Teflon	0.33 ± 0.22
SO_4^{-2}	Teflon	0.10 ± 0.09
NH_4^+	Teflon	0.06 ± 0.09
Cl^-	Teflon	2.36 ± 1.77
Ca^{+2}	Teflon	0.23 ± 0.18
Ca^{+2}	dew	0.46 ± 0.36
Mg^{+2}	Teflon	0.15 ± 0.12
Mg^{+2}	dew	0.41 ± 0.20
Na^+	Teflon	0.33 ± 0.34
Na^+	dew	0.42 ± 0.67
K^+	Teflon	0.69 ± 0.72
K^+	dew	0.88 ± 0.89

concentrations in the dew than the rain and much lower H^+ concentrations. Thus, it was concluded that the acidity of dew at this site is controlled more by the deposition of large basic particles than by the deposition of acids and acid precursors. This suggests a potential for higher dew acidity in areas where Ca^{+2} deposition is lower or under conditions which minimize the deposition of large particles. This effect was demonstrated previously when it was shown that the acidity of dew was much higher on downward-facing surfaces than upward-facing surfaces (10). It is also seen in this study with the artificial dew samples which had much lower Ca^{+2} concentrations than the natural dew samples and thus more H^+.

Deposition rates of all species except K^+ and Na^+ were 1.4 to 7.2 times greater to the artificial dew than to the natural dew. Apparently the moisture increased the retention of particles and permitted the deposition of water soluble gases. This latter effect was particularly important for the deposition of NO_3^-, NH_4^+, S(IV) and SO_4^{-2}.

The deposition velocities to the artificial dew and the dry plates appear reasonable under the experimental conditions of this study. The particulate SO_4^{-2} v_d of 0.10 cm/s is somewhat lower than the weekly summertime average SO_4^{-2} v_d of 0.29 cm/s to a dry deposition bucket reported by Dasch and Cadle (15) for this site. This difference is reasonable since their estimate did not account for any SO_2 deposition which may have occurred when the bucket was damp. Also, average atmospheric stability conditions may have been different. The particulate NH_4^+ v_d to the Teflon plate, 0.06 cm/s, was in good agreement with the SO_4^{-2} v_d. Both these species are present as small particles at this site. The particulate NO_3^-, Cl^-, and K^+ v_d's were in good agreement with those reported by Dasch and Cadle (15), while the Ca^{+2} and Mg^{+2} v_d's to the dry plate were lower by factors of 9 and 7, respectively. The v_d's of gaseous species to the artificial dew were calculated after correcting the total deposition rates for the deposition of particles. Particulate deposition rates were assumed to be close to those observed to the dry plate. The estimated SO_2 v_d was 0.15 cm/s as compared to the ~0.04 cm/s recently reported by Pierson et al. (9). The difference between these results reflects the less turbulent nighttime conditions during their study and the decreased SO_2 solubility due to the low pH of their samples. A higher estimate for SO_2 v_d of 0.69 cm/s to a deionized water surface has been reported by Dasch and Cadle (19) for spring days at this site. The HNO_3 v_d, 0.39 cm/s, was also higher than the 0.24 cm/s reported by Pierson et al. (9). The HCl v_d, 0.73 cm/s, is in reasonable agreement with the HNO_3 v_d. However, the NH_3 v_d, 1.9 cm/s, appears to be inconsistent with our other results.

Overall, it is concluded that acid dew is not a significant environmental concern at this site. However, the ability of dew to increase the deposition rate of water soluble gases to some surfaces, and thus increase the acidity of the dew, may be important at other locations. The deposition velocities reported above can be used to estimate dew concentrations at other sites as long as differences in particle sizes and atmospheric conditions are taken into consideration.

Acknowledgments

The authors wish to acknowledge Kenneth Kennedy of the Environmental Science Department for the analytical assistance he provided and Sudarshan Kumar for his helpful discussions.

Literature Cited

1. Anderson, E. A. In "Atmospheric Corrosion of Non-ferrous Metals"; ASTM STP 175, American Society for Testing and Materials, Philadelphia, PA, 1955.
2. Fassina, V. Atmos. Environ. 1978, 12, 2205-2211.
3. Bangay, G. E.; Riordan C. United States-Canada Memorandum of Intent on Transboundary Air Pollution, Impact Assessment. 4-39, 1983.
4. Yaalon, D. H.; Ganor, E. Nature 1968, 217, 1139-1140.
5. Brimblecombe, P.; Todd, I. J. Atmos. Environ. 1977, 11, 649-650.

6. Anderson, E. A.; Landsberg, H. E. Environ. Sci. Technol. 1979, 13, 992-994.
7. Smith, B. E.; Friedman, E. J. "The Chemistry of Dew as Influenced by Dry Deposition: Results of Sterling, Virginia and Champaign, Illinois Experiments," Mitre Corporation Working Paper WP82W00141, 1982.
8. Wisniewski, J. Water, Air, Soil Pollution 1982, 17, 361-377.
9. Pierson, W. R.; Brachaczek, W. W.; Gorse, R. A., Jr.; Japar, S. M.; Norbeck, J. M. "On the Acidity of Dew," Presented at the 78th Annual Meeting of the Air Pollution Control Association, Detroit, MI, 1985, Paper No. 85-7.4.
10. Cadle, S. H.; Groblicki, P. J. In "Transactions of the APCA Specialty Conference, The Meteorology of Acid Deposition"; Samson, P. J., Ed., 1983, pp. 17-29.
11. Dasch, J. M.; Cadle, S. H. Atmos. Environ. 1984, 18, 1009-1015.
12. Shaw, R. W.; Bowermaster, J.; Tesch, J. W.; Tew, E. Atmos. Environ. 1982, 16, 845-853.
13. Okita, T.; Kaneda, K.; Yanaka, T.; Sugai, R. Atmos. Environ. 1974, 8, 927-936.
14. Shendrikar, A. D.; Lodge, J. P., Jr. Atmos. Environ. 1975, 9, 431-435.
15. Dasch, J. M.; Cadle, S. H. Atmos. Environ., in press.
16. Dasch, J. M. "Measurement of Dry Deposition to a Deciduous Canopy," General Motors Research Publication GMR-5019, 1985.
17. Dasch, J. M. In "Precipitation Scavenging, Dry Deposition, and Resuspension; Pruppacher, H. R.; Semonin, R. G.; Slinn, W. G. N., Eds.; Elsevier Publishing: New York, Amsterdam, Oxford, 1983, Vol. 2, 883-902.
18. Cadle, S. H. Atmos. Environ. 1985, 19, 181-188.
19. Dasch, J. M.; Cadle, S. H. "Dry Deposition to Snow in an Urban Area," Presented at the 78th Annual Meeting of the Air Pollution Control Association, Detroit, MI, 1985, Paper No. 85-6B.3.

RECEIVED January 13, 1986

METALLIC CORROSION

6

Influence of Acid Deposition on Atmospheric Corrosion of Metals: A Review

Vladimir Kucera

Swedish Corrosion Institute, Box 5607, S-114 86 Stockholm, Sweden

In outdoor atmospheres dry deposition of S-pollutants and especially of SO_2 is of greatest importance. Dose response functions describing corrosion as function of SO_2 and time of wetness are available for some materials as steel and zinc. NO_x in combination with SO_2 has a synergistic corrosion effect especially indoors on electrical contact materials, copper and steel. The influence of acid precipitation may differ for different metals and depends also on the pollution level. The atmospheric corrosion of metals due to acid deposition is in most regions mainly a local problem restricted to areas close to the pollution source. A reduction of losses due to corrosion has been achieved in many Swedish cities by reduction of SO_2 emissions and by introduction of district heating systems.

The increased combustion of fossil fuels has created the problem of acidification of the environment, which today is considered to be one of the most serious environmental problems. In Scandinavia, where scientists first gave warning of the problem, the long distance transport of pollutants from Europe and unfavourable geological conditions in big areas make the situation especially serious. Among the effects caused by acidification the corrosion of materials belongs to the first observed and best documented. It has thus been known for a long time that S-pollutants originating from burning of fossil fuels accelerates in the first place the atmospheric corrosion of certain metals. Their influence was described e.g. in Sweden's case study for the U.N. conference on human environment in 1972, which was the first effort to describe the general picture of the acidification problem (1). More recently also the influence of N-pollutants on the atmospheric corrosion has attracted attention. This may be seen in the light of the fact that during the 1970s the wet deposition of sulphate in Sweden remained by and large at a constant level. The wet

0097-6156/86/0318-0104$06.00/0

deposition of nitrate by contrast increased at a rate of 3-4 per cent a year (2).

The corrosion effect on constructions in the atmosphere due to acidification is in most areas mainly of local nature. It should be stressed, however, that acidification of soil and water can lead to increased corrosion of buried installations and of installations in water including water pipes (3, 4). This is on the other hand mainly a regional problem, where the long-range transport of air pollutants plays an important role. The following subdivision may thus be used.

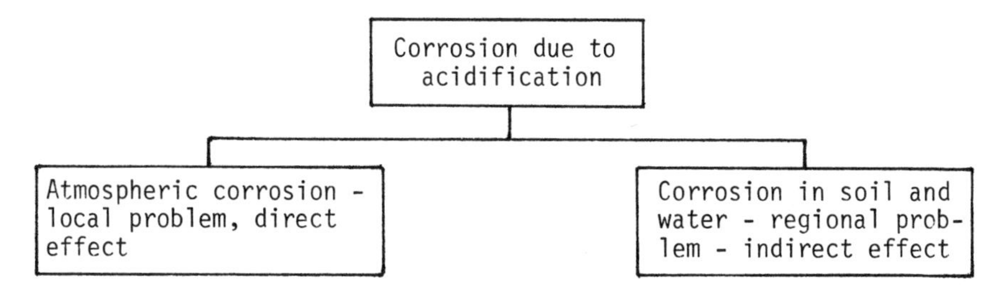

In the present lecture a brief review will be given of the influence of acidifying air pollutants on the atmospheric corrosion of metals based mainly on more recent results from Europe and especially from Scandinavia.

The deposition process

Both sulphur and nitrogen pollutants are emitted into the air in the form of oxides, SO_2 and NO respectively, which may then further be oxidized during the transport in the air. The processes of deposition of air pollutants are:

- absorption of gas on material surfaces } - dry deposition
- impaction of particles }
- removal of gas and aerosols by precipitation - wet deposition

The relative significance of the individual deposition mechanisms for corrosion of materials may vary in different areas depending e.g. on the distance from the emission source, and also for different materials depending on differences in corrosion mechanisms and nature of protective layers of corrosion products.

The influence of sulphur pollutants

The influence of mainly SO_2 on the corrosion rate of several materials has been shown in numerous national exposure programs. During the last decades a number of empirical relations have been derived from measurements of atmospheric corrosion rates of the most important structural metals and from measurements of environmental factors. The results are usually presented in form of equations including pollution and meteorological parameters (5).

Carbon steel

Several investigations have shown the dominating effect of SO_2 on the corrosion rate, recent results from an inter-Nordic and from a Swedish-Czechoslovak exposure may serve as examples (6, 7). The following dose-response functions have been obtained (FIG. 1):

1 year: $K_{Fe} = 4.0\ SO_2 + 58$ $r = 0.98$

4 years: $K_{Fe} = 8.0\ SO_2 + 160$ $r = 0.94$

where K = weight loss for steel (g/m^2), SO_2 = deposition rate ($mg/m^2.d$) and r = correlation coefficient. These investigations show that in a broad range of the temperate climatic zone and in SO_2 pollution levels representative for Western and Central Europe as well as for Southern Scandinavia most of the variations in the corrosion rate may be explained solely by the SO_2 deposition rate. In this region the average wetness conditions do not differ to a greater extent. This explains the fact that time of wetness is not included in the dose response relationship.

It should be pointed out, however, that during shorter time periods, the variations of wetness conditions are very extensive. Predictions of one or three months steel corrosion needs apart from the SO_2 pollution data also at least values of the time of wetness as illustrated by the following equation (10):

1 month: $K_{Fe} = 0.51\ SO_2 + 0.11\ TOW - 9$ $r = 0.82$

where K = weight loss for steel (g/m^2), SO_2 = deposition rate ($mg/m^2.d$), TOW = time of wetness (hours with RH >80% and T >^{0}C), r = correlation coefficient.

The results from lengthy exposure periods are of course more relevant for practical purposes e.g. for classification of the corrosivity of atmosphere on a given location or for cost-benefit analysis. In this case the corrosion rate may be assessed from yearly mean values of the concentration of pollutants and from the time-of-wetness class estimated from meteorological measurements.

Zinc

Zinc belongs to the materials that exerts a strong dependence of the corrosion rate on the concentration of sulphur pollutants. In several investigations the corrosion rate in urban atmospheres was found to be 2-6 times higher than in rural atmospheres (5, 8). Also for zinc dose-response functions obtained in the temperate climatic zone show the dominating influence of SO_2 on inland sites. The inclusion of time of wetness in the linear model did not further significantly improve the correlation. The following dose-response functions have been obtained in the previously mentioned investigation (FIG. 2):

1 year: $K_{Zn} = 0.13\ SO_2 + 7.21$ $r = 0.81$

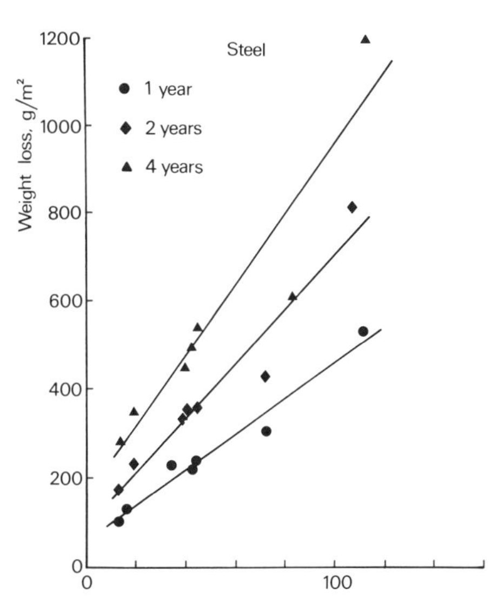

Figure 1. The weight loss of steel after 1, 2, and 4 years' exposure as a function of SO_2 deposition rate at 7 test sites in Sweden and Czechoslovakia (7).

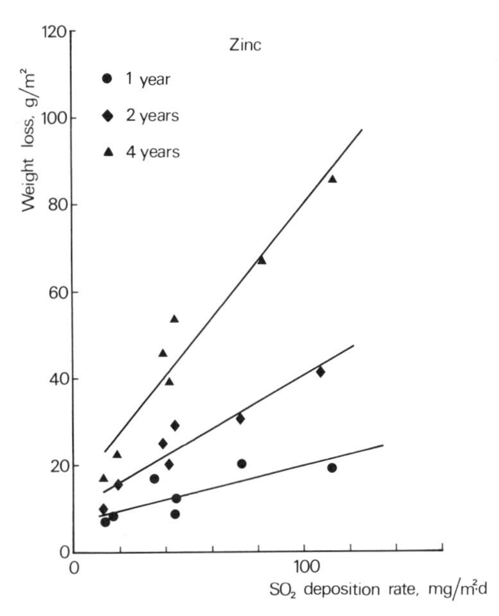

Figure 2. The weight loss of zinc after 1, 2, and 4 years' exposure as a function of SO_2 deposition rate at 7 test sites in Sweden and Czechoslovakia (7).

4 years: $K_{Zn} = 0.66\ SO_2 + 14.20$ $r = 0.97$

The corrosion rate of zinc is strongly dependent on the wetness conditions during the early days or weeks of exposure (9). Therefore often no useful equations for corrosion can be found in terms of monthly averages of environmental factors (10). The corrosion rate of zinc may be elevated during the first year of exposure especially at rural sites with large amount of precipitation. This so called memory effect may exert its influence during the first years of exposure. After some years of exposure the corrosion rate has diminished and in the long run it seems to reach a value corresponding to the pollution level (6).

Copper

It is a well known fact that formation of the green patina takes a substantially shorter time in urban than in rural atmosphere, where often very long time elapses before the surface is covered by patina or in very pure atmospheres the surface remains covered by a black oxide layer. Also the corrosion rate in rural atmosphere is usually lower ($<1\ \mu m$/year) than in urban or industrial atmospheres (1-3 μm/year) (8).

As may be seen from FIG. 3 the correlation of the corrosion and the SO_2 deposition rate is not as good as for steel and zinc. The following dose-response functions have been obtained:

1 year $K_{Cu} = 0.22\ SO_2 + 3.55$ $r = 0.75$

4 years: $K_{Cu} = 0.69\ SO_2 + 5.98$ $r = 0.85$

where K_{Cu} = weight loss for copper (g/m^2), SO_2 = deposition rate ($mg/m^2.d$) and r = correlation coefficient.

Aluminum

Aluminum is a material possessing a very good corrosion resistance in S-polluted atmospheres. Nevertheless also for aluminum a higher corrosion is obtained at more polluted sites, FIG. 4. A linear regression analysis gave the following equations:

1 year: $K_{Al} = 0.008\ SO_2 + 0.022$ $r = 0.63$

4 years: $K_{Al} = 0.136\ SO_2 + 0.013$ $r = 0.70$

where K_{Al} = weight loss for aluminum (g/m^2), SO_2 = deposition rate ($mg/m^2.d$) and r = correlation coefficient.

Even if there is a significant correlation between corrosion rates of copper and aluminum and SO_2 pollution there are other factors than SO_2 that influence significantly the corrosion rate of the two metals also at exposure within the temperate climate zone.

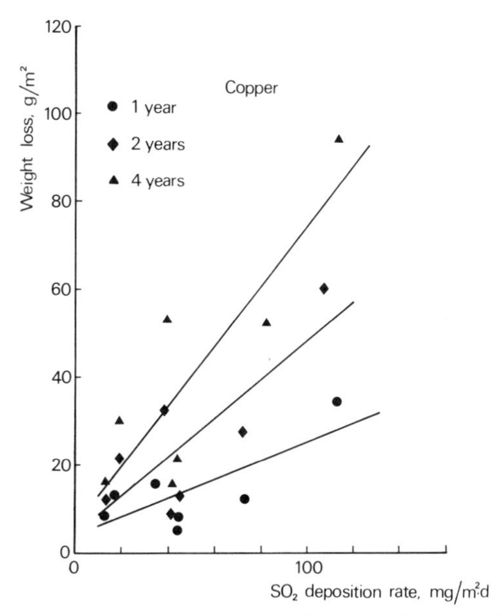

Figure 3. The weight loss of copper after 1, 2, and 4 years' exposure as a function of SO_2 deposition rate at 7 test sites in Sweden and Czechoslovakia (7).

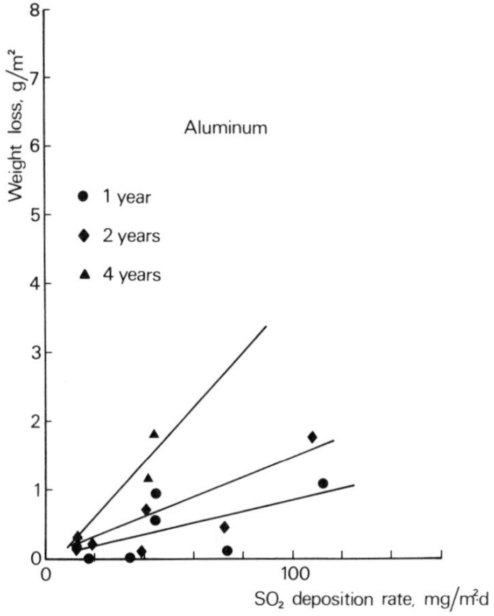

Figure 4. The weight loss of aluminum after 1, 2, and 4 years' exposure as a function of SO_2 deposition rate at 7 test sites in Sweden and Czechoslovakia (7).

Nickel

Also nickel, often used in decorative Ni-Cr coatings, is a metal sensitive to the influence of sulphur compounds in the atmosphere. This may be illustrated by results from a field exposure giving the corrosion rate of 0.4 μm/year in rural and 2.7 μm/year in urban atmospheres (11). In other investigations as high corrosion rates as 6 μm/year has been found in industrial atmospheres (12). Also the life of decorative Ni-Cr coatings is substantially shorter in urban/industrial than in rural areas (13).

The influence of nitrogen pollutants

On combustion most of the nitrogen oxides are emitted as NO. In the atmosphere oxidation takes place successively to NO_2, which usually may be considered the main nitrogen pollutant near the emission source. NO_2 is then further oxidized to HNO_3. This reaction has a very low rate especially at low contents of NO_x (<0.1 ppm). Therefore in the vicinity of the emission source the contents of HNO_3 and nitrates are very low (14, 15).

The influence of NO_x on atmospheric corrosion, especially under outdoor conditions, has so far been investigated only on a limited scale. A number of studies is, however, at present ongoing or planned.

Steel

A systematic chamber study (16) of weathering steel and zinc has revealed that 0.05 and 0.5 ppm NO_2 had no significant effect on the corrosion rate, whereas SO_2 in the same concentrations has a very strong effect. A recent Swedish laboratory study (17) has, however, shown that whereas 3.0 ppm NO_2 or 1.3 ppm SO_2 at 50% RH alone caused only insignificant corrosion on carbon steel, a combination of both gases increased the corrosion rate by a factor 30, FIG. 5. This effect was not obtained at 90% RH. The synergistic effect has in this investigation been explained by formation of hygroscopic corrosion products containing nitrates. This influence of a mixture of NO_2 and SO_2 may proof to be of practical importance e.g. at storage indoors in polluted atmospheres.

Outdoors the influence of NO_x has been studied in a Japanese exposure program at 4 test sites in the Niigata district (18). During a 4 years' exposure no significant effect of NO_x was found, whereas both SO_2 and chlorides showed a significant correlation with the corrosion rate. It may also be stressed that the amount of NO_x deposited on the surface was 10 to 100 times lower than the amount of SO_2 which indicates a low deposition rate of NO_x.

Zinc

Except the above mentioned chamber study (16) which showed no direct or synergistic effect of NO_2 a laboratory study has indicated some increase of corrosion rate when NO_2 is added to a SO_2

containing atmosphere (19). For outdoor conditions so far no results seems to be reported.

Copper

A recent laboratory study (19) has shown that a combination of NO_2 and SO_2 at 90% RH causes rapid corrosion of copper compared to the influence of the two gases when present alone, see FIG. 6. These results indicate that the synergistic effect of NO_2 and SO_2 may be of importance for atmospheric corrosion of copper outdoors.

Aluminium

For aluminium the experimental evidence in this field is very limited. A laboratory and a field study have, however, shown that the degree of hydration of the aluminium oxide layer shows the best correlation with the NO_x content (20). It seems, however, at present not be possible to judge whether nitrogen oxides are of any practical importance for corrosion of aluminium.

Contact materials for electronics

Air pollutants affect not only corrosion of constructions outdoors but they may have a harmful effect even indoors. To the most important problems in this context belongs the corrosion on electrical and electronic equipment. This type of equipment is very sensitive and even restricted corrosion attack may cause interruption in service. The indoor climate contains in principle the same gaseous pollutants as outside, though usually in lower concentrations. During the last years the research efforts in this field have been intensified. In field tests both in the U.S.A. (21) and in Norway (22) relatively high levels of NO_2 has been found in the indoor atmosphere and nitrogen compounds have been detected in corrosion products. Elevated amounts of nitrates have been found on electric equipment working indoors and have caused stress corrosion cracking of wire springs of Ni brass (23).

The influence of NO_x on electric contact materials seems often to be due to synergistic effects with other pollutants. Whereas exposure of e.g. copper and gold contacts in an atmosphere of NO_2 shows no corrosion effect the synergistic effect of SO_2 and NO_2 is very pronounced (24, 25). A possible explanation is that NO_2 oxidizes SO_2 to sulfuric acid according to the following reaction:

$$SO_2 + NO_2 + H_2O \quad H_2SO_4(l) + NO$$

This reaction may proceed under humidity conditions when otherwise SO_2 is not deposited and oxidized on the surface. The thin layer of sulphuric acid then causes corrosion in the underlaying metal in pores of the gold layer. It should also be mentioned that the chemical composition of corrosion products which form on gold contacts at accelerated testing in a mixture of SO_2 and NO_2 have shown the best correspondence with specimens exposed in telephone central office environments (22).

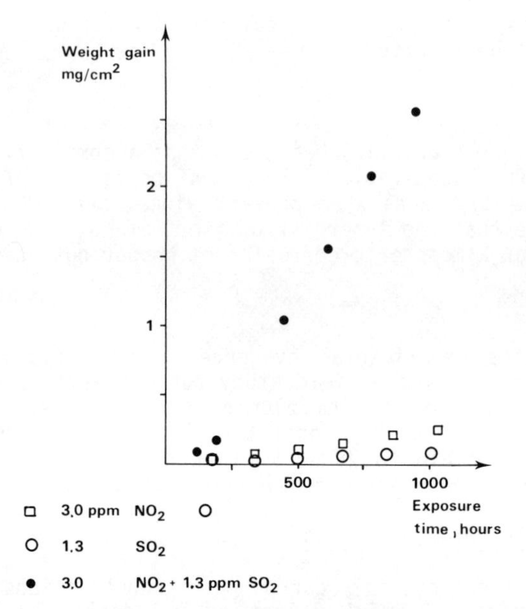

Figure 5. Corrosion of mild steel in air containing SO_2 and/or NO_2 at 50% relative humidity (17).

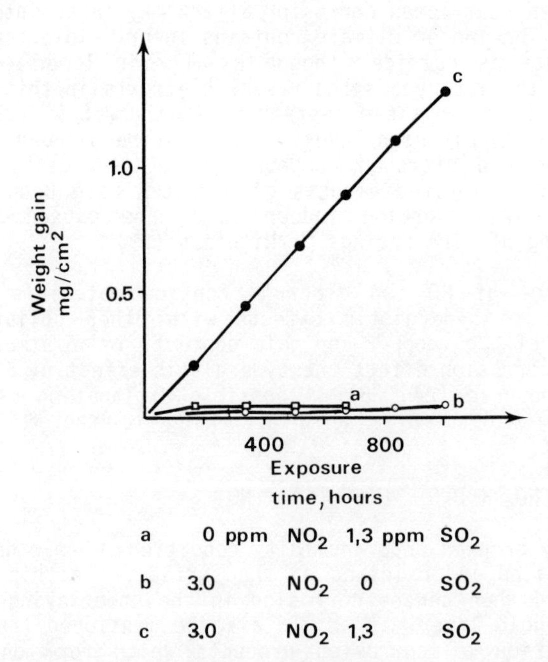

Figure 6. Corrosion of copper in air containing SO_2 and/or NO_2 at 90% relative humidity (19).

The influence of rain and its composition

The total effect of acid rain on corrosion cannot be unequivocally determined. In principle rain exerts an influence on corrosion in the following ways:

- by wetting the surface
- by transport of stimulators of corrosion e.g. H^+ and SO_4^{2-} ions to the surface
- by washing away pollutants from the surface deposited during the preceding dry period.

The relative influence of those three effects depends i.a. on the degree of pollution, on the corrosion mechanism and on the nature of corrosion products of the metal.

At exposure of steel in heavily polluted industrial atmosphere the corrosion rate on the upper side of steel panels exposed at 45° inclination was only 37 per cent of the total corrosion. In clean air, by contrast, the corrosion effect of rain was predominant and the upper sides of the test panels corroded faster than the undersides (26). The atmospheric corrosion of steel proceeds in local cells, where the sulphate nests acts as anodes. This may be the explanation why the washing effect of rain prevails in polluted atmospheres, as rain water may wash away sulphates from the nests.

For zinc and copper which are metals whose corrosion resistance may be ascribed to a protective layer of basic carbonates and basic sulphates the pH value of rain seems to be of significance. If the pH of rainwater falls to values close to 4 or even lower, as may be seen in a potential- pH diagram for copper in FIG. 7, this may lead to dissolution of the protective coatings.

For aluminum the corrosion rate especially in polluted atmosphere is usually lower at open outdoor compared to sheltered exposure. This indicates that the composition of the surface layer of moisture which in sheltered positions is created by dry deposition of pollutants, is more corrosive than precipitaiton.

The local nature of atmospheric corrosion

From the practical and economic point of view atmospheric corrosion is closely associated with centers of population. Three factors here coincide: high pollution level, high density of population, which in turn means great use of materials. The rate of atmospheric corroion decreases sharply with increasing distance from the emission source. This may be illustrated by the corrosion of carbon steel as function of the distance from the stack of a polluting industry in Kvarntorp, see FIG.8 (26).

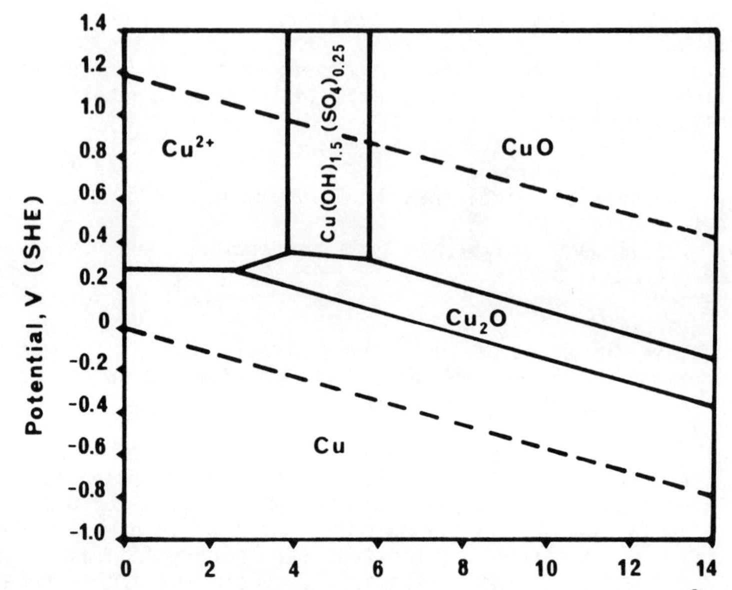

Figure 7. Potential-pH-diagram for the system Cu - SO_4^{2-} H_2O; 25°C, 10^{-1}M Cu, 10^{-3}M SO_4^{2-} (27).

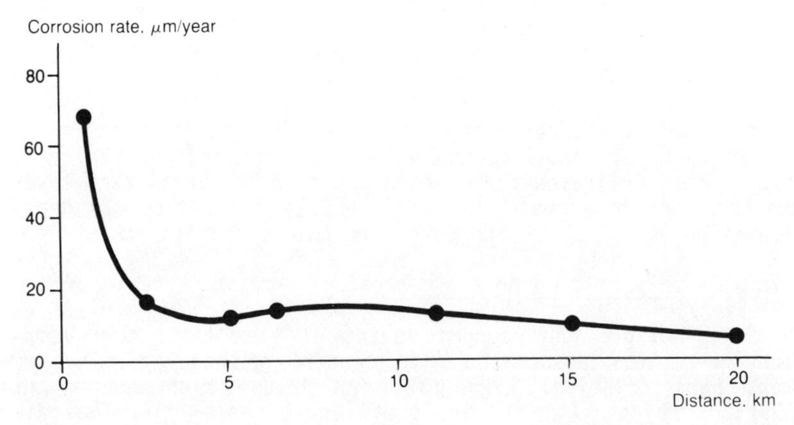

Figure 8. Corrosion rate of carbon steel as a function of the distance from the emission source - a chimney in Kvarntorp (26).

The great differences of the corrosion rate in restricted geografical areas have also been demonstrated by construction of corrosion maps for cities or whole countries. The corrosion map of zinc for UK (28), the corrosion map of several metals for the Sarpsborg/Fredrikstad area in Norway (29) and the corrosion map of steel for Madrid (30) may serve as examples. The very strong local variations of atmospheric corrosion of metals implies also the major role of dry deposition of pollutants. The wet deposition does not by far exhibit such strong variations as the corrosion rate.

Atmospheric corrosion is thus, at least in Scandinavia, a local effect mainly caused by the country's own emissions and not affected by long-distance transport of pollutants. The situation may, however, be different in densly populated areas of e.g. Western or Central Europe, where also transport of pollutants over the national boundaries may cause appreciable corrosion damage.

TABLE 1. Corrosion rate of zinc exposed in Central Stockholm (Vanadis).

Period	Corrosion rate µm/year
1938-1953	2.0
1958-1959	5.0
1958-1963	4.3
1967-1968	4.0
1967-1970	3.0
1975	2.9
1975-1976	2.3
1977-1978	2.1
1974-1978	2.0
1979-1980	1.7
1984	1.4

A reduction of the losses due to atmospheric corrosion may thus be achieved by reducing emissions, which is of course the preferable solution when bearing in mind the problem of "acid rain" in its complexity. But also building high stacks and discharging the flue gases high up in the air may be a solution of the local effects caused by acid deposition. This has been done or is planned in most Swedish conurbations in pursuance of legislation concerning a low maximum sulphur level in oil for heating and the expansion of district heating systems. It has already had very positive effects, leading among other things to a substantial decrease in the corrosion rate of zinc at points in the inner city of Stockholm at the test site Stockholm Vanadis, (TABLE 1). A similar development has occurred in a number of Swedish towns and brought important economic savings as consequence of less corrosion.

Conclusions

- The atmospheric corrosion of metals caused by acid deposition is mainly a local problem restricted to areas close to the pollution source.

- The main corrosive effect in this respect is caused by dry deposition of air pollutants. The influence of acid precipitation may differ for different materials and depends also on the pollution level.

- The effect of S-pollutants and especially of SO_2 are of greatest importance in this respect.

- Dose-response functions describing the corrosion as function of SO_2 and time of wetness are today available for steel and zinc. In extensive areas of the temperate climate zone the corrosion loss may be described by SO_2 pollution solely.

- According to the present knowledge NO_x has only limited influence on corrosion of steel and zinc in outdoor atmospheres, there are however indications that NO_x in combination with SO_2 promote corrosion of e.g. copper.

- Indoors NO_2 in combination with SO_2 has a synergistic corrosive effect on electrical contact materials like gold plating and copper and may also be important for other materials like carbon steel at storage in industrial atmospheres.

- A reduction of losses due to atmospheric corrosion has been achieved in many Swedish cities by reduction of SO_2 emissions and by introduction of district heating systems.

References

1. "Air pollution across national boundaries. The impact on the environment of sulfur in air and precipitation." Sweden's case-study for the U.N. Conference on the human environment 1972. Ministry of Agriculture, Stockholm 1972.

2. "Acidification Today and Tomorrow." A Swedish Study prepared for the 1982 Stockholm Conference on the Acidification of the Environment. Ministry of Agriculture, Stockholm 1982

3. Kucera, V. Proc. 9th Scandinavian Corrosion Congress, 1983, p. 153-170.

4. Kucera, V. Proc. Electrochemical Soc. Fall Meeting, 1985 (To be published).

5. "Airborne Sulphur Pollution. Effects and Control." ECE Air Pollution Studies 1. United Nations, New York 1984.

6. Haagenrud, S.; Kucera, V.; Atteraas, L. Proc. 9th Scand. Corrosion Congress, 1983, p. 257.

7. Knotkova, D.; Gullman, J.; Holler, P.; Kucera, V. Proc. 9th Int. Corr. Congress, 1984, Vol. 3, p. 198.

8. Kucera, V; Mattsson, E. In "Corrosion Mechanisms" Ed. by Mansfeld, F., Marcel Dekker Inc., N. York (in print).

9. Ellis, D. B: Proc. ASTM 48 (1948) 152.

10. Gullman, J.; Holler, P.; Kucera, V.; Knotkova, D. Proc. 7th European Congress of Corrosion, 1985.

11. Kucera, V.; & Mattsson, E. In "Atmospheric Corrosion" Ed. by W. H. Ailor, John Wiley & Sons Inc., N. York 1982, p. 561-574.

12. Evans, T. E. Proc. 4th Int. Congress on Metallic Corrosion, 1969, NACE, 1972, p. 408-418.

13. Fintling, E.; Hasselbohm, B. Proc. of 5th Scandinavian Corrosion Congress, 1968.

14. Grennfelt, P. "Nitrogen oxides in the atmospheres." Report B 585, IVL, (in Swedish).

15. Grennfelt, P. "Nitrogen oxides in Swedish and Norwegian conurbations." IVL report for project KHM 705, 1982 (in Swedish).

16. Haynie, F. H.; Spence, J. W.; Upham, J. B. In "Atmospheric Factors Affecting the Corrosion of Engineering Metals"; ASTM STP 646, ASTM 1978, p. 30.

17. Johansson, L-G. Proc. 9th Int. Corrosion Congress, 1984 , Vol. 1, p. 407.

18. Katoh, K. et al. Boshoku Gjiutsu 30 (1981) 337.

19. Johansson, L-G. Proc. Electrochemical Soc. Fall Mtg, 1985 (to be published).

20. Byrne, S. C.; Miller, A. C. In "Atmospheric Corrosion of Metals"; ASTM STP 767, ASTM 1982, p. 359.

21. Rice, D. W. et al. J. Electrochem. Soc. 1980, 127, p. 891.

22. Zakipour, S.; Leygraf, C. J. Electrochemical Soc. (submitted for publication).

23. McKinney, N.; Hermance, H. W. ASTM STP 452, 1967, p. 274.

24. Svendung, O.; Johansson, G. Proc. 9th Scand. Corrosion Congress, 1983 p. 337.

25. Rice, D. W. et al. J. Electrochem. Soc., 128, 1981, 275.

26. Kucera, V. Ambio 5, 1976, 243.

27. Mattsson, E. Materials Performance 21, 1982, 928.

28. Shaw, T. R. In "Atmospheric Factors Affecting the Corrosion of Engineering Metals"; ASTM SPP 646, 1978.

29. Haagenrud, S. E.; Henriksen, J. F.; Gram, F. Proc. Electrochemical Soc. Fall Mtg, 1985 (to be published).

30. Morcillo, M.; Feliu, S. Rev. Metal. CENIM 13, 1977, 212.

RECEIVED January 21, 1986

7

Environmental Effects on Metallic Corrosion Products Formed in Short-Term Atmospheric Exposures

D. R. Flinn, S. D. Cramer, J. P. Carter, D. M. Hurwitz, and P. J. Linstrom

Bureau of Mines, U.S. Department of the Interior, Avondale, MD 20782-3393

The Bureau of Mines has measured short- and long-term atmospheric corrosion damage on five metals and two metal-coated steel products at four sites in the east and northeast United States as part of the National Acid Precipitation Assessment Program to evaluate the effects of acid deposition on materials. The composition of the corrosion product on carbon steel, weathering steel, copper, zinc, and galvanized steel is relatively unchanged in 1- and 3-month exposures over a wide variety of environmental conditions. Spalling and runoff losses are observed on all metals. Massive reorganization of the corrosion film by a mechanism of cyclic dissolution and precipitation was observed on carbon steel, Cor-Ten A, zinc, and galvanized steel. Loss of corrosion product from zinc in runoff was a function of both dissolution in rain water and neutralization by hydrogen ion loading, with dissolution contributing the greater portion of the loss.

The corrosion of metallic materials in the atmosphere has been studied extensively (1). The majority of the work in this area has been to determine the performance of materials and to evaluate mitigation techniques in environments of interest. With only a few exceptions (see for example references 2, 3), attempts have not been made in studies conducted in the United States to fully characterize the environment and to determine the relationships between components of the environment and the performance of the material of interest (see reference 4 for a recent assessment of this area). Adherent corrosion products are often characterized, but no attempts have been made, except in laboratory studies (5), to quantitatively relate the corrosion film chemistry to environmental parameters.

In 1981, a field study was initiated by the Bureau of Mines to determine the effects of the environment, including acid deposition, on the corrosion of a number of commonly used metallic materials of construction. This study, which is fully described in a recent paper (6), is being conducted at field sites where continuous air

Published 1986, American Chemical Society

quality, rain chemistry, and meteorology measurements are being made. Of equal importance, the corrosion products are characterized by both wet chemical and instrumental methods in order to provide additional information regarding environmental effects on the materials. In this paper, the results of the corrosion product characterizations for short-term (generally 1 year or less) exposures are described.

Experimental

The field study is being conducted at five sites: Research Triangle Park, NC; Washington, DC; Chester, NJ; Newcomb, NY; and a recently added site at Steubenville, OH. These sites encompass the range of gaseous pollutant levels and meteorology variables typically observed in most rural, urban, and suburban locations in the United States. Hydrogen ion loading in the rainfall varies by a factor of 2 between the four original sites, and annual average rainfall pH ranges from 4.18 to 4.41.

The metals exposed in the program are 1010 carbon steel, Cor-Ten A[1] weathering steel, 3003-H14 aluminum, 110 copper, rolled zinc (alloy 191), G-90 galvanized steel, and Galvalume (a 55Al-45Zn coating on steel). Weight-loss samples, 4 by 6 inches, are installed (in triplicate) and removed following preselected exposure periods of 1 month, 3 months, 1 year, and 3 years; the 3-month and 1-year exposures are installed every 3 months, with the winter season defined as December through February. The sample exposure racks are inclined at 30° to the horizon and face south. After the exposed samples are weighed, the corrosion product is chemically removed from the weight-loss samples, and these stripping solutions are analyzed by standard chemical techniques (7). The cleaned samples are weighed again to determine metal loss during exposure. Smaller, 1 by 1.75 inch, samples ("microanalysis samples") are exposed concurrently with the larger samples.

The microanalysis samples are used in studies of the corrosion film by a number of instrumental techniques. X-ray diffraction (XRD) was performed on a Phillips X-ray diffraction unit at 40 KV using CuKα radiation. Samples were examined on an ISI scanning electron microscope (SEM) using backscattered electrons. Elemental depth profiles for the corrosion films were measured by a combination of ion scattering spectroscopy (ISS) and ion etching with $^{3}He^{+}$ at a rate of 0.3 to 0.5 nm per minute. Chemical information from the corrosion film surfaces was obtained with a Surface Science Laboratory SSX-100 X-ray photoelectron spectrometer (XPS) using an AlKα monochromatic X-ray source. Thermogravimetric analysis (TGA) was conducted using a Cahn RH vacuum electrobalance system and a Cahn Mark II time derivative computer to determine the amount of water in the corrosion films and the presence of compounds that may decompose at temperatures up to about 600° C. TGA was done in an argon atmosphere.

For complete description of the site environments, materials studied, and experimental procedures, see reference 6. Complete computer files of the air quality and meteorology data are being prepared so that it will soon be possible to evaluate correlations between these data and the metallic corrosion rates. With the exception of Steubenville, OH, more than 2 years of corrosion rate data (25 one-month exposure periods, 11 three-month periods, and 7 year-long exposures) have been measured at each site.

Results and Discussion

Table I shows the average corrosion rates observed for Cor-Ten A, zinc, and copper in 1-month, 3-month, and 1-year exposures. The corrosion rates for copper and zinc were lowest at the Washington, DC, site, while the Cor-Ten A corrosion rate was lowest at the New York site. The corrosion rates for galvanized steel and for Galvalume were less than 1 μm/y in exposures up to 1 year. The corrosion rate for aluminum, as expected, was low (less than 0.1 μm/y).

Table I. Average corrosion rates at four field sites

Metal	Exposure[1] Months	Corrosion Rate, μm/y			
		NC	DC	NJ	NY
Cor-Ten A	1	33.8	50.7	43.4	18.6
	3	27.1	36.3	34.5	14.7
	12	16.1	15.1	20.1	10.6
191 Zinc	1	6.5	1.0	3.5	4.6
	3	2.6	0.9	2.0	2.4
	12	1.4	1.1	1.4	1.0
110 Copper	1	6.2	1.9	5.4	5.8
	3	4.3	1.4	3.6	3.8
	12	2.4	1.1	2.2	2.1

[1]One-month exposures for period May 1982 through May 1984; 3- and 12-month exposures begun in period March 1982-December 1983.

In a general way, the growth of a corrosion film on a metal can be viewed schematically as a process with three distinct steps. The first is the reaction of metal A with environmental species B

$$A + B \rightarrow AB \quad (1)$$

The second step is the interaction of this corrosion product with environmental factors such as the weather, gaseous pollutants, and wet and dry deposition to produce, while conserving metal in the film, a modified or weathered corrosion product AB*

$$AB + \text{environment} \rightarrow AB^* \quad (2)$$

The third involves the loss of mass from the corrosion product through mechanical and chemical processes

$$AB^* + \text{environment} = AB^{**} + R_L + S_L \tag{3}$$

where R_L represents the runoff losses produced by the dissolution and removal of corrosion product in precipitation, and S_L represents the spalling losses caused by thermal and volume distortion stresses. In a well-developed corrosion film, all three steps may be occurring at the same time.

If m_A is the mass of A lost due to corrosion, S_L and R_L are the mass of A lost from m_A in spalling and runoff, and $m_{AB^{**}}$ is the mass of weathered corrosion product retained on the metal in Equation 3, then a mass balance for the corrosion film yields:

$$m_{AB^{**}} = \left(m_A - R_L - S_L\right)\left[\sum_{i=1}^{N} X_i\left(\frac{M_i}{M_A}\right)\right] \tag{4}$$

where X_i is the fraction of A in the i-th constituent of $m_{AB^{**}}$, compared to the total mass of A retained in $m_{AB^{**}}$. M_A and M_i are the molecular weight of A and the i-th constituent, respectively, N is the number of constituents present in AB**, and

$$\sum_{i=1}^{N} X_i = 1 \tag{5}$$

For some materials, runoff and spalling losses are negligible in short-term exposures, e.g., $R_L = S_L = 0$, so that m_A replaces $\left(m_A - R_L - S_L\right)$ in Equation 4. If, furthermore, the system of oxides, hydroxides, etc., which comprise the corrosion product are thermodynamically stable, i.e., relatively insensitive to changes in the environment so that the composition does not vary greatly, a plot of $m_{AB^{**}}$ as a function of m_A, using the gravimetrically determined film weight and weight-loss data, should yield a straight line with slope

$$b = m_{AB^{**}}/m_A = \sum_{i=1}^{N} X_i\left(\frac{M_i}{M_A}\right) \tag{6}$$

such as that shown in Figure 1 for Cor-Ten A. The concentration, X_i^*, of the i-th constituent in the corrosion product expressed as a fraction of $m_{AB^{**}}$ is then

$$X_i^* = \frac{X_i m_A}{m_{AB^{**}}}\left(\frac{M_i}{M_A}\right) = \left(\frac{X_i}{b}\right)\left(\frac{M_i}{M_A}\right) \tag{7}$$

Data presented as in Figure 1 has the advantage of accentuating differences due to spalling and runoff losses. These differences result in points which lie below the line describing the short-term data.

In cases where runoff or spalling losses are significant, $\left(m_A - R_L - S_L\right)$ must be estimated from an analysis of the corrosion product retained on the corroding metal. This was done by wet chemical analysis of the stripping solutions that contain the corrosion product removed from exposed samples. Again, if the corrosion product is thermodynamically stable for a wide range of environmental

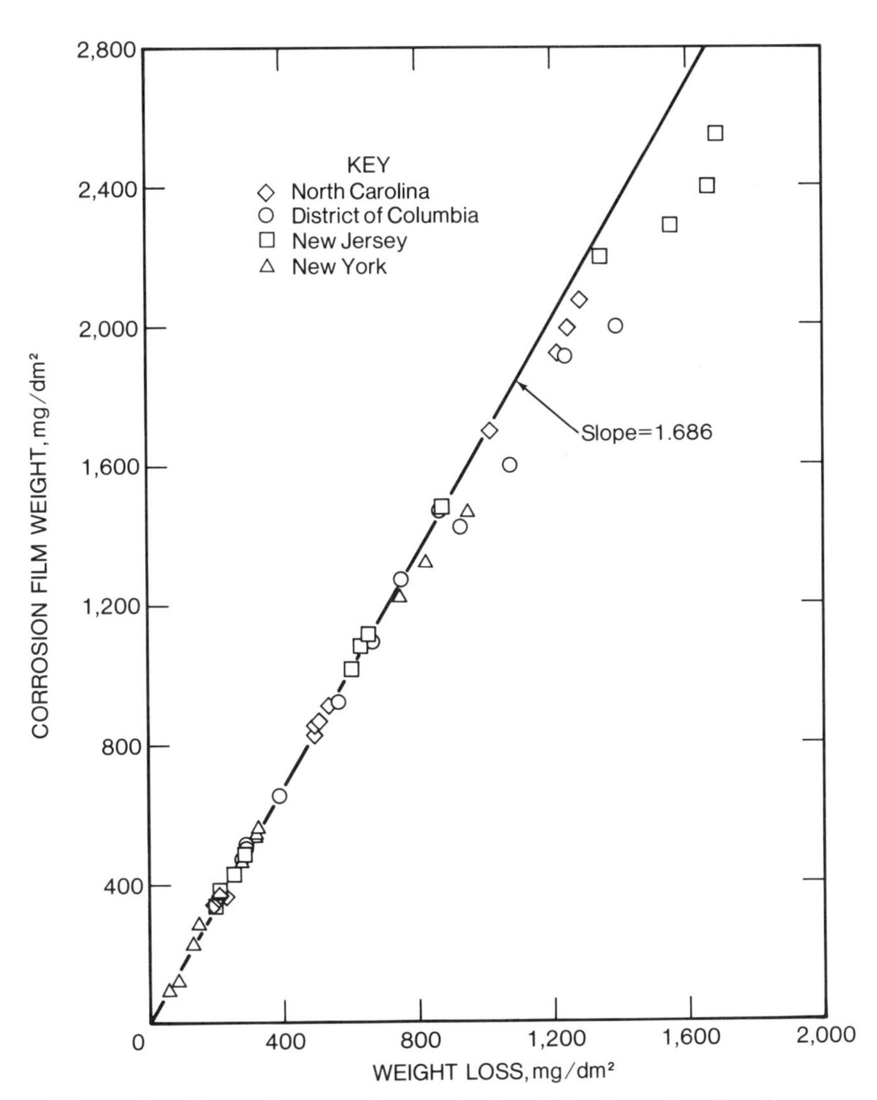

Figure 1. Mass of corrosion product retained on Cor-Ten A weight-loss panels in exposures of 1, 3, and 12 months at 4 sites. Exposures were initiated in period May 1982 - May 1983.

conditions, Equation 4 shows that a plot of m_{AB**} as function of $(m_A-R_L-S_L)$ should yield a straight line with a slope

$$b = \frac{m_{AB**}}{(m_A-R_L-S_L)} \quad (8)$$

With this approach, spalling and runoff losses do not affect the computation of the slope. On the other hand, the precision of the wet chemical measurements is lower than for the gravimetric measurements and slopes computed by Equation 8 are not as well defined as those from Equation 6. In the limit when $S_L=R_L=0$, these two approaches to computing the slopes give similar results.

Experimental estimates of b are given in Table II. Except for galvanized steel, these values were calculated using gravimetric results and Equation 6. For galvanized steel, b was calculated from wet chemical results for the amount of zinc in the corrosion film, i.e., $(m_A-R_L-S_L)$ and Equation 8. The number of data points, N, used in the computation are also given to show that the slopes are based on a large data set. Experimental values of the relative uncertainty in b, i.e., $\Delta b/b$, are given in Table II, column 5, as $2tS/b$, where S is the standard error in the estimate of b, and t is the measure of the dispersion about b, at a given confidence level. One- and 3-month exposure gravimetric data were used to compute b for Cor-Ten A and carbon steel since 1-year data showed evidence of spalling losses (confirmed by SEM examination) and to a lesser extent runoff losses. One- and 3-month gravimetric data were also used for copper. The zinc and galvanized data showed considerable scatter due to runoff losses. No spalling of these materials was observed using SEM examination of their surfaces. The 1- and 3-month exposure zinc gravimetric data define an upper limit to the amount of corrosion film retained on the zinc and were used to compute the slope b. On the other hand, the galvanized steel data had much more scatter than the zinc data and defining the limiting slope was more difficult. As noted above, this was overcome by using the wet chemical estimate for $(m_A-R_L-S_L)$ in Equation 8.

Table II. Relative sensitivity factor and experimental slopes

Metal	$(1/b)\partial b/\partial x$	Constituents	b	$2tS/b$[1]	N
Cor-Ten A	1	FeOOH or Fe_2O_3, H_2O	1.686	0.005	420
1010 Carbon Steel	1	FeOOH or Fe_2O_3, H_2O	1.677	0.009	425
191 Zinc	0.2	$Zn(OH)_2$, $ZnCO_3$	1.645	0.014	280
Galvanized Steel	0.4	ZnO, $ZnCO_3$	1.512	0.030	136
110 Copper	0.1	CuO, Cu_2O	1.115	0.007	429

[1]S evaluated at 99 pct confidence level for all metals except galvanized steel; galvanized steel evaluated at 90 pct confidence level.

The sensitivity of the slope b to changes in the composition of the corrosion film is characterized by the relative sensitivity factor, $(1/b)\partial b/\partial x$. Values of this factor are given in Table II for the case where the corrosion film consists of the 2 major constituents shown in column 3. Cor-Ten A and 1010 carbon steel had the highest sensitivity factor; a 1 pct change in composition would result in a 1 pct change in slope. Copper had the lowest sensitivity factor and a 1 pct change in composition would result in only a 0.1 pct change in slope. Comparing values of the relative uncertainty in b, column 5, with the sensitivity factors suggests that the composition of most of the corrosion films in short exposures is independent of environmental influences. In fact, using the 2-constituent corrosion film model, the experimental slopes define the amount of each constituent in the corrosion product to within several weight percent. Of course, the corrosion films on these materials are more complicated than the 2-constituent model. Nevertheless, these results indicate that the composition of the corrosion films is quite stable and that, regardless of site-to-site differences in the important environmental factors, the relative amounts of each constituent are fairly constant. The major influence of the environmental factors then is not to alter the thermodynamic stability of the corrosion products but to influence the kinetics of the rate limiting corrosion reaction.

Consider now Equation 3 describing losses of corrosion product. Assuming that spalling does not occur, evidently true for most materials in this study except Cor-Ten A and carbon steel, and that dry deposition effects can be neglected, the principal effects leading to loss of corrosion product in runoff are: (1) dissolution of the corrosion product in rainwater; and (2) reaction of corrosion product with hydrogen ion in the rain to form a more soluble species. Let the flux of hydrogen ion, i.e., the hydrogen ion load, to the skyward side of the weight loss panels be x (in mg H^+/m^2) and the flux of precipitation be y (in L H_2O/m^2). The loss of metal A from the corrosion product AB* to runoff is R_L (in mg A/dm^2). For simplicity, assume that the AB* is a single compound. Let H^+ react with this product according to

$$AB^* + nH^+ \rightarrow A^{n+} + H_nB^* \tag{9}$$

The amount of AB consumed in runoff due to this reaction is (xM_A/nM_H), where M_H and M_A are the molecular weights of H and A, respectively. In addition, the amount of A dissolved as AB* in H_2O is $fS_{AB^*}y(M_A/M_{AB^*})$, where S_{AB^*} is the solubility of AB* in water (in mg AB*/L), M_{AB^*} is the molecular weight of AB*, and f is the fraction of saturation that is achieved. This fraction is likely to be below 1.0 since the residence time for precipitation on the inclined surface of the panels is short. Combining terms, rationalizing units, and correcting areas for the incline of the panel leads to:

$$R_L = \frac{\cos 30°}{200}\left[fS_{AB^*}\,y\left(\frac{M_A}{M_{AB^*}}\right) + \left(\frac{M_A}{M_H}\right)\frac{x}{n}\right] \tag{10}$$

This equation includes a factor of 2 because runoff is assumed to occur only from the skyward side but m_A, $m_{AB^{**}}$, S_L, and R_L are computed based on the total sample surface area (skyward and groundward). This equation has the same form as that used to interpret wet deposition damage to stone (4, 8, 9). It predicts that total runoff losses involve contributions from both solubility and hydrogen ion effects.

Carbon Steel and Cor-Ten A Steel. The corrosion rates for Cor-Ten A steel at the four sites are shown in Table I. Corrosion rates for carbon steel at the North Carolina, District of Columbia, and New Jersey sites are lower than those for Cor-ten A during 1-month exposures, equal to Cor-ten A during 3-month exposures, and higher than Cor-ten A for 1-year exposures. Corrosion rates for carbon steel at the New York site are lower than Cor-ten A during the 1- and 3-month exposures and equal to Cor-ten A after 1 year of exposure. These results indicate, relative to carbon steel, that Cor-ten A must corrode a definite amount before producing a corrosion film of sufficient thickness to exhibit protective properties.

The formation and composition of atmospheric rust films on carbon steel and weathering steels has been discussed extensively in the literature (10-22). In general, the rust which forms on steels can be viewed as a 2-layer structure generated by an electrochemical corrosion cycle which involves the oxidation of metal and reduction of oxide corrosion products in the inner layer and oxidation of reduced corrosion products in the outer layer. The corrosion film is basic in nature and comprised essentially of a series of oxides and oxyhydroxides representing partially dehydrated forms of ferric hydroxide. The corrosion product in the outer layer dissolves during wet periods and reprecipitates as amorphous $FeO_x(OH)_{3-2x}$ upon drying. The differences in the corrosion of carbon steel and weathering steel have been attributed to the enrichment of alloying elements in the inner rust layer which (1) favors the formation of amorphous subtances which strengthen the rust coating, (2) increases the amount of water found in the corrosion film and slows drying to produce a more compact film, and (3) decreases the formation of intermediate substances that can form crystalline magnetite and improve the conductivity of the rust coating.

A plot of metal weight loss (m_A) for 1-, 3-, and 12-month exposures of Cor-Ten A at the 4 sites versus film weight, m_{AB**}, is shown in Figure 1. The slope of the line for 1- and 3-month data was 1.686. The narrow confidence limits for the slope, Table II, indicate that the corrosion film composition is similar at all sites for short-term exposures. This means that the thermodynamically favored corrosion products are the same regardless of site. Carbon steel gave similar results. Deviations from the straight line at higher weight-loss values correspond to 1-year data. Based on an SEM examination of the corrosion films, much of this loss is the result of spalling. Rust staining of the corrosion test racks indicates that runoff losses also occur. The slope of 1.686 for Cor-Ten A suggests that the corrosion film is largely FeOOH and hydrated Fe_2O_3, Table III. It is unlikely that FeO, Fe_3O_4, or Fe_2O_3 is present in large amounts as these products would bring the slope down significantly. Conversely, $Fe(OH)_3$ and $FeSO_4$ would raise it considerably. Carbon steel gave a similar slope and, at least in short-term exposures, the chemistry of the corrosion film on carbon steel and Cor-Ten A with respect to the major constituents apppears the same. If the corrosion film is assumed to be FeOOH, the computed

Table III. Corrosion film mass balance results from wet chemical and gravimetric estimates of metal retained in corrosion film

Metal	b Confidence Interval	Major Constituents		Comment
		Likely	Not Likely	
Cor-Ten A	1.68-1.69*	FeOOH $Fe_2O_3 \cdot nH_2O$	FeO Fe_3O_4 Fe_2O_3 $FeSO_4$ $Fe(OH)_3$	Contains 5 wt pct water of hydration
1010 Carbon Steel	1.67-1.68*	Same as Cor-Ten A		Same as Cor-Ten A
191 Zinc	1.633-1.657*	$ZnCO_3$ $Zn(OH)_2$	ZnO $ZnSO_4$ ZnS $ZnS \cdot H_2O$	Corrosion film 30 pct $ZnCO_3$ and 70 pct $Zn(OH)_2$ in short exposures
	1.617-1.945**	$ZnCO_3$ $Zn(OH)_2$		Corrosion film at 3 sites 65 wt pct $ZnCO_3$ and 35 wt pct $Zn(OH)_2$
		$ZnCO_3$ $Zn(OH)_2$ ZnO		DC site (b=1.436)
Galvanized Steel	1.156-1.200*	ZnO	$Zn(OH)_2$ $ZnCO_3$ $ZnSO_4$ ZnS $ZnS \cdot H_2O$	Low b for DC site
	1.489-1.535**	$ZnCO_3$ $Zn(OH)_2$ ZnO		
110 Copper	1.111-1.119*	Cu_2O CuO	Cu_2S CuS $CuSO_4$ $CuSO_4 \cdot 3Cu(OH)_2$ $Cu(OH)_2$ Cu_2CO_3 $CuCO_3 \cdot Cu(OH)_2$ $2CuCO_3 \cdot Cu(OH)_2$ $CuCl_2 \cdot 3Cu(OH)_2$	Low b for DC site. One-month DC data agrees with other sites; 3- and 12-month data fall below curve
	1.047-1.225**	Cu_2O CuO		DC data higher than other sites corresponding to 60 wt pct CuO and 40 wt pct Cu_2O

*Metal loss determined gravimetrically; 99 pct confidence level.
*Metal loss determined from wet chemical analysis of corrosion product 90 pct confidence level.

slopes would suggest that there is approximately 5 wt pct water present in the film.

The results from TGA, Table IV, indicate that the corrosion film on Cor-Ten A contains from 30 to 70 wt pct FeOOH. The

Table IV. Thermogravimetric analysis (TGA) of corrosion film on microanalysis samples

Metal	Major Constituents		Description	Mass Balance Results
	Likely	Not Likely		
Cor-Ten A	FeOOH H_2O	$FeSO_4 \cdot H_2O$	Weight loss during bake-out, <70° C	4-12 wt pct, "free" H_2O
			Weight loss from 110°-160°C	1.7-4.4 wt pct water of hydration
			Large weight loss from 160°-260° C, e.g., "goethite terrace"	30-70 wt pct FeOOH
191 Zinc	$ZnCO_3$	$ZnSO_4 \cdot H_2O$ $Zn(OH)_2$	Small weight loss from 110° C to 150°-200° C	$Zn(OH)_2$ decomposition
			Large weight loss from 200°-250° C	40-100 wt pct $ZnCO_3$
110 Copper	—	$CuCO_3$ $CuCl_2 \cdot 2H_2O$ $CuSO_4$ CuS $CuCl_2 \cdot Cu(OH)_2$	Weight loss builds at small but steady rate from 110°-350° C	—

decomposition at 160°-260° C clearly matched the decomposition of goethite reported in the literature (23). There is 4 to 12 pct loosely bound water absorbed in the film, and 1.7 to 4.4 pct water more tightly bound as waters of hydration presumably as $Fe_2O_3 \cdot n\ H_2O$ or the amorphous iron oxyhydroxide. The only crystalline phase identified in the corrosion film was goethite, Table V. After TGA analysis, XRD analysis showed both magnetite and hematite present. This would indicate that decomposition of the corrosion film to Fe_3O_4, the expected final thermal decomposition product (24), is incomplete.

XPS analysis of the corrosion film on Cor-Ten A, Table VI, gives results similar to those obtained by corrosion film mass balance, TGA, and XRD. The principal film constituents in the outer 10 nm of the film are Fe_2O_3 and FeOOH, with there being somewhat more iron present as Fe_2O_3 than FeOOH. This would be expected in a film that dehydrates. However, if this is true, then the 2:1 atomic ratio of oxygen to iron indicates that even the outermost surface of the film contains substantial water. ISS depth profiles for the

Table V. X-ray diffraction analysis of corrosion film on microanalysis samples

Metal	Phases Present	
	3-year Exposure, Avondale, MD	After TGA
Cor-Ten A	Goethite (FeOOH)	Magnetite (Fe_3O_4) Hematite (Fe_2O_3)
1010 Carbon Steel	Goethite (FeOOH)	(1)
191 Zinc	Unknown phases	Zincite (ZnO)
110 Copper	Cuprite (Cu_2O)	Cuprite (Cu_2O)

(1) Not determined.

corrosion films on Cor-Ten A, Table VII, show sulfur (or Cl) present only at the surface, with concentrations dropping to zero in the first nm. Normalizing the data to Fe_2O_3 for the layer at 1-5 nm in depth, the corrosion film is rich in iron at the surface (0-5 nm), probably due to more extensive drying of the film. At depths greater than 15 nm, the film contains relatively more oxygen, corresponding to a compound such as FeOOH.

SEM examination of the corrosion films on Cor-Ten A and carbon steel from exposures of 1, 12, and 36 months showed massive reorganization of the corrosion product. In short-term exposures the corrosion film on carbon steel and Cor-Ten A is characterized by mounds of corrosion product, Figure 2A, produced by intense localized corrosion cells which lie beneath them. The displacement of this corrosion product outward, away from the metal-film interface, can cause severe fracturing of the corrosion film and give moisture easy access to the inner layer and the corroding surface. Superimposed on this general structure are 3 features which suggest the massive reorganization just noted. First, in longer exposures the mounds become an increasingly less distinct feature of the corroding surface. Their tops are flattened and the large space which had existed earlier between them is filled. Secondly, there is little evidence of healed cracks in the corrosion film, although there are numerous fresh cracks and the well-defined features of these cracks could not survive even moderate solution alteration of the surface. These cracks were also observed by light microscopy on samples which had not been exposed to the vacuum of the SEM. Third, there is abundant evidence of relocation by precipitation of large amounts of corrosion film. Figures 2B and 2C show just one such example. A reef-like feature in the top right corner of 2B is spreading laterally from its attachment to a mound as new material precipitated along its edges. In doing so, it is spreading outward over material which is in the "valley" between mounds, Figure 2C, and evidently will fill in this space. This is one mechanism whereby the mounds coalesce, as is occurring in the lower left quarter of Figure 2B, to

Figure 2. Groundward side of Cor-Ten A microsample E231 exposed 1 year at NC site showing dissolution and precipitation features. Magnifications: (a) 80X; (b) 240X; (c) 1600X.

Table VI. X-ray photoelectron spectroscopic analysis of corrosion film on microanalysis samples

Metal/Exposure	XPS Peak	Observation[1]
Cor-Ten A	$C1S_{1/2}$	C-C, C-H: major C-O, C=O: minor
	$O1S_{1/2}$	Fe_2O_3: major hydroxide [FeOOH], water: major C-O,C=O: minor $O(as\ FeOOH) > O(as\ Fe_2O_3)$
	$Fe2P_{3/2}$	Fe_2O_3: major FeOOH: major $Fe(as\ FeOOH) < Fe(as\ Fe_2O_3)$ Note: O/Fe~2
Galvanized Steel	$C1S_{1/2}$	C-C, C-H: major CO_3: major
	$O1S_{1/2}$	hydroxide [$Zn(OH)_2$], water: major carbonate [$ZnCO_3$]: major oxide [Cr_2O_3 or CrO_3]: minor $O(as\ Zn(OH)_2) > O(as\ ZnCO_3)$
	$Zn2P_{3/2}$	2 peaks [$Zn(OH)_2$ and $ZnCO_3$]
	$Pb4f_{7/2}$	1 peak [PbO or Pb_3O_4]
	$Cr2P_{3/2}$	Cr_2O_3 CrO_3 } $Cr^{3+} \leqslant 6\ Cr^{6+}$

[1]Phases in brackets indicate probable identities of peaks.

Table VII. Ion scattering spectroscopic analysis of corrosion films on microanalysis samples

Metal	Distance From Outer Surface, nm	Description
Cor-Ten A	0-0.2	2-10 a/o Cl/S O rich layer (O/Fe>>1)
	0.2-1	1 a/o Cl/S Fe rich layer (O/Fe ~ 1)
	1-5	Fe_2O_3[1]
	>15	FeOOH (O/Fe~2.2) 0 a/o Cl/S
Zinc	0-1	10-55 a/o Cl/S ZnO
	1-15	0.5-5 a/o Cl/S ZnO, with 2-4 a/o excess Zn
	>15	O>Zn, composition shifting towards carbonate or hydroxide
Galvanized Steel	0-1	3-13 a/o Cr .2-2 a/o Al 16-29 a/o Zn 55-80 a/o O
	>1	ZnO 9-13 a/o Cr 2-7 a/o Al
Copper	0-0.3	3-8 a/o Cl/S 30-85 a/o Cu 10-62 a/o O
	0.3-2	CuO[1] 1-3 a/o Cl/S
	>2	Cu rich phase (Cu/O~1.6)

[1]Data normalized to this value for computational purposes.

give the smooth flat surface characteristic of the corrosion film formed in long exposures. For such a reef-like feature to grow, the "valley" or depressed area between mounds must be filled with a concentrated solution of dissolved rust. This of course could not have occurred during a rain because the solution would have washed away and the features seen in Figure 2 obliterated. Instead, it is proposed that the thin layer of water which remains after a rainfall precipitation event absorbs acidic gases from the air before the surface dries and results in an acid electrolyte that dissolves rust. Then, on drying the dissolved material precipitates in the "valleys" and at the edge of the "reef-like" structures to gradually fill in the "valley" areas and level the corrosion film. The numerous small mounds in the valleys appear to be precipitate nucleation sites. The sites are fewer and larger at the rim of the solution filled areas, indicating a lower rate of precipitation when the solution volume is larger. But, as electrolyte evaporates and the solution draws down into the valley, the rate of precipitate nucleation increases and numerous small mounds consistent with accelerated drying are formed.

Cracks from an earlier drying cycle would not survive such extensive alteration of the surface. This evidently accounts for the absence of healed or partially healed cracks in the surface. The cracks apparently form as the last step in the drying cycle. New cracks, of course, can form along the path of earlier cracks.

The overall leveling that occurs in longer exposures suggests that the crests of the mounds are the most active sites for dissolution when wet. The water layer is thinnest here, and exchange with acidic gases from the atmosphere to produce a solution which will dissolve the rust should occur most rapidly at such sites. If the crests are sites of most active dissolution, then the "valleys" are the sites of most active precipitation and one follows the other as the volume of water retained on the surface decreases during drying. Horton (25) has observed that dust particles are found exclusively in the outer layer of the corrosion film. This is consistent with the massive reorganization of the film proposed here by a cycle of dissolution and precipitation during the latter stages of drying. The outer layer then corresponds to a solution altered structure modified intimately by contact with the environment, while the inner layer is modified basically by diffusion and conduction processes and is relatively isolated, except along crack lines, to massive intrusions of water.

In a few cases precipitation of material along cracks was observed, Figure 3. This was more likely to be seen on the groundward side than the skyward side, more often for exposures at the North Carolina site than New Jersey, and in longer exposures than in short exposures. The precipitated material is seen in Figure 3B to be densest along that portion of the crack lying in the depressions above and below the rim in the center of this figure. This is consistent with the observation that the higher areas of the film are sites of dissolution and the lower areas are sites for preciptation. Evidently, in the case shown in Figure 3, dissolved material from within the crack has moved to the surface where it precipitated as the corrosion film dried. Further drying then reformed the crack.

The three features described here suggest that the outer part of the corrosion films on carbon steel and Cor-Ten A undergo massive reorganization by a mechanism of solution and precipitation during the latter stages of drying. In many ways this process is similar for both materials and only differs in small but significant details related to the presence of alloying elements in the weathering steel. The ability to heal and self-repair the corrosion film is an essential aspect of forming a more protective corrosion film and would appear linked not only to the chemistry of the alloy but also to the chemistry of the interacting environment.

191 Zinc. In contrast to the steels, zinc does not yield a simple linear relationship between the amount of corrosion film retained on the surface and the zinc loss due to corrosion, Figure 4. The data fans out considerably with increasing weight loss. A mass balance on the 1-and 3-month data which define an upper limit to the amount of corrosion film retained on the surface, using Equation 6, yields a line with a slope b = 1.645. The small confidence interval for b, Table III, and the large number of points (N = 280) used in computing b would suggest that runoff and spalling of the zinc corrosion film are not a significant factor in the short-term exposures and that b is a consequence only of Equations 1 and 2. Values of b computed for the individual sites agree well with that shown in Figure 4 except for the District of Columbia site, where b = 1.41. Comparison of b with the theoretical values for various compounds, Table III, suggests that the corrosion film at the North Carolina, New Jersey, and New York sites is largely a combination of $Zn(OH)_2$ and $ZnCO_3$. Applying Equation 7 for a film composed of two main constituents, the difference between the experimental slope 1.645 and the theoretical slopes, 1.52 for $Zn(OH)_2$ and 1.92 for $ZnCO_3$, indicates a film consisting of about 30 wt pct $ZnCO_3$ and 70 wt pct $Zn(OH)_2$ in short-term exposures. The lower slope for the District of Columbia site strongly indicates the presence of ZnO. Slopes were also computed from Equation 8 using the results from the wet chemical analyses of the stripping solutions. The results agree with those from gravimetric determinations and lead to the same set of major constituents for the corrosion film, Table III. The computed slopes were, however, somewhat higher, as would be expected if there were some runoff losses, and recomputation of the corrosion film composition based on these slopes gave values nearer to 65 wt pct $ZnCO_3$ and 35 wt pct $Zn(OH)_2$ for short-term exposures at the North Carolina, New Jersey, and New York sites.

Corrosion film constituents identified by TGA in an inert atmosphere, Table IV, agree well with those determined by the mass balance on the corrosion film. Corrosion films on samples exposed at the District of Columbia and North Carolina sites for 1 to 12 months gave similar results. Pure $ZnCO_3$ was run as a standard. The large weight loss at 200°-250° C coincided exactly with the decomposition of $ZnCO_3$ and results reported by Anderson (26). Mass balance calculations indicate that the weight loss corresponds to between 40 and 100 wt pct $ZnCO_3$. No weight loss of the $ZnCO_3$ standard was observed below 200° C. Therefore, a small loss of weight from the corrosion film that occurs in the range 110°-200° C was interpreted as $Zn(OH)_2$ decomposition. X-ray diffraction spectra from the

a

b

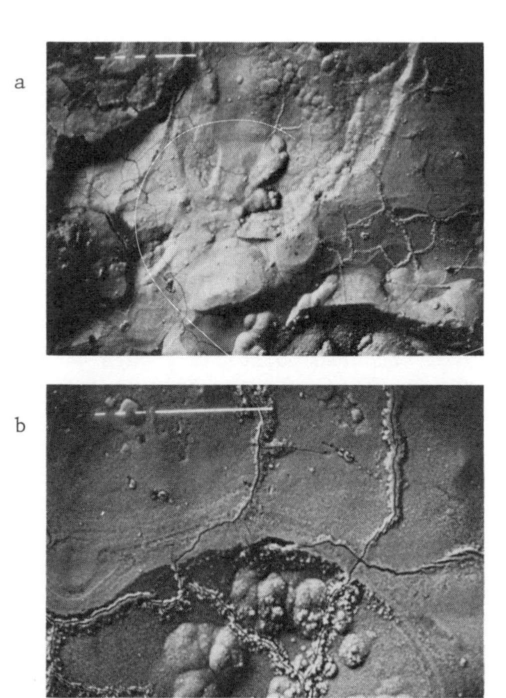

Figure 3. Groundward side of 1010 carbon steel microsample A46 exposed 3 years at NC site showing dissolution and precipitation features. Magnifications: (a) 80X; (b) 240X.

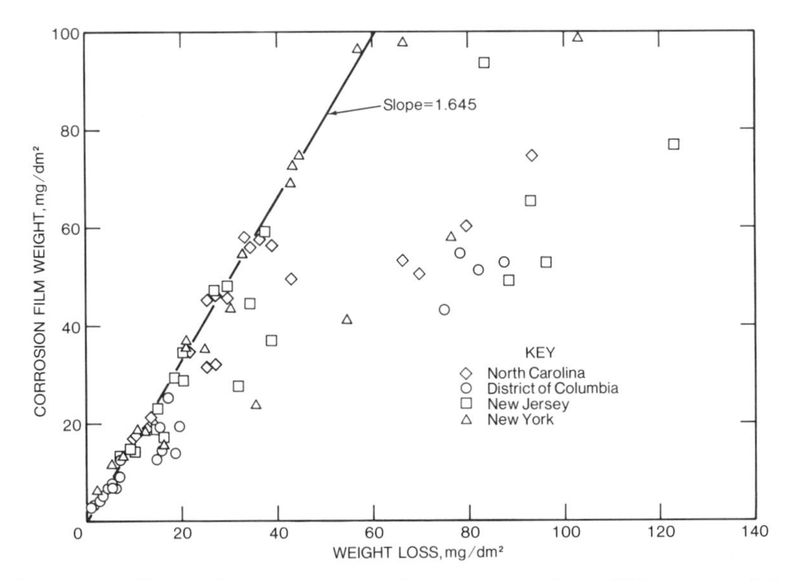

Figure 4. Mass of corrosion product retained on 191 zinc weight-loss panels in exposures of 1, 3, and 12 months at 4 sites. Exposures were initiated in period May 1982 - May 1983.

corrosion film on samples exposed for 3 years, Table V, contained several peaks which could not be matched with known phases. X-ray diffraction of the samples following TGA showed the corrosion film to be ZnO, the normal thermal decomposition product of $Zn(OH)_2$ and $ZnCO_3$.

Ion scattering spectroscopy of the corrosion film formed in 1-month exposures at the New York site gave an O/Zn atomic ratio consistent with ZnO in the outer 15 nm of the surface and which trended towards $ZnCO_3$ or $Zn(OH)_2$ deeper into the film, Table VII. The very outer 1 nm was rich in the surface contaminants Cl or S. Some Zn enrichment of the outer 15 nm of the film is observed which could be associated with the surface contaminants or related to drying effects.

Although many of the data points in Figure 4 lie on the mass balance line defined by the short-term data, a significant number of 3- and 12-month points lie in the fan-shaped area to the right of the line. Given the chemistry of long-term zinc corrosion products in the atmosphere (27), this shift cannot be due to the transformation of short-term corrosion products by weathering into some lower molecular weight compound. Instead, it represents a substantial loss of zinc corrosion product from the surface in runoff or possibly by spalling. SEM examination of the corrosion film on zinc samples exposed up to 36 months showed no evidence of spalling and, hence, it is assumed that this material loss is due entirely to runoff. The vertical difference between the data points in Figure 4 and the mass balance line roughly approximates the amount of this loss. However, a better estimate of the zinc loss to the environment in runoff is given by the difference in the gravimetric zinc weight loss and the zinc retained in the corrosion film as determined from the wet chemical analyses, i.e., (m_A-R_L-S_L). This quantity is plotted in Figure 5 for 1-month exposures and in Figure 6 for 3- and 12-month exposures as a function of the total hydrogen ion load during the exposure period. The 1-month data in Figure 5 show that little, if any, of the corrosion film is lost in runoff due to hydrogen ion load or through dissolution in water. The zinc losses exhibit a roughly normal distribution about zero, and the wide distribution in values, particularly the negative values, are due to low precision in the wet chemical analysis of the corrosion film chemistry. This observation, that essentially no runoff losses occur in short-term exposures, is consistent with the earlier interpretation of the mass balance results for corrosion films formed in 1- and, in some cases, 3-month exposures.

The zinc losses for longer exposures, Figure 6, show a decidedly different result. Here there is an increasing trend of zinc loss from the corrosion film as the hydrogen ion load increases. A least squares fit of the data yields the following relationship between the zinc runoff loss, R_L(in mg Zn/dm^2), and the hydrogen ion load, x (mg H^+/m^2):

$$R_L = 0.746\ x \tag{11}$$

Assume, for simplicity, that the corrosion film is $ZnCO_3$ and that hydrogen ion reacts with the corrosion film to form bicarbonate. Equation 9 then becomes

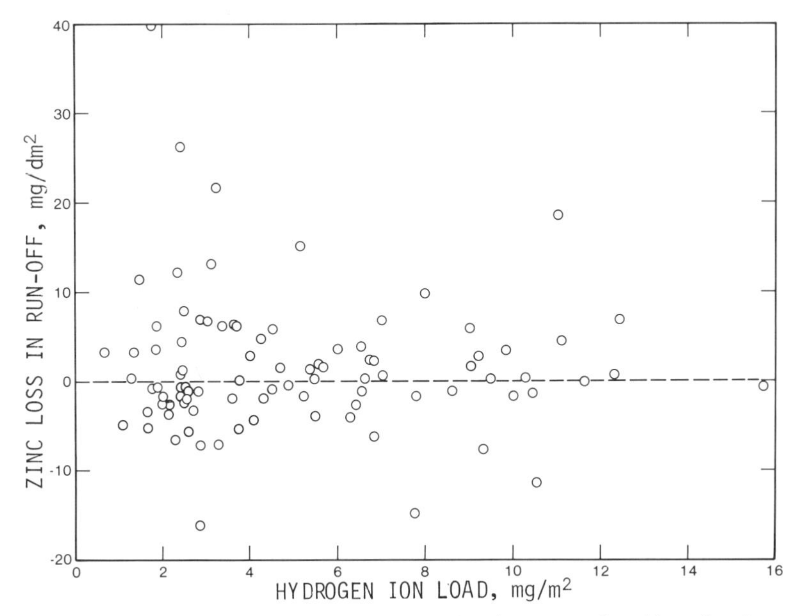

Figure 5. Influence of hydrogen ion loading on the dissolution of corrosion product during one-month exposures of 191 zinc at 4 sites.

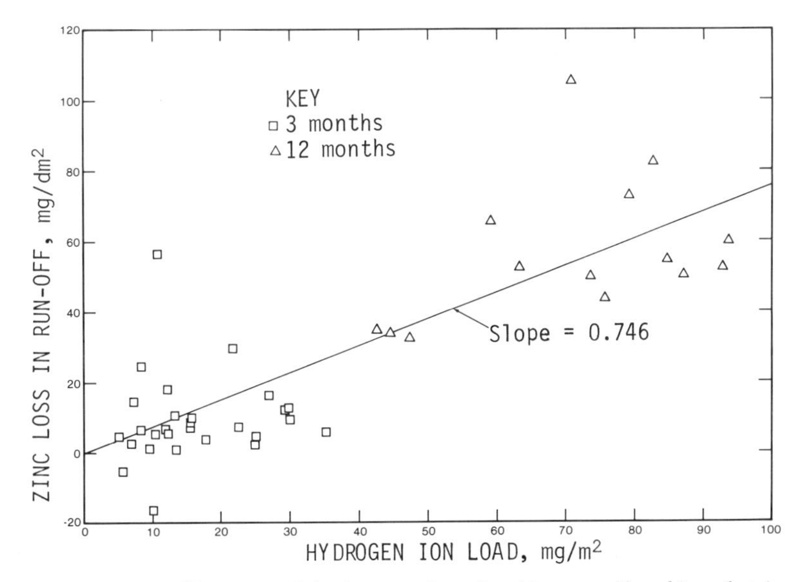

Figure 6. Influence of hydrogen ion loading on the dissolution of corrosion product during 3- and 12-month exposures of 191 zinc at 4 sites.

$$ZnCO_3 + H^+ \rightarrow Zn^{2+} + HCO_3^- \tag{12}$$

Using the annual average pH and the hydrogen ion load to estimate rainfall volume

$$y = \frac{(10^{-3+pH})X}{M_H} \tag{13}$$

Equation 10 can then be rewritten in form of Equation 11 to give

$$R_L = 0.0043 \left[fS_{ZnCO_3} \, 10^{-3+pH} \left(\frac{M_{Zn}}{M_{ZnCO_3}}\right) + M_{Zn} \right] \frac{X}{M_H} \tag{14}$$

The solubility of $ZnCO_3$ in neutral water is given as 10 mg/L (28). (The often quoted value given by Ageno and Valla (29), 206 mg/L, is evidently in error by a factor of 10 since they also report their value as 1.64×10^{-4} mole/L.) Substituting into Equation 14, using 4.2 as a reasonable annual average pH for the sites (6), and assuming that the runoff from the sample is saturated with $ZnCO_3$, i.e., f=1 gives

$$\begin{aligned} R_L &= (0.356 + 0.281)X \\ &= 0.637X \end{aligned} \tag{15}$$

This is in good agreement with Equation 11 and indicates that runoff losses from zinc are due to a combination of dissolution of corrosion product in water, the principal effect, and hydrogen ion loading. Equilibria involving CO_2, H_2CO_3, carbonate, and bicarbonate are also involved in the dissolution reaction and rough calculations have indicated that the runoff will be decidedly more basic than the incident rain. However, this does not alter the fact that hydrogen ion load is a significant factor in the dissolution of the zinc corrosion product.

Zinc corrosion rates decrease substantially with exposure time, Table I, indicating that the corrosion film is becoming increasingly more protective and retarding further corrosion. Moreover, corrosion rates at the four sites are converging to similar values even though there was a 6-fold difference in their initial rates. Thus, it would appear that a certain mass of corrosion product must form on the metal surface to provide the degree of protection observed in longer exposures. If runoff losses are significant, as suggested by Figure 6 and Equation 11, then zinc must corrode at a rate sufficient to replace the corrosion product lost by dissolution in long exposure times when roughly steady state conditions exist. In this way hydrogen ion load and acid deposition have a very specific and definable role in accelerating the corrosion of zinc.

Scanning electron microscopic examinations of zinc samples exposed 1, 12, and 36 months showed extensive reorganization of the corrosion film through repeated cycles of dissolution and precipitation not unlike that which occurs on carbon steel and Cor-Ten A. The corrosion product on zinc exposed 3 years at the New Jersey site, where short-term zinc corrosion rates are low, is uniform, somewhat nodular, and fine grained on both the skyward and groundward sites, Figure 7. The same was true for samples from the District of Columbia site. On the other hand, the corrosion film on

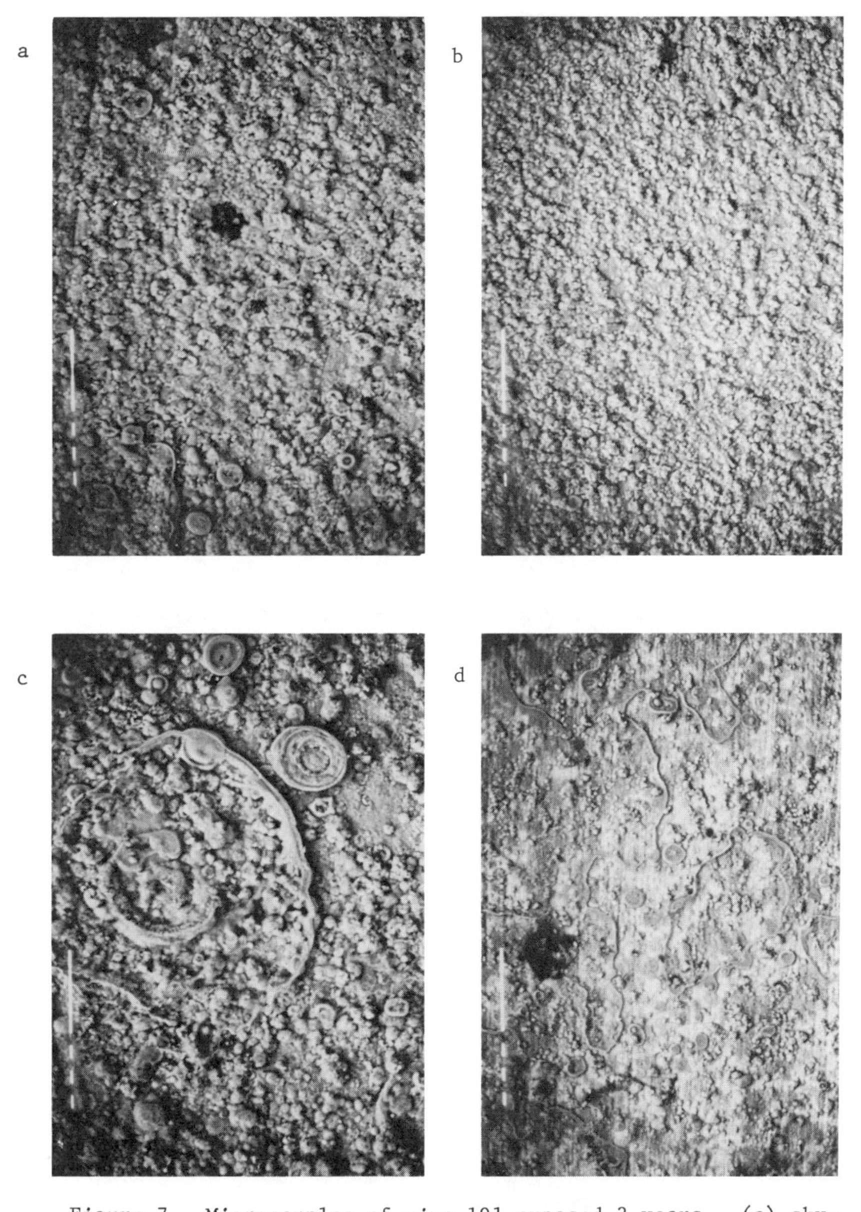

Figure 7. Microsamples of zinc 191 exposed 3 years. (a) skyward side and (b) groundward side, sample C43, from NJ site; (c) skyward side and (d) groundward side, sample C31, from NC site. Magnification: 136X.

zinc exposed 3 years at the North Carolina site, where short-term corrosion rates were highest, was very different from those for New Jersey and skyward and groundward sites were dissimilar. Massive dissolution and precipitation features are superimposed on the skyward side over a uniform, nodular corrosion product similar to that observed at New Jersey. These features, the enlarged and filled oval shapes and the curving solid barrier crossing dozens of the smaller nodular shapes, are evidence for a process of dissolution, concentration, redistribution, and precipitation that builds larger structures on the metal surface. Were this to happen during rainfall the dissolved material would wash away. Instead, as for the steels, it would appear to happen when the surface is drying and involves absorption of acidic gases from the atmosphere. The groundward side, Figure 7D, shows broad open areas covered by only a thin, porous corrosion product and bound by looping barriers of precipitated material. It would appear that corrosion on the groundward side of zinc exposed at the North Carolina site is substantially less than on the skyward side. Single-sided experiments are now in progress to confirm this possibility (6).

Figure 8 shows in greater detail the corrosion product that forms on the skyward side of zinc exposed at the North Carolina and New Jersey sites for 1 and 3 years. The massive features that develop at North Carolina in 3 years, Figure 8B, are not yet apparent after 1 year and the corrosion product that uniformly covers the surface, and on which is superimposed the nodular shapes, appears porous, Figure 8A. In contrast, the uniform layer covering the surface at New Jersey after 1 year, while cracked, is dense and shows evidence of dissolution and redistribution of material to give a corrosion film of more uniform thickness, Figure 8C. The corrosion product formed in 3 years exhibits several types of structures due to still unidentified phases, Figure 8D.

The SEM photomicrographs in Figures 7 and 8 show that reorganization of the zinc corrosion product on weathering is a complex phenomena involving competing processes. For the highly structured features to develop and be preserved over long periods, it is clear that, while dissolution does occur, the corrosion products are not readily soluble and accumulate rather than wash from the surface.

Galvanized Steel. The relationship between corrosion film weight and zinc weight loss for galvanized steel was similar to that shown in Figure 4 for zinc. There were, however, fewer points defining the line for the upper limit to the amount of corrosion product retained on the surface. There was also much more scatter in the data below this line indicating that substantial corrosion product was lost in runoff. The slope for the limiting line was 1.178, which would indicate that ZnO was a major constituent of the corrosion film. However, the slope was recomputed using the zinc present in the corrosion film as determined by wet chemical analysis of the film stripping solution, Equation 8. This produced a linear relationship with very little scatter between film weight and zinc in the corrosion film. Slopes computed from this data for each of the sites, including the Washington, DC, site, were in good agreement and ranged from 1.45 to 1.55. These values are consistent with a corrosion film containing $ZnCO_3$, ZnO, and $Zn(OH)_2$, as reported by

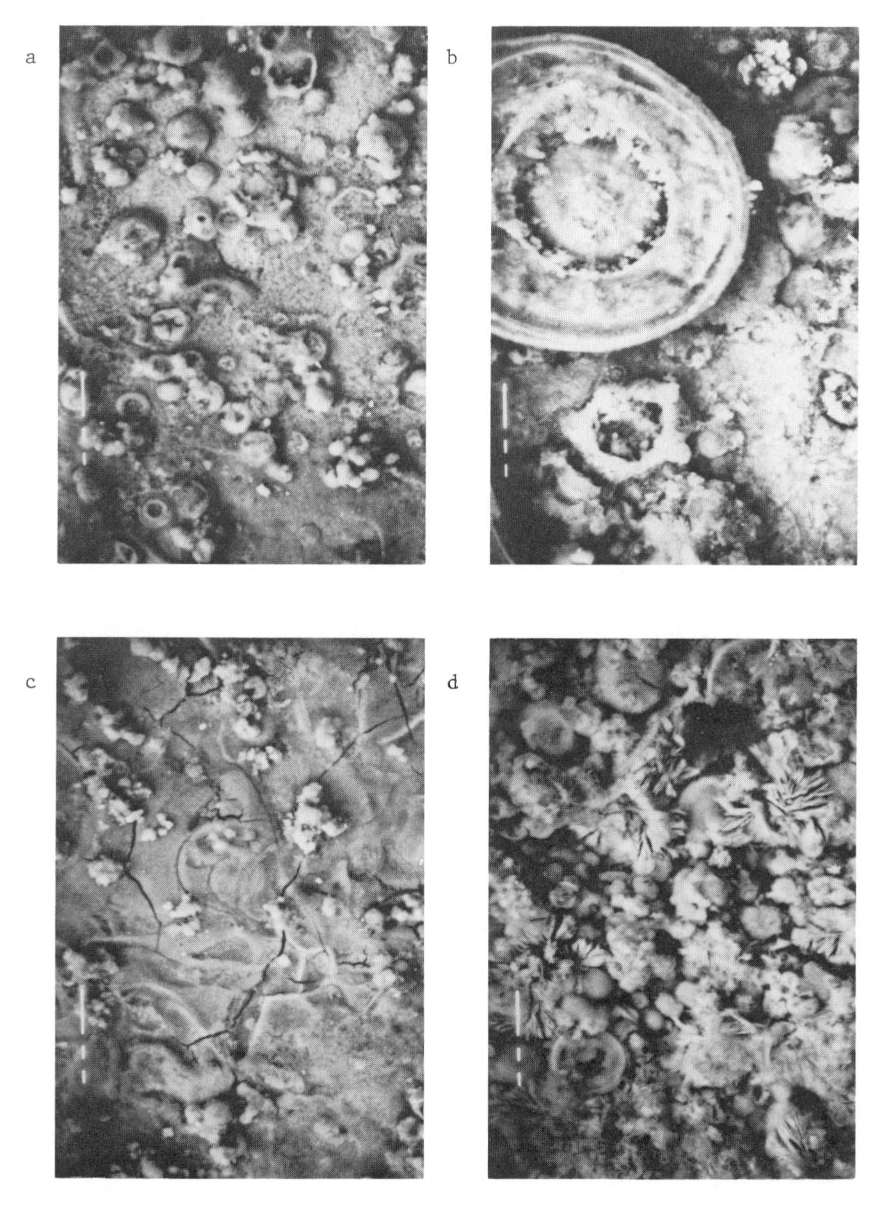

Figure 8. Skyward side of 191 zinc microsamples. (a) 1-year exposure and (b) 3-year exposure at NJ site; (c) 1-year and (d) 3-year exposure at NC site. Samples are (a) C229; (b) C31; (c) C231; (d) C43. Magnification: 720X.

Biestek (27) and Barton (30). Assuming a 2-constituent corrosion film of $ZnCO_3$ and ZnO, a slope of 1.5 corresponds roughly to a corrosion film consisting of 60 wt pct $ZnCO_3$ and 40 wt pct ZnO.

The slopes and the constituents associated with those slopes are different for zinc and galvanized steel. This was unexpected since, to the atmosphere, they should appear the same. However, there was one significant difference between the two materials. The surface of the galvanized steel received a chromate treatment to improve its resistance to "white rusting" during storage (31, 32).

X-ray photoelectron spectroscopy, and ISS analysis of the corrosion film on galvanized steel has shown that the surface is rich in chromium from this chromate treatment, Tables VI and VII. XPS analysis shows both Cr^{3+} and Cr^{6+} present in the corrosion film following a 1-month exposure at the New York site; Cr^{3+} was as much as 6 times more abundant in the corrosion film than Cr^{6+}. Lead was a minor constituent of the corrosion film, probably as PbO or Pb_3O_4. Three peaks were found for oxygen; a relatively small one for the chromium oxides; one for $ZnCO_3$; and a third identified as $Zn(OH)_2$. ISS shows as much as 13 at pct Cr in the outer 1 nm of the corrosion films and small amounts of aluminum. Beyond 1 nm into the corrosion film, the oxygen concentration, after correcting for the amount combined with Cr in Cr_2O_3 and Al in Al_2O_3, was stoichiometrically equivalent to ZnO. The corrosion film on galvanized steel may be similar to that on zinc where there appears to be a ZnO layer on the outer surface, perhaps as a consequence of drying, and $Zn(OH)_2$ is a constituent deeper into the corrosion film. These results clearly show that Cr is present in substantial amounts on the surface of the galvanized steel corrosion film and evidently serves to inhibit the corrosion of the zinc coating.

SEM examination of galvanized steel exposed 3 years at the North Carolina site shows very different corrosion films on the groundward and skyward sites, Figure 9. The large solution-altered features on the skyward side look not unlike those for zinc exposed at the same site. However, the small nodular structure observed on zinc was not present in the corrosion film on galvanized steel and there were areas (one appears as a smooth white area in the lower right corner of Figure 9A) which are uncorroded. These areas are presumed to be due to the continued passivation of the surface by the residual chromate coating. The groundward side, Figure 9B, is characterized by large areas of uncorroded surface. Small areas of localized breakdown are distributed over this surface including areas where corrosion has proceeded along scratch marks. It is probable that these scratches occurred after the chromate treatment and penetrate the protective chromate layer. By removing the chromate in these areas the scratches have made the zinc coating more susceptible to corrosion.

Given the nonuniform distribution of corrosion on the surface of the galvanized steel it is possible that the ISS and XPS results concerning ZnO and $Zn(OH)_2$ are not in conflict; two distinctly different areas may have been examined. Perhaps a largely uncorroded surface, as seen in Figure 9B, was examined by ISS. This area would then be characterized as passivated by Cr and consisting of a thin ZnO layer. On the other hand, corrosion product from a more actively corroding surface, as in Figure 9A, may have been examined by

XPS. This area would be characterized as containing substantial water and $Zn(OH)_2$ and $ZnCO_3$ as well as ZnO.

Three-year corrosion films on galvanized steel from the New Jersey site did not possess solution-altered features on the skyward side. This was true also for the corresponding zinc samples. There were, however, substantial areas of uncorroded surface on both the skyward and groundward sides. While short-term corrosion rates are different between zinc and galvanized steel, in long-term exposures there are close parallels in the development of the weathered corrosion film on both materials.

Galvalume. A plot of film weight versus weight loss showed a scatter of points with a poorly defined upper limit like that for galvanized steel. Corrosion film weights were often less than the weight of material which had dissolved to form the corrosion film. This indicates that Galvalume, like the other zinc containing materials, loses corrosion product to runoff during exposure.

SEM photomicrographs of an unexposed Galvalume blank and Galvalume exposed 3 years at the New Jersey site are shown in Figure 10. Galvalume is a 2-phase AlZn coating which consists of an Al-rich dendritic phase and a Zn-rich interdendritic phase (33). These structures are clearly evident in Figure 10A for the blank, with most of the surface covered by the oxide on the Al-rich phase, and with isolated windows through this dendritic phase exposing the Zn-rich phase. Corrosion appears to be confined largely to those areas where the Zn-rich phase is exposed, Figure 10B. Average corrosion rates were less than 1 μm/y in 1-month exposures at the New Jersey site from May 1982 through May 1984 (6). Actual corrosion rates would be substantially higher than this if based on the area that is corroding, i.e., the interdendritic phase. The voids that are produced in this area leave a surface which, on a microscopic scale, is quite rough and can readily trap fine dust particles, aerosols, and corrosion product.

110 Copper. Plots of film weight versus metal loss for 1- and 3-month exposures of 110 copper generally show a linear relationship between film weight and metal loss. If 1-year exposures are included in the plot they will be scattered and fall below the line established by the 1- and 3-month exposures. This may be an indication of runoff losses in longer exposures. The lines based on short-term data had slopes of 1.12, 1.19, 1.10, and 1.08 for the North Carolina, New Jersey, New York, and District of Columbia sites, respectively, when based on gravimetric data, Equation 6. The District of Columbia data points exhibited somewhat greater scatter then the other sites, which could represent increased runoff loss for these samples.

Plots of film weight versus the copper determined by wet chemistry from the stripping solutions, Equation 8, are linear. The slopes--1.12, 1.15, 1.14, and 1.20 for the North Carolina, New Jersey, New York, and District of Columbia sites--generally confirm the results from the gravimetric data. No significant site-to-site differences were apparent in the slopes. The District of Columbia site had a slightly higher slope using the wet chemical analyses, but it also had the lowest slope by gravimetric methods. The slopes fell between the theoretical values for pure CuO and Cu_2O,

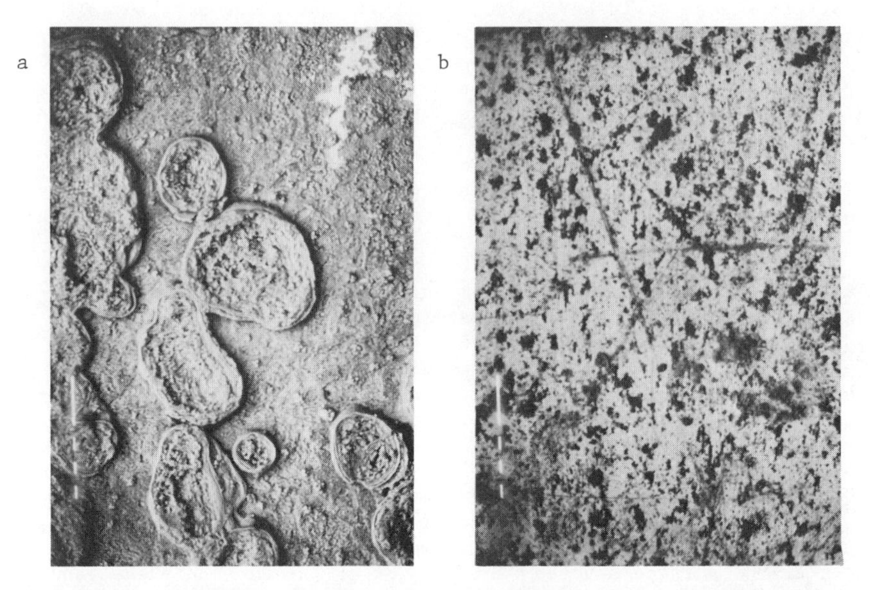

Figure 9. Skyward (a) and groundward (b) sides of galvanized steel microsample G51 exposed 3 years at NC site. Magnification: 80X.

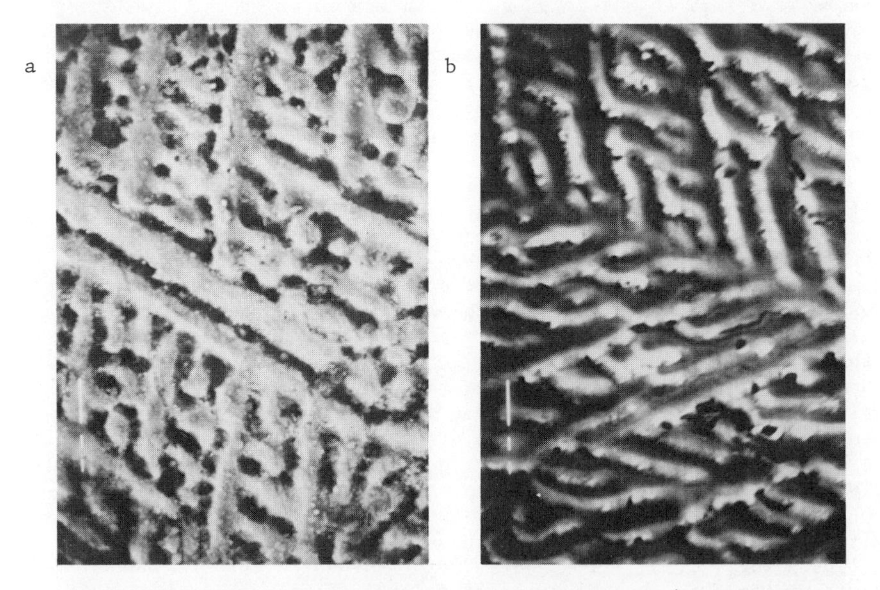

Figure 10. Galvalume blank (a) and skyward side of microsample F38 (b) exposed 3 years at NJ site. Magnification: 720X.

Table III, indicating that the film is composed primarily of these two constituents. These results are consistent with observations of Mattson (34), who indicated that Cu_2O was the primary film constituent for short-term exposures.

ISS analysis of copper films confirmed that the film is composed mostly of oxygen and copper. The atomic percentage of chlorine or sulfur is relatively small and diminishes rapidly with distance into the film. Thermal analysis results for 1-month and 1-year exposures, Table IV, rule out the possibility that large amounts of sulfur, chlorine, or carbonate containing compounds are present in the film. A slow, constant rate of weight loss was observed at temperatures between 110° and 350° C by TGA which was not typical of the thermal decomposition reported for any inorganic copper compounds (23). In measurements on pure CuO, no weight change was observed in this temperature range. Interaction of carbonaceous material deposited on the film with CuO to produce Cu_2O and CO_2 could possibly explain the slow weight loss from samples during TGA. X-ray diffraction results, Table V, show the presence of Cu_2O in samples both before and after TGA.

The SEM photomicrographs in Figure 11 show that copper corrodes locally and nonuniformly. Figure 11B shows a sample, exposed for 1 year at the North Carolina site, which has had its corrosion film stripped. Flat uncorroded areas, identified by the rolling marks from the original copper surface, are present. Figure 11A shows a similar sample with the corrosion film intact. Thick mounds formed in the corrosion film appear to be associated with the areas of localized metal attack seen in Figure 11B.

XPS was used to study films on copper samples from several sites for different exposure periods. As indicated in Table VIII major changes in the composition of the corrosion film surface appear to be occurring with time. Different sites gave similar results for the same exposure period, although there was some variation in the $C1S_{1/2}$ peaks among the sites. This could indicate that the nature of the carbon compounds being deposited on the film varies from site to site. The $Cu2P_{3/2}$ peak was interpreted using data from Schon (35) to distinguish CuO from Cu and Cu_2O. It was not possible to separate the contributions of Cu and Cu_2O. The 1-month samples showed large amounts of both CuO and Cu or Cu_2O. At longer exposures the amount of CuO increased while the amount of the less oxidized species decreased. The groundward sides of the samples tended to have less CuO and more Cu_2O or Cu than the skyward. The results indicate that a layer of CuO is produced in the outer 10 nm of the film surface with increasing exposure time and that this process proceeds faster on the skyward side than the groundward side.

3003-H14 Aluminum. The Al-Mn 3003 alloy corrodes at rates less than 0.1 μm/y in exposures of 1, 3, and 12 months. Rates at the New Jersey site are 2 to 3 times those at the other sites. SEM examination showed pitting in exposures of 12 and 36 months, Figure 12. The pits were in the early stages of development with a small, deeper central area surrounded by shallow localized attack radiating outwards, particularly along grain boundaries. This type of attack is similar to that described by Sowinski (36). Pits were more fully developed at the New Jersey site, encompassing a larger area of

Figure 11. Skyward side of 110 copper exposed 1 year at NC site. (a) microsample D231 with corrosion film intact; (b) wieght-loss panel D1-91 after removing corrosion film. Magnification: 240X.

Figure 12. Skyward side of aluminum 3003-H14 weight-loss panel B2-73 exposed 1 year at NJ site after removing corrosion film. Magnification: 720X.

Table VIII. X-ray photoelectron spectroscopic analysis of corrosion film on microanalysis samples

Metal/Exposure	XPS Peak	Observation[1]
110 Copper/1 month	$Cu2P_{3/2}$	Cu, Cu_2O: major CuO: major Less Cu as CuO on groundward side than skyward side.
	$C1S_{1/2}$	C-H, C-C: major C-O, C=O: minor
	$O1S_{1/2}$	Oxide [CuO, Cu_2O]: major hydroxide, water: minor C=O, C-O: minor
110 Copper/1 year	$Cu2P_{3/2}$	Cu, Cu_2O: major CuO: major CuO $>$ Cu, Cu_2O
	$C1S_{1/2}$	C=O, C-O: major C-H, C-C: minor
110 Copper/3 year	$Cu1P_{3/2}$	skyward side-CuO: major groundward side-CuO: major Cu, Cu_2O: minor
	$C1S_{1/2}$	C=O, C-O: major C-H, C-C: minor
	$O1S_{1/2}$	C=O, C-O: major oxide [CuO, Cu_2O]: minor

[1]Phases in brackets indicate probable identities of peaks.

localized breakdown and involving more intense dissolution of grains and grain boundaries. Pitting was more evident on the skyward side than the groundward side of the samples. Also present on the skyward side was a finely structured nonuniform general attack which textured the surface in exposures of 3 years and obscured original surface detail such as rolling marks and fine scratches. This attack may be associated with small intermetallic $FeAl_3$ and $FeMnAl_6$ precipitates dispersed in the 3003 alloy. Similar texturing of the groundward side had not occurred in 3-year exposures and much of the surface detail was retained.

Conclusions

For purposes of analysis, growth of an atmospheric corrosion film was viewed schematically as a three step process consisting of: (1) formation of a corrosion product; (2) weathering of that product while conserving mass of the metal; and (3) corrosion product losses through runoff and spalling. One or more of these steps may be occurring during any stage in film growth depending upon the material and the environment. Corrosion films formed on metals and metal-coated steel products exposed by the Bureau of Mines at field sites in the eastern U.S. for times from 1 to 36 months had: (1) chemistries that were independent of environment in short-term exposures; (2) morphologies that were, in some cases, highly altered by solution processes; and (3) runoff losses for zinc that were dependent upon hydrogen ion loading.

Corrosion Film Chemistry. A linear relationship exists between the mass of corrosion product formed on carbon steel, Cor-Ten A, zinc, galvanized steel, and copper and the mass of metal in the corrosion film. This relationship is independent of site and the wide variation in environmental parameters between the sites in short-term exposures of 1 and 3 months. The ratio of the two masses is relatively sensitive to the composition of the corrosion film. The independence of this ratio from substantial variations in air quality, meteorology, and rain chemistry is interpreted as indicating, at least for the major constituents, that the composition of the corrosion film is independent of the environment in short-term exposures.

The corrosion film on Cor-Ten A contains between 30 and 70 wt pct FeOOH, about 4-12 wt pct loosely-bound water, about 1-5 wt pct water of hydration, and the balance hydrated Fe_2O_3. No significant concentrations of FeO, Fe_2O_3, Fe_3O_4, $FeSO_4$, and $Fe(OH)_2$ are present. The corrosion film on carbon steel is similar to that on Cor-Ten A in short exposures.

The corrosion film on zinc contains about 65 wt pct $ZnCO_3$ and 35 wt pct $Zn(OH)_2$. The corrosion film from the Washington, DC, site may also contain ZnO. The corrosion film on galvanized steel consists of $ZnCO_3$, $Zn(OH)_2$, and ZnO. The presence of the ZnO is probably due to the stabilization of an initial passive ZnO film by Cr^{3+} which persists over parts of the surface in exposures up to 3 years.

The corrosion film on copper contains mostly Cu_2O and some CuO. The corrosion films from the Washington, DC, site contain a higher percentage of CuO, i.e., 40 wt pct Cu_2O and 60 wt pct CuO. The films do not contain large amounts of the copper sulfides, sulfates, hydroxides, carbonates, chlorides, or combinations of these compounds.

Corrosion Film Weathering. Significant spalling and runoff losses occur for all of the metals in exposures of 1 year. Zinc and galvanized steel exhibit runoff losses in 3- and 12-month exposures. Massive reorganization of the corrosion films on carbon steel, Cor-Ten A, zinc, and galvanized steel occurs which produces new morphologies in the weathered film. These morphologies are dependent on site, exposure time, and orientation (skyward and groundward). They develop over extended periods of time, indicating that the solution and precipitation processes producing them do not occur primarily during precipitation events, when washing would totally remove dissolved corrosion products, but during the drying phase of precipitation events and when moisture collects on the surface with little or no runoff, as with dew. The solution present on the surface during the drying phase is a powerful solvent, altering features of the corrosion film which appear unaffected by the continuous washing of rain. Absorption of acidic gases from the atmosphere is probably important to the formation of this solvent. With continuing evaporation of the solution, dissolution of corrosion product diminishes and precipitation begins. Differences exist in the details of this process for the individual metals, e.g., Cor-Ten A and zinc. Corrosion rates for all of the metals except Galvalume and aluminum decrease with increasing time, indicating the formation of a more protective corrosion film in the longer exposures. Conditions which affect the weathering processes described here will have a marked effect on the corrosion of the metals themselves, as their ability to achieve a stable, low corrosion rate in long-term exposures depends entirely on the development of the corrosion film.

Runoff Losses Due to Hydrogen Ion Loading. Runoff losses from zinc, and presumably galvanized steel, are dependent upon hydrogen ion load in 3- and 12-month exposure. Such an effect is not apparent in 1-month exposures. Hydrogen ion dissolves zinc carbonate, perhaps the major constituent of the corrosion film, by the reaction

$$ZnCO_3 + H^+ \rightarrow Zn^{2+} + HCO_3^-$$

Additional corrosion film is lost in the runoff due to the limited solubility of the corrosion product in rain. The relative contribution of these effects to the runoff in 3- and 12-month exposures was 55 pct dissolution and 45 pct hydrogen ion loading. To maintain the stable corrosion film that develops on zinc in long-term exposures, it is evident that zinc must corrode at a rate sufficient to replace the corrosion product lost in runoff.

Acknowledgment

This research has been funded by the Bureau of Mines and the National Acid Precipitation Assessment Program through a cost-sharing Interagency Agreement between the Bureau and the Environmental Protection Agency. Use of trade names or company names does not imply endorsement by the Bureau of Mines.

Literature Cited

1. Flinn, D. R.; Cramer, S. D.; Carter, J. P.; Lee, P. K; Sherwood, S. I. "Acidic Deposition and the Corrosion and Deterioration of Materials in the Atmosphere: A Bibliography. 1880-1982"; PB83-126091; National Technical Information Service, July 1983.
2. Haynie, F. H.; Upham, J. B. In "Corrosion in Natural Environments"; STP 558; American Society for Testing and Materials: Philadelphia, 1974; 33-51.
3. Mansfeld, F. B. "Effects of Airborne Sulfur Pollutants on Materials"; PB81-126351; National Information Service, January 1980.
4. Lipfert, F. W.; Benarie, M.; Daum, M. L. Derivation of Metallic Corrosion Damage Functions For Use in Environmental Assessments. Draft Report, available from senior author, Dept. of Energy and Environment, Brookhaven National Laboratory, Upton, NY 11973.
5. Franey, J. P.; Graedel, T. E.; Kammlott, G. W. In "Atmospheric Corrosion"; Ailor, W. H., Ed.; John Wiley and Sons: New York, 1982; 383-392.
6. Flinn, D. R.; Cramer, S. D.; Carter, J. P.; Spence, J. W. Durability of Building Materials, 1985, 3(2), 147-175.
7. Bureau of Mines. Quality Assurance Project Plan. Interagency Agreement AD-14-F-1-452-0 between Bureau of Mines and Environmental Protection Agency, October 1983. (Available from D. R. Flinn, Bureau of Mines, Avondale Research Center, Avondale, MD 20782-3393.)
8. Reddy, M. M.; Sherwood, S. I. Limestone and Marble Dissolution by Acid Rain. In this book.
9. Youngdahl, C. A.; Doe, B. R.; Sherwood, S. I. Roughening Recession and Chemical Alteration of Marble and Limestone Sample Surfaces After Atmospheric Exposure in the Northeastern United States. In this book.
10. Cohen, M.; Hashimoto, K. J. Electrochem. Soc., 1974, 121(1), 42-45.
11. Inouye, K.; Ichimura, K.; Kaneko, K.; Ishikawa, T. Corrosion Sci., 1976, 16, 507-517.
12. Kameko, K.; Inouye, K. Bull. Chem. Soc. of Japan, 1976, 49(12), 3689-3690.
13. Suzuki, I.; Masuko, N.; Hisamatsu, Y. Corrosion Sci., 1979, 19, 521-535.
14. Cramer, S. D.; Carter, J. P.; Covino, B. S., Jr. "Atmospheric Corrosion Resistance of Steels Prepared from the Magnetic Fraction of Urban Refuse"; U.S. Bureau of Mines, RI 8477, 1980.
15. Suzuki, I.; Hisamatsu, Y.; Masuko, N. J. Electrochem. Soc., 1980, 127(10), 2210-2215.
16. Matijevic, E. Pure and Applied Chemistry, 1980, 52(5), 1179-1193.
17. Spedding, D. J.; Sprott, A. J. Proc. 8th Intern. Congr. Met. Corr.; Dechema: Frankfurt, 1981; Vol. 1, 329-335.
18. de Meybaum, B. R.; Ayllon, E. S.; Bonard, R. T.; Granesse, S. L.; Ikeha, J. L. Proc. 8th Intern. Congr. Met. Corr.; Dechema: Frankfurt, 1981; Vol. 1, 317-322.

19. Keiser, J. T.; Brown, C. W.; Heidersbach, R. H. Corrosion, 1982, 38(7), 357-360.
20. Leidheiser, H., Jr.; Czako-Nagy, I. Corrosion Sci., 1984, 24(7) 569-577.
21. Cleary, H. J. Corrosion, 1984, 40(11), 606-608.
22. Albrecht, P.; Naeemi, A. H. "Performance of Weathering Steel in Bridges"; National Cooperative Highway Research Program Report 272, Transportation Research Board, National Research Council: Washington, DC, 1984.
23. Liptay, G. "Atlas of Thermoanalytical Curves"; Heyden and Sons: New York, 1971; Vol. 2, Section 89.
24. Evans, U. R. "The Corrosion and Oxidation of Metals"; E. Arnold: London, 1960; p. 25.
25. Horton, J. B. Proc. San Francisco Regional Technical Meeting; American Iron and Steel Institute: Washington, DC, November 18, 1965; 171-195.
26. Anderson, E. A.; Fuller, M. L. Metals and Alloys, 1939; Vol. 10, pp. 292-287.
27. Biestek, T. In "Atmospheric Corrosion"; Ailor, W. H., Ed.; John Wiley and Sons: New York, 1982; 631-643.
28. Handbook of Chemistry and Physics; 60th Ed.; CRC Press: Boca Raton, FL, 1979; B-142.
29. Ageno, F.; Valla, E. Hydrolysis. Atti Accad. Lincei, 1911; Vol. 20, Part II, 706-712.
30. Barton, K. "Protection Against Atmospheric Corrosion"; Wiley and Sons: New York, 1973; 49.
31. Williams, L. F. G. Plating, 1971, 59(10), 931-938.
32. Duncan, J. R. Surface Technology, 1982, 16, 163-173.
33. Zoccola, J. C.; Townsend, H. E.; Borzillo, A. R.; Horton, J. B. In "Atmospheric Factors Affecting the Corrosion of Engineering Materials"; STP 646; Coburn, S. K., Ed.; American Society for Testing Materials: Philadelphia, 1978; 165-184.
34. Mattson, E.; Holm, R. In "Atmospheric Corrosion"; Ailor, W. H., Ed.; John Wiley and Sons: New York, 1982; 365-381.
35. Schon, G. Surface Sci., 1973, 35, 96-108.
36. Sowinski, G.; Sprowls, D. O. In "Atmospheric Corrosion"; Ailor, W. H., Ed.; John Wiley and Sons: New York, 1982; 297-328.

RECEIVED January 2, 1986

8

Bronze, Zinc, Aluminum, and Galvanized Steel: Corrosion Rates as a Function of Space and Time over the United States

D. E. Patterson[1,3], R. B. Husar[1], and E. Escalante[2]

[1] **Washington University, St. Louis, MO 63130**
[2] **National Bureau of Standards, Gaithersburg, MD 20899**

The corrosion of metals exposed to the atmosphere is known to be caused by a mixture of natural and anthropogenic factors. To apportion the cause of metal corrosion, one may conduct controlled laboratory experiments or well designed field exposure experiments. A complicating factor may arise in the case that the progress of corrosion is not constant in time, but may change significantly after an initial period, e.g. due to erosion or formation of protective surface layers. Such erratic behavior has been documented by Guttman (1) for variability of zinc corrosion with season of first exposure. Therefore it is also desirable to examine the records of long term exposure studies to better reflect the actual fate exposed metals. Unfortunately, the environmental data associated with such exposures are generally insufficient to document the meteorological and chemical causes of the corrosion.

The usefulness of existing long term exposure metals corrosion data thus depends upon reconstruction of the meteorological and chemical histories which are relevant to corrosion. To do so involves analysis of data on meteorology and pollutant emissions in conjunction with data interpolation tools, i.e. pollutant dispersion models. This report discusses the current status of such an effort at Washington University, and examines the existing exposure data for evidence of key features which may clarify the likely importance of manmade pollutants in metals corrosion.

Metal Corrosion Data

The existing long term exposure metal corrosion data has been reported in a number of papers in journals and proceedings over the years. The considerable task of locating, verifying, standardizing and assembling the numerous bits of information into a coherent whole was undertaken by E. Escalante of the National Bureau of Standards. Thus, the final data set contains only those experiments which were conducted according to standard ASTM procedures and had adequate documentation of the site characteristics and period of exposure.

[3] Current address: 3073 Andover, St. Louis, MO 63121

0097-6156/86/0318-0152$06.00/0

The corrosion data includes periods from 1936 to 1978, with exposure durations of 1 to 20 years. Most data refer to the period of the 1960s, and reflect durations of 1 to 7 years. The sites are predominantly in the eastern United State or in California.

Six metals are included in the assembled records: carbon steel, weathering steel, galvanized steel, zinc, aluminum and bronze. A total of 63 sites are found in the data base, with many sites reporting values for numerous metals. Sites were classified as marine, industrial, or urban. The site distribution is given below.

Metal	Industrial	Rural	Marine
Aluminum	7	5	6
Bronze	3	5	4
Galvanized Steel	5	9	2
Weathering Steel	2	2	1
Zinc	17	8	10
Carbon Steel	22	10	16

The parameters for the corrosion data include site location and classification, duration and time period of exposure, source reference, and a series of corrosion measures: corrosion rate ($mg/dm^2/da$, or mdd), percentage change in strength and elongation, pit depth, years to first rust, and others which were rarely used. The change in strength and in elongation are more directly important as measures of structural integrity than is the more commonly measured weight loss measure of mdd.

In the process of examining the metal corrosion data, care was taken to average only over narrow time ranges. A graphical presentation of the averaged data value is contained in the final report associated with the metal corrosion study (2). The reader is referred to that report for the details of the data set. In this paper we present some key results.

Paradigm

The retrospective study has been formulated with a clear paradigm of the parameters of the corrosion system. It is assumed that the significant contributors to the corrosion of metals exposed to the atmosphere are water, sea salts, sulfur oxides, nitrogen oxides, and the acidity of precipitated water.

In the absence of water, it is presumed that no significant corrosion activity will take place. Surfaces may be moist due to precipitation, dew, hygroscopic action, and other causes. Precipitation is presumed to act in three modes: by cleansing the surface of accumulated dry deposited matter, by moistening the surface, and (at low pH) by direct chemical attack upon the metal and corrosion products.

In the presence of moisture, it is assumed that the key chemical reactions are initiated by SO_2 and perhaps by NO_x gases. Other mechanisms, such as dry deposition of acidic aerosol or deposition of HNO_3 to dry surfaces, are not considered. Synergistic effects are similarly disregarded.

With this outlook, the key factors appear to be the coincident occurrence of SO_2 and moisture. The controlled metal exposures from the ASTM studies were of flat panels in standard orientation. The simple geometry avoids many complications from moisture retained in surface texture features. Therefore, as a first approximation, we take the relevant time of wetness to be short periods following precipitation and periods of dew formation. The formation of a film of water is most likely at high relative humidities; high humidities are most prevalent during the night hours. In the eastern U.S., summer predawn humidity is typically 80-90% compared to afternoon humidities of 50-60%; winter patterns are less pronounced in areas north of the Ohio river, but also show afternoon minima (3).

The impact of near surface and elevated emission sources of SO_2 on surface concentrations has been shown to differ very strongly (4). Elevated emission sources contribute little to night time SO_2 concentrations, with the plume typically touching down in midmorning during summer conditions. Near surface emissions, however, cause high concentrations of SO_2 (and NO_x) overnight. These surface source emissions are trapped within a shallow nocturnal mixing layer with relatively little dispersion, thus making SO_2 available at the same time that moisture is most likely to be present. The "corrosion potential" for nearby surface sources is thus disproportionately large for its emissions.

Retrospective Reconstruction of Environmental Histories

The reconstruction of the pollutant concentrations to which a material is likely to have been exposed requires both emission information and diffusion models to relate emission trends to ambient concentrations downwind.

Using historical archives of fuel use, Husar (5) has developed and applied a methodology for estimating the historical SO_2 emissions on a state by state basis back to 1900. No such central record exists, unfortunately, for emissions on the urban scale.

The CAPITA Monte Carlo regional model of pollutant transmission through the atmosphere was developed as a diagnostic tool to simulate the formation and transport of sulfate aerosol. It has subsequently been modified and applied to simulation of sulfur wet deposition as well as sulfate concentration (6,7). For the current purposes,the model was substantially modified in order to adequately model the diurnal patterns of SO_2 impact from near surface and elevated sources. The simulation is appropriate for the regional scale, with time step of 3 hours and spatial resolution of about 100 km in a grid of roughly 12000 km^2. This scale, however, is inadequate for simulation of primary pollutants such as SO_2 within the first 100-200 km of transport. The average over the grid, even if correct,does not reveal the orders of magnitude higher concentrations experienced in small areas nearest the sources of emissions.

Therefore a local model was developed to better simulate the near field concentration and deposition impacts of primary emissions. This simulation utilizes much of the same formulation in terms of meteorology as the regional model (with improved initial vertical dispersion simulation), and with the identical kinetics for transformation of SO_2 to $SO_4^{=}$, dry deposition of sulfur, and wet deposition of sulfur. This model was used to estimate correction factors to be applied to the regional model results.

The full methodology of retrospective analysis was applied for the case of marble tombstone deterioration in an urban and a remote cemetery near New York City (8,9). In this exercise it was found that our technique may estimate rural SO_2 concentrations acceptably well, but it does not reproduce detailed structure of urban concentrations adequately. The main reasons are uncertainty about the effective release heights of sources and a lack of detailed local emission information. For the purposes of metal corrosion, the current state of the retrospective reconstruction of environmental histories is not sufficiently quantitative to warrant extraction of damage functions.

The data set of metal corrosion provides a valuable insight as to which parameters dominate the corrosion process. The general meteorological features are well established in North America (although micrometeorological factors may dominate at any single exposure site). The pollutant related questions may be simply stated: (1) what are the magnitudes of marine and industrial corrosion rates compared to the rural values; (2) what are the relative rates in the rural eastern United States compared to the western rural values; (3) what are the broad trends in corrosion rate over the years?

Roughly speaking, the first question addresses the importance of SO_2, while the second and third relate to the regional scale impacts of SO_2, $SO_4^{=}$, and acidic deposition. SO_2 in urban areas is a matter for local control efforts, whereas the regional impacts (which are due to long range transport of emitted material) require larger scale controls and fall into the category of "acid rain" associated damage.

Observations

This section discusses a few of the major patterns evident in the metals corrosion database. In many respects, the results are restatements of the findings of earlier research by those associate with the original projects. For the purposes of the retrospective reconstruction project, it is nevertheless useful to determine whether the initial assumptions appear reasonable and to examine the relative importance of the meteorological and chemical factors as revealed in the data. This will tell where to place additional effort for the best improvement. More specifically, if the aim of the study is to explain the observed variations in corrosion rate, we first wish to know what variations are strong enough to warrant further study.

First the data were examined for variability. The convenient measure of noise to signal (standard deviation divided by the mean) yielded values which were quite large for a large number of carbon steel exposure tests. Due to this variability and the typically short exposure times, carbon steel was not examined in greater detail. Single year exposures of zinc plates indicated that the relative variability was 30% for marine sites, 20% for rural sites, and 10% for industrial sites. The marine sites had the highest corrosion rates as well as the highest relative variability, while the industrial sites indicated rather consistent behavior even over this short time period. At the other extreme, bronze triplicate panels yielded relative variability of below 10% after 20 year exposures. The overall result of such considerations was the decision to average data wherever multiple values were available with nearly identical exposure times. Similarly, it indicated that small differences in corrosion rates (particularly during the first few years) should not be overinterpreted.

The near coastal effect of wetness and sea salt is evident in the data, particularly for 1 and 2 year exposures. Zinc corrosion rates at Cape Kennedy, FL, dropped by a factor of 4 as distance from the coast increased from 60 to 880 m. At Kure Beach,NC, Zn rates dropped by a factor of 3 as distance increased from 25m to 250m from sea; corresponding Al corrosion rates fell by a factor of 2. Clearly sites within a few hundred meters of the sea must be segregated from other sites in any further analysis.

The general meteorological environment in the continental U.S. is well known, ranging from the desert Southwest to the roughly 1m annual rainfall over much of the country east of the Mississippi river. The number of days with measurable precipitation is about 50% higher in the northeastern quadrant of the nation than in the Great Plains. The number of days with heavy fog is higher by a factor of 2 or more in the mountainous portions of the eastern U.S. and in coastal areas than in the Great Plains or lower Ohio River valley (3). Therefore some variation in corrosion rates within the eastern U.S. may be expected just from variation in time of wetness.

Four sites with multiple alloys exposed over the years recur in most of the metal corrosion data sets. These sites include rural marine exposures at Point Reyes, CA and Kure Beach, NC; industrial exposures at Newark, NJ; and an eastern U.S rural inland location at State College, PA.

Aluminum alloys indicate generally low corrosion rates, except in the Chicago area; for several alloys, corrosion rates there were nearly an order of magnitude higher than at other sites. For AL 6061 (T6), the industrial environment caused substantial corrosion and loss of strength. For three other alloys (AL 2014 (T4), AL 2024 (T81) and AL 7075 (T6)), the higher rates were noted at the marine sites, with less industrial effect. For all these alloys, the rural Pennsylvania corrosion rates were lower than the corresponding rates at industrial sites, by a factor of 2 to 7 after 7 years exposure. The rural site exposures induced no significant strength loss. Therefore it may be tentatively concluded that the regional scale causes of aluminum corrosion are not sufficiently pronounced to warrant further research at this time.

Weathering steel data indicated rather constant corrosion rates after 8 year exposures (1967-1975) at marine, urban and rural sites in the eastern U.S. Although this data is limited, it suggests that the corrosion history of this weathering steel alloy must be relatively insensitive to both SO_2 concentration and to the pH of rainfall. An alternative explanation is that pollutant levels above some threshold value are sufficient to induce the observed corrosion. Even so, the data do not offer clues as to the pollutant effect.

Galvanized steel data is available from 20 year exposures (1936-1956) and from 2 year exposures (1971-1973). The twenty year exposures show that corrosion rates in the Great Plains range from 0.2 to 0.4 mdd, while rural rates in the Ohio River valley and the northeastern U.S. range 0.5 to 0.6 mdd. Such rural differences are likely attributable to variations in time of wetness. Industrial exposures are associated with rates of 1-3 mdd.

The more recent short term exposures at the rural State College site indicate lower corrosion rates than were registered during the earlier 20 year exposures. Although it is difficult to make such comparisons, the suggestion that rural eastern U.S. corrosion rates have increased over the past several decades is not supported by these measurements.

Zinc provides the most extensive set of measurements among the metals. Data are available from 7 year exposures ending in 1956 and in 1965, as well as dual sets of 20 year exposures ending in 1951 and 1978. The data show strong dependence on presumed time of wetness. The two distinct 20 year exposure periods include use of the same site at State College, PA. At this rural site, the long term corrosion rates over the 1931-1951 period were identical to those of the 1958=1978 period. Thus the idea of increasing rural corrosion rates over the past 5 decades is not supported. Industrial corrosion rates are substantially higher than rural rates; during the earlier exposure study, neighboring sites indicated industrial rates 3 to 6 times higher than those recorded at rural sites (Figure 1).

These data do not offer enough spatial coverage to assess the differences between eastern and noneastern rural corrosion rates. The Interstate Surveillance Project (10) provided measures of zinc corrosion rates at a number of sites in the Great Plains south to Texas as well as in the eastern U.S. This data represents the effects after single year exposures, so that large variability is expected. A clear overall pattern is evident, however, which suggests that eastern U.S. rural corrosion rates for zinc are higher than the corresponding rates in the Great Plains by a factor of 2 to 3. This is more than can be accounted for by meteorological factors alone; it is presumed, therefore, that rural zinc corrosion rates in the eastern U.S. are affected to a significant degree by regional pollution. The regional effect is not likely to be a new phenomena.

Bronze data is sparse. An alloy of 92% Cu exposed from 1931-1951 showed no corrosion in the desert environment and very little at the rural sites. Coincident zinc data indicates that over the 20 year period the average corrosion rates were higher for bronze at marine sites but much lower for bronze than for zinc at rural sites (Figure 2).

At the rural State College, PA, site the corrosion rate for two different bronze alloys during the 1958-1978 period was nearly twice that of the earlier bronze exposure, while the industrial Newark area corrosion rates appeared to have declined.

A bronze with 99% Cu in alloy with 1.25% Sn and P was exposed during the 1958-1978 period at the four standard ASTM sites. The corrosion rates at the rural site and the western marine site were less than half those at the eastern marine and industrial sites.

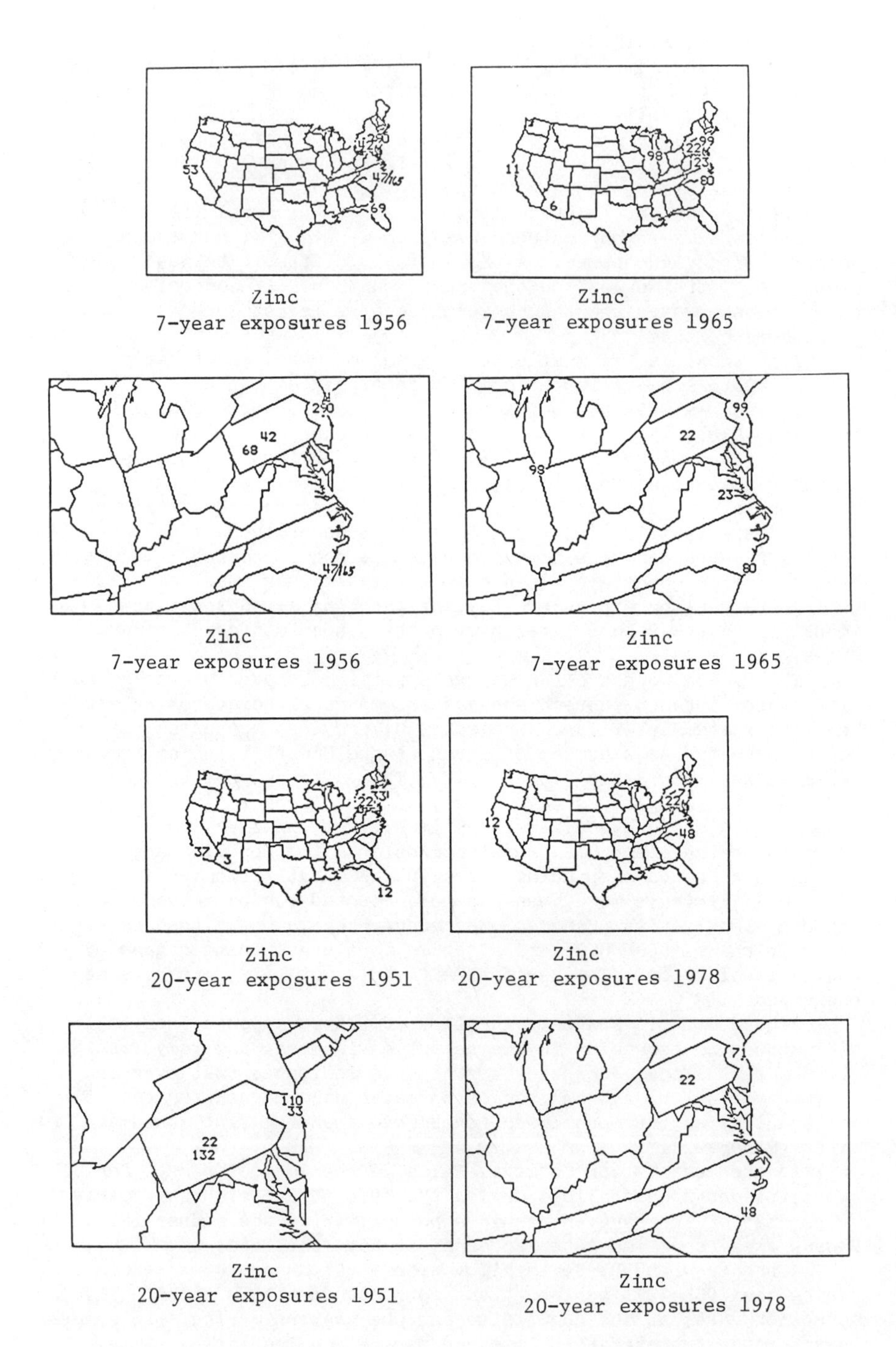

Figure 1. Corrosion rates (mg/dm^2,da) for zinc.

Bronze Corrosion
1.25% Sn, P: 1958-1978
% Strength (7)

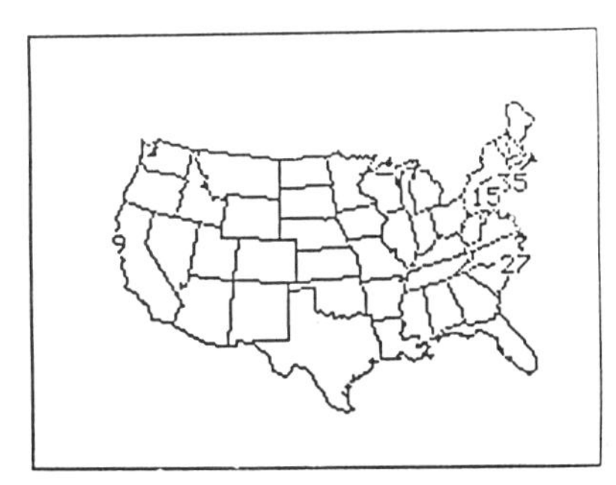

Bronze Corrosion
1.25% Sn, P: 1958-1978
20 yr

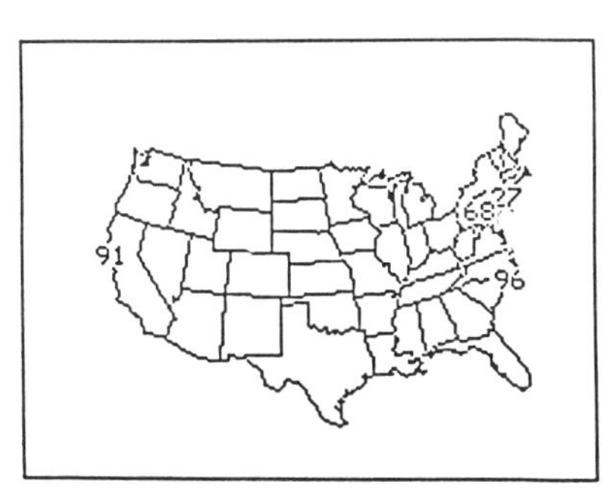

Bronze Corrosion
1.25% Sn, P: 1958-1978
% Loss attach

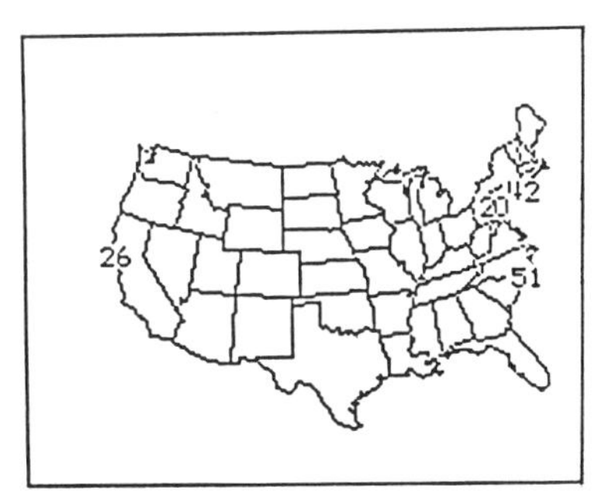

Bronze Corrosion
1.25% Sn, P: 1958-1978
2 yr

Bronze Corrosion
92% Cu: 1931-1951
mg/dm^2, da

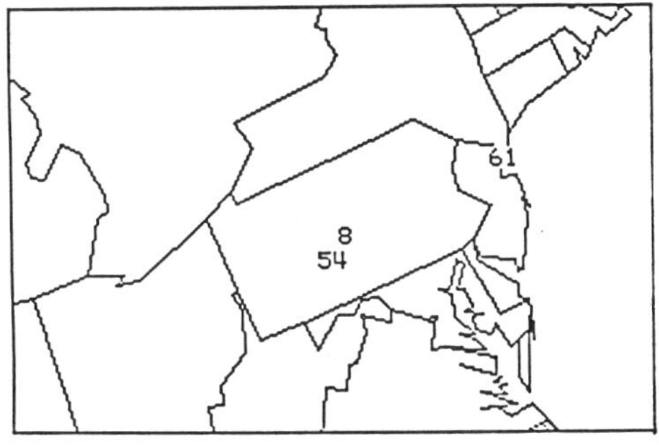

Bronze Corrosion
92% Cu: 1931-1951
mg/dm^2, da

Figure 2. Corrosion rates (mg/dm^2,da) of bronze (92% Cu).

Chemical analyses of the corrosion product (11) indicated that no sulfur compounds were present at either of the marine sites; apparently pollutants did not cause the marine bronze deterioration. Over 90% of the corrosion product was still adhering at these marine sites after 20 years. At the rural site, 67% remained, while only 27% of the corrosion product remained at the industrial site.

A bronze alloy more representative of statuary bronzes (5% Sn and P) was also exposed at the four standard sites. The rural and Western marine sites again show corrosion rates less than half those of the eastern marine and the industrial site. The corrosion rates at the industrial and the marine site increase with exposure time for this alloy, in marked contrast to the usual pattern of declining corrosion rate with time. It is also curious that the measured loss of strength at the rural PA site is disproportionately large for its corrosion rate (1.4% vs 2.0% at the industrial site).

The bronze alloys were exposed at the same time and place as the zinc during the period 1958-1978, with measurements at 2, 7 and 20 year exposures. Examination of the corrosion rates revealed two separate patterns: a marine trend of high initial corrosion rate with sharp reduction in rate after the first few years, and an industrial-rural trend correspond to different chemical mechanisms at work, with only the industrial-rural corrosion being related to sulfur compounds.

The corrosion rates over distinct time periods are given below. The three sets of rates correspond to the initial two years of exposure, the 5 years covering the period from 2 to 7 years after exposure, and the final 13 years for the period of 7 to 20 years after exposure.

Site	Zinc			1% Bronze			5% Bronze		
	2	5	13	2	5	13	2	5	13
Kure Beach, NC	.77	.81	.30	.51	.40	.18	.92	.49	.30
Point Reyes, CA	.26	.05	.13	.26	.04	.09	.55	--	.13
Newark, NJ	.68	1.11	.56	.42	.38	.33	.34	.51	.51
State College, PA	.18	.24	.23	.20	.14	.15	.14	.14	.20

Although ratios of the accumulated weight losses do not show a consistent pattern, the discrete time periods do show a sudden agreement after the initial 7 year exposure period. During the last 13 years, the ratio of bronze to zinc corrosion rates (in mdd) falls into the range 0.65 ± .05 for both the marine and nonmarine sites for the 1% alloy, with marine values of 1.00 and nonmarine values of 0.90 ± .03 for the 5% alloy. These ratios correspond to factors of the coefficients for Cl^- and SO_2, respectively, in damage functions such as proposed by Benarie (12). An additional bronze alloy containing 7% Al was examined, and no such relationship was found. Clearly the behavior of a bronze is strongly dependent upon the alloy; it appears that the phosphor bronzes exhibit corrosion behavior quantitatively similar to that of zinc after the first few years of exposure.

Summary

A project was undertaken to perform retrospective reconstruction of environmental histories at the sites of previous long term atmospheric metal exposures. The effort required development of appropriate emission information and dispersion modeling capabilities on both the regional and urban spatial scales. The development of useful urban scale emission inventories dating back several decades proved to be a limiting factor. At present, therefore, the retrospective reconstruction of environmental histories is not possible for the large number of sites in the metals corrosion data base. This precludes derivation of damage functions at this time.

Examination of the metal corrosion data base itself has indicated that the reported corrosion values for zinc and bronze, at least, may be associated with regional pollution as well as local SO_2 sources. It also was found that zinc and the two tin-phosphor bronzes behave similarly after an initial period of exposure to the atmosphere.

Literature Cited

1. Guttman, H. (1968) Effects of atmospheric factors on rolled zinc. Atmospheric Corrosion of Metals, ASTM STP 767, American Society for Testing and Materials, 286-308.
2. Patterson, D.E. and Husar, R.B. (1985) Final Report on Metals Damage. Submitted to National Park Service, Preservation Assistance Division, Washington, DC.
3. Baldwin, J.L.(1973) Climate of the United States. Environmental Data Service, NOAA, U.S. Dept. of Commerce, UDC 551.582(73).
4. Husar, R.B., Patterson, D.E., Husar, J.D., Gillani, N.V. and Wilson, W.E. (1978). Sulfur budget of a power plant plume. Atmos. Environ. 12, 549-568.
5. Husar, R.B. (1985). Manmade SO_x and NO_x Emission Patterns and Trends. In National Academy of Sciences Trend Report.
6. Husar, R.B., Patterson, D.E. and Wilson, W.E. (1985a) Monte Carlo simulation of regional air pollution: transport dynamics. Submitted to Journal of Climate and Applied Meteorology, June 1985.
7. Husar, R.B., Patterson, D.E. and Wilson, W.E. (1985b) Monte Carlo simulation of regional air pollution: semiempirical regional scale source-receptor relationship. In internal EPA review.
8. Husar, R.D., Patterson, D.E.(1985c) SO_2 concentration estimates for New York City,1880-1980. Report to EPA Atmospheric Sciences Research Laboratory, Research Triangle Park, NC.
9. Husar, R.B., Patterson, D.E. (1985d). Marble tombstone deterioration in New York City area.Report to EPA Atmospheric Research Laboratory, Research Triangle Park, NC.
10. Cavendar, J., Cox, W., Georgevich, M., Huey, N., Yutze, G. and Zimmer C. (1971). Interstate Surveillance Project. US EPA, GPO 5503-0006, Washington, DC.
11. Costas, L.(1982) Atmospheric corrosion of copper alloys exposed for 15 to 20 years. Atmospheric Corrosion of Metals, ASTM STP 767, American Society for Testing and Materials, 106-115.

12. Benarie, M. (1984) Metallic corrosion functions of atmospheric pollutant concentrations and rain pH. Report under contract DE-AC02-76CH0016 to Department of Energy.

RECEIVED March 19, 1986

9

Environmental Factors Affecting Corrosion of Weathering Steel

Fred H. Haynie

Atmospheric Sciences Research Laboratory, U.S. Environmental Protection Agency, Research Triangle Park, NC 27711

Weathering steel samples were exposed for periods of up to 30 months at nine air monitoring sites in the St. Louis, Missouri area. Climatic and air quality data were recorded during the exposure period and subjected to a rigorous evaluation to eliminate recording errors and to estimate missing values. Weight loss was used as the measure of steel corrosion. Corrosion rate was evaluated with respect to, 1) flux of pollutants (sulfur oxides, nitrogen oxides, oxidants, and particles) to the steel during both wet and dry periods, 2) temperature, and 3) exposure history. Different definitions of when the steel was wet were evaluated to determine the most likely "critical relative humidity." Non-linear multiple regression techniques were used to determine the statistical significance of each factor and develop a theoretically consistent environmental damage function.

From the fall of 1974 to the spring of 1977, EPA conducted an air pollution modeling study in St. Louis, Missouri (1). Nine of 25 continuous air monitoring sites were selected for studying the effects of pollutants on eight types of materials (2). Weathering steel was one of the materials. This paper presents the results of analyzing the corrosion of weathering steel with respect to environmental data.

Theoretical Considerations

Many metals form corrosion product films as they corrode. These films tend to restrict the rate of corrosion. In general, the rate of corrosion is inversely proportional to the thickness of the corrosion product film (rate controlled by diffusion through the film). When the film is insoluble and does not change structure with time, the corrosion-time function is parabolic ($c = \alpha\sqrt{t}$). Many metal corrosion products have solubilities that are proportional to acidity. Thus, in very acid solutions, films are very thin and the resulting corrosion-time function is almost linear. In atmospheric exposures, there are many sets of environmental conditions where these two mechanisms are competing. The results is the empirically observed relationships, $c = At^n$, where the exponent, n, most often has a value between 0.5 and 1.0 (2-4).

Theoretically, the amount of corrosion (c) is the sum of the metal accumulated in the corrosion product film (T) and the amount of metal solubilized from that film with time (βt_w).

$$c = T + \beta t_w \tag{1}$$

where t_w is time-of-wetness and β is a solubilization rate which is a function of acidic pollutant fluxes. The rate of film thickness growth in units of metal corroded is;

$$dT/dt_w = \alpha/T - \beta \tag{2}$$

where α is a function of diffusivities through the film. Pollutants may also affect this coefficient. Under constant conditions, integration of equation (2) yields;

$$-\beta^2 t_w = \beta T + \alpha \ln(1 - \beta T/\alpha) \tag{3}$$

substituting equation (1) into equation (3) and rearranging produces the transcendental equation:

$$c = \beta t_w + \alpha(1-\exp(-\beta C/\alpha))/\beta \tag{4}$$

which is equivalent to:

$$c = \beta t_w + \alpha/(dc/dt_w) \tag{5}$$

A least squares fit of equation (4) cannot be used to determine the effects of enviornmental factors on the coefficients α and β because as α/β becomes large the equation approaches being an identity. When $\beta c/\alpha$ is large, $\exp(-\beta c/\alpha)$ approaches zero and equation (4) becomes;

$$c = \beta t_w + \alpha/\beta \tag{6}$$

At early stages of corrosion, however, from the empirically observed relationship, $dc/dt_w \simeq nc/t_w$ which yields;

$$c = \beta t_w + \alpha t_w/nc \tag{7}$$

During the first few years of corrosion of weathering steel, nearly all of the corrosion is accumulated in the film and α/nc is much greater than β, which would indicate that the corrosion-time function should be nearly parabolic. The corrosion product film on weathering steel, however, has the property that, generally, a given film thickness becomes more protective with time. An exception is when the film stays wet. In that case, the corrosion behavior of weathering steel is not much better than carbon steels. This means that n in the empirical equation At^n can be less than

0.5 and is expected to increase with fraction of time-of-wetness (f).

Measurement and Data Analysis

Environmental data. Hern, et al., (5) describes the data collection and evaluation system for the environmental parameters at the nine sites. Subsequently, the resulting Regional Air Pollution Study (RAPS) data base was revised several times using better validation techniques. The portions of that data base used in this study are hourly averages of temperature, dew point, windspeed, wind direction, total oxidant, NO_x, SO_x, and 24-hour total suspended particulate matter (TSP) samples. The validated data base contains a lot of missing hourly averages and the system was not operating during the first month or the last fifteen days in which weathering steel samples were exposed. A methodology was developed (6) to estimate these missing values using total system relationships and relationships between the system and climatological data from Lambert Field (airport meteorological station located about 16 kilometers northwest from the St. Louis central business district). Rainfall was not recorded at the exposure sites. With rainfall expected to be an important parameter, data from Lambert Field were used in this study.

Weathering Steel Corrosion. Triplicate specimens were exposed for periods varying from three to thirty months with exposures started at each of the four seasons during the first year. The experimental procedure and exposure schedule are documented in an EPA report (2). The results are 153 sets of triplicate weight loss data with site, exposure time and initial exposure season as primary variables.

Analysis of Environmental Data. Although the methodology for analyzing the data has been previously reported (6) there are some differences that should be noted. First, a later version of the RAPS data base was used as an initial sourse. Second, time-of-wetness in this paper is defined differently, thus, a relationship to calculate relative humidity from temperature and dew point is based on data for dew points greater than 0°C. Third, deposition velocities are calculated from boundary layer theory rather than empirical relationships.

This later version of RAPS data base eliminated many errors but produced more missing data.

Time-of-wetness as previously defined was the time exceeding some critical relative humidity (6). In this paper time-of-wetness is the time a critical relative humidity is exceeded and the dew point is greater than 0°C, plus any time the critical humidity is not exceeded and it is raining. For that reason a regression relating relative humidity to dew points above 0°C and temperature was used to calculate relative humidity for each hour. This relationship is:

$$RH = 100 \exp\{[-0.0722 + 0.00025(T + DP)][T-DP]\} \quad (8)$$

where RH is relative humidity, T is temperature and DP is dew point, the later both in °C.

Deposition velocities were calculated from windspeed data. Windspeeds at tower height can be used to calculate windspeeds at specimen rack height using the relationship for rough surfaces (7):

$$V^{+} = 8.5 + 2.5 \ln(Z/e) \quad (9)$$

where V^{+} is the dimensionless velocity and z and e are measuring height and roughness height respectively. The rack height was about three meters from the ground. At three of the sites the meteorological towers were 10 meters while at the others they were 30 meters. The average ground roughness height was assumed to be 0.1 m. Thus, the rack height windspeed is 0.75 times the windspeed at 30 meters or 0.85 times the windspeed at 10 meters.

From anology with momentum transport, gases with a Schmidt number of approximately one that readily react at a surface, have a deposition velocity of:

$$u = V^{*2}/V \quad (10)$$

where u is the deposition velocity, V* is the friction velocity and V is the average windspeed. The friction velocity is equal to $V\sqrt{f/2}$ where f is the friction factor. From boundary layer theory for smooth flat plates (7):

$$f = 0.03/(RE_{L})^{1/7} \quad (11)$$

where $RE_L = LV/\nu$, L = length of surface over which the air flows, and ν is the kinematic viscosity of air (0.15 cm^2/sec). L is assumed to be the geometric mean of the panel dimensions ($\sqrt{10.2 \times 15.2}$ = 12.45 cm). Thus:

$$u = 0.35\, V_{10}^{.86} \quad (12a)$$

$$u = 0.31\, V_{30}^{.86} \quad (12b)$$

with u in cm/sec and V_{10} and V_{30} in m/sec.

Deposition velocities were calculated on an hourly basis and averaged over exposure periods. Because windspeeds are not normally distributed, the average deposition velocity is about 91% of the deposition velocity calculated from average windspeed for an exposure period.

Pollutant Fluxes. Hourly deposition velocities were multiplied by hourly pollutant concentrations to get hourly pollutant fluxes. These were summed over exposure periods for hours of wetness with different critical relative humidity criteria (75 to 90% in 5% intervals). Average fluxes were then calculated by dividing by the time-of-wetness. The results were compared with fluxes calculated by multiplying average deposition velocities for a period by the average pollutant concentration during times of wetness. The values by the two methods were fairly consistent.

Fluxes of TSP were calculated by multiplying average TSP by two-tenths of the average deposition velocity for gases (actual deposition velocities vary considerably with the size distribution of particles). Rain fluxes were calculated by dividing the amount of rain for an exposure period by time of wetness.

Statistical Analysis of Data

An initial linear regression was performed on the data to determine the relative significance of each of the factors. The form of the model was:

$$\ln(C)=\alpha_0 \ln(t)+\Sigma\alpha_i \ln f_i+\beta Q+\Sigma\gamma_i P_i \tag{13}$$

where C is amount of steel corrosion, t is total time of exposure, f is fraction of time of wetness for different critical relative humidities, Q is a temperature factor (1000[1/(273.16+T°C)-1/(273.16+T°C avg]) (6), and the P_is are the fluxes of pollutants and rain. Both forward and backward stepwise regressions were done.

The results indicated that fraction of time of wetness (f) for a critical relative humidity of 85% was the most significant of the f values, the temperature factor was somewhat less significant, and only ozone flux was not a significant factor. TSP and SO_2 had positive coefficients while the NO_2 and rain coefficients were negative.

Because the amount of corrosion of steel is expected to follow the empirical form $C = At_w{}^n$, where both A and n are variables with changing environmental factors, a linear regression on the ln-ln form was performed on the following model:

$$\ln(C)=\alpha_0+\alpha_1\ln(t_w)+\alpha_2 f\ln(t_w)+\alpha_3\ln(f)+\beta Q+ \Sigma\gamma_i P_i \tag{14}$$

where t_w and f are time-of-wetness and fraction of time-of wetness respectively, for a critical relative humidity of 85%. The results are given in Table I.

Table I. Results of the Linear Regression of Equation:

$$\ln(C)=\alpha_0+\alpha_1\ln(t_w)+\alpha_2 f\ln(t_w)+\alpha_3\ln(f)+\beta Q+ \Sigma\gamma_i P_i$$

Variable	Units	Coefficient	Standard error	Partial F**
ln(C)	ln(μ)	α_0 = 4.245		
$\ln(t_w)$	ln(years)	α_1 = 0.3116	0.0444	49.19
$f \ln(t_w)$	ln(years)	α_2 = 0.8583	0.1561	30.23
ln(f)		α_3 = 0.4399	0.0819	28.82
Q	*	β = -0.7023	0.1809	15.07
TSP Flux	mg/cm^2 year	γ_1 = 0.8729	0.1004	75.62
SO_2 Flux	mg/cm^2 year	γ_2 = 0.1152	0.0299	14.87
NO_2 Flux	mg/cm^2 year	γ_3 = -0.0370	0.0128	8.33
Rain Flux	cm/year	γ_4 = -3.18×10^{-4}	1.24×10^{-4}	6.58

* Q = 1000(1/273.16+T)-1/(273.16+Tavg) where T is in °C and Tavg is overall average of average temperatures during times-of-wetness (13.55°C).

**A measure of the statistical significance of adding the specific variable to all of the others.

The percent of variability explained by regression is 95.85%.

The fraction of time-of-wetness affects the value of n in the empirical equation $C = At_w^n$, (n=0.312+0.858f), which means that the film becomes more protective with time when f is low, but less protective with time when f is high. All of the other terms in equation 14 are considered to be a part of the A coefficient.

A pseudo steel corrosion rate was calculated by dividing steel corrosion by $f^{0.44}\ t_w^{(0.312+0.858f)}\ \exp(-0.702Q)$. This value was regressed against the pollutant and rain fluxes to determine non-exponential coefficients that make up the A term. The results are given in Table II.

Table II. Results of Regression of Pseudo Corrosion Rate Against Pollutant Fluxes

Variable	Units	Coefficient	Standard error	Partial F
Pseudo Corrosion Rate		A_0=65.63		
TSP Flux	mg/cm^2 year	A_1=85.13	9.39	82.26
SO_2 Flux	mg/cm^2 year	A_2=10.67	2.46	18.88
NO_2 Flux	mg/cm^2 year	A_3=-3.51	1.04	11.36
Rain Flux	cm/year	A_4=-0.0302	0.0085	12.51

The resulting empirical damage function, $C=At_w^n$, has the coefficients:

$$A = (65.63+85.13TSP+10.67SO_2-3.51NO_2-.03Rain)f^{0.44}\exp(-.7Q) \quad (15a)$$

$$n = 0.312+0.858f \quad (15b)$$

where the pollutant fluxes have the units in Table II. This equation accounts for 95.93 percent of the variability.

Evaluation of data with respect to theory. The significant factors in the empirical equations 15a and 15b may affect either or both α and β coefficients in equation 7. The relative effects of each parameter on α and β were determined by regressing steel corrosion/time-of-wetness (c/tw) against all of the fluxes, $1/nC_p$ (where $C_p = At_w^n$ with the coefficients calculated using equations 15a and 15b), $\exp(-.7Q)/nC_p$, f/nC_p, and all of the products of fluxes and $1/nC_p$. Stepwise regression was used in the order of most significant to least significant variable and including only those variables with a 0.95 probability of significance. A total of 13 independent variables were considered. Table III gives the results.

Table III. Regression Coefficients for Theoretical Model of Weathering Steel Corrosion $C/t_w = \Sigma\beta_i P_i + \Sigma\alpha_i P_i/nC_p$

Variable	Units	Coefficient	Standard error	Partial F
C/t_w	μ/year	β_0=39.12		
TSP/nC_p		α_1=2292	175	171.10
f/nC_p		α_2=1573	262	36.05
$1/nC_p$		α_0=1381	308	20.15
NO_2 Flux	mg/cm^2 year	β_1=-8.85	1.85	22.88
SO_2 Flux	mg/cm^2 year	β_2=18.32	4.62	15.71
Rain Flux*	cm/year	β_3=-0.0798	0.0227	12.37
$EXP(-.7Q)/nC_p$		α_3=1104	335	10.85

* Rain Flux is expressed in cubic centimeters of rain per square centimeter of surface per year of wet time.

Variables are listed in the order in which they were entered into the regression. The resulting damage function ($C=(\Sigma\beta_i P_i+\Sigma\alpha_i P_i/nC_p)t_w$) can account for 95.65% of the variability.

Discussion of Results

The empirical damage function with coefficients calculated using equations 15a and 15b provides the best fit of the data (graphically presented in Figure 1). The gaseous pollutant fluxes are based on hourly concentrations and deposition velocities during periods of wetness. These do not differ dramatically from fluxes calculated from exposure period averages of deposition velocities and concentrations. The TSP fluxes are calculated from exposure period averages of deposition velocities and concentrations. The rain fluxes are the amounts of rain divided by times of wetness. Fraction of time when wet (f) is for a critical relative humidity of 85%. This equation can be used to predict weathering steel corrosion as a function of environmental conditions.

The coefficients in Table 3 provide a better theoretical understanding of how the different factors affect the corrosion of weathering steel. The β coefficients affect the solubility of the protective oxide layer and the α coefficients affect the diffusivity through the layer. The large ratio of α/β confirms the relative insolubility of the rust on weathering steel in most environments. Sulfur dioxide increases the solubility of the film while NO_2 and rain decrease the solubility. Rain apparently washes away acidic components (deposited during dew formation) that increase the solubility.

The α coefficient for TSP is highly significant. Accumulation of particles in the oxide layer appears to increase the diffusivity of ions or the electrical conductivity of the film. TSP does not appear to enhance the solubility of the film. Increasing the fraction of time-of-wetness appears to incrase the diffusivity through the layer. This could actually be a time function of a matter of hours. The diurnal cycle suggests that while the film is wet it becomes less protective with time; when it is dry it becomes

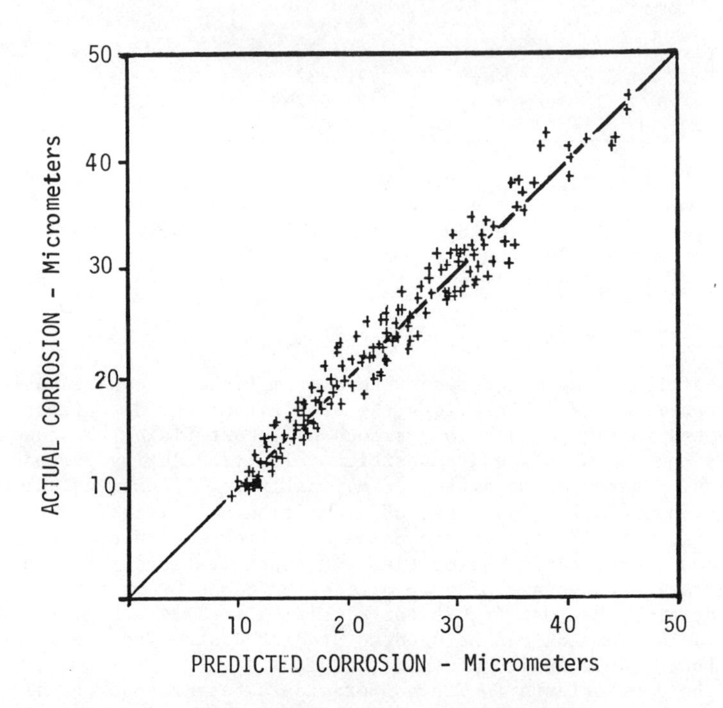

Figure 1. Fit of weathering steel corrosion data to At_w^n model, where A and n are functions of environmental parameters

more protective with time. When f is large the film apparently is not dry long enough to reach a desired level of protectivity. The temperature effect is as expected; diffusion increases as temperature increases.

Conclusions

Weathering steel corrosion can be described as competing mechanisms of formation and dissolution of a protective oxide layer during periods of wetness.

Empirically, the best fit of the corrosion data suggests that time-of-wetness of the steel is best defined as the time when the dew point exceeds 0°C and the relative humidity exceeds 85% plus the time during rain when the relative humidity does not exceed 85%.

Three variables increase the diffusivity through the oxide film; 1) fraction of time when wet, 2) temperature, and 3) TSP flux.

Solubility of the oxide film is increased by increasing the flux of sulfur oxides during periods of wetness.

Dissolution of the oxide film is reduced by increasing the fluxes of nitrogen oxides and rain.

Ozone appears to have no significant effect on the corrosion of weathering steel.

Literature Cited

1. Schiermeier, F. A. Environmental Sci. Technol. 1978, 12, 644.
2. Mansfeld, F. "Regional Air Pollution Study: Effects of Airborne Sulfur Pollutants on Materials"; EPA-600/4-80-007, 1980.
3. Haynie, F. H.; Upham, J. B. Materials Protection and Performance 1971, 10, 18.
4. Mattsson, E. Materials Performance 1982, 21, 9.
5. Hern, D. H.; Taterka, M. H. "Regional Air Monitoring System Flow and Procedures Manual"; EPA Contract 68-02-2093, Rockwell International, Creve Coeur, Mo. 1977.
6. Haynie, F. H. Durability of Building Materials 1982/1983, 1, 241.
7. Knudson, J. G.; Katz, D. L. "Fluid Dynamics and Heat Transfer"; University of Michigan, Ann Arbor, Michigan, 1954, p. 38 and p. 149.

RECEIVED January 2, 1986

10

A Laboratory Study to Evaluate the Impact of NO_x, SO_x, and Oxidants on Atmospheric Corrosion of Galvanized Steel

Edward O. Edney[1], David C. Stiles[1], John W. Spence[2], Fred H. Haynie[2], and William E. Wilson[2]

[1]Northrop Services, Inc., Research Triangle Park, NC 27709
[2]Atmospheric Sciences Research Laboratory, U.S. Environmental Protection Agency, Research Triangle Park, NC 27711

A series of laboratory experiments was conducted in which galvanized steel samples were exposed to NO_2 in air and irradiated propylene/nitrogen oxides/air mixtures in the absence and presence of SO_2. Dew was produced periodically on the test panels, and, at the end and/or during the experiments, panels were sprayed with either deionized water or an ammonium bisulfate solution (pH of 3.5). Gas phase concentrations were monitored, and dew and rain rinse samples were analyzed for nitrite, nitrate, sulfite, sulfate, formaldehyde, and zinc.

The average deposition velocities measured during periods of wetness were sulfur dioxide, 0.8 cm/s; formaldehyde, 0.6 cm/s; nitric acid, 0.7 cm/s; and nitrogen dioxide, 0.03 cm/s. Analysis of dew samples suggests that the dry deposition of sulfur dioxide, nitric acid, formaldehyde, and possibly nitrogen dioxide accelerates the atmospheric corrosion of galvanized steel. Wet deposition of acidic species accelerates corrosion. A general model for the atmospheric corrosion of galvanized steel is presented.

Galvanizing steel is a well-known technique for protecting steel against corrosion. Galvanized steel consists of a zinc (Zn) coating strongly bonded to a steel substrate. Zn is anodic with respect to iron and will react electrochemically before iron in the presence of an electrolytic solution. Because it is one of the least expensive methods for protecting steel against atmospheric corrosion, a number of field studies have been conducted to measure the atmospheric corrosion rate of galvanized steel and/or Zn (1-11). Many of the studies have focused on the effect of SO_2 and time of wetness on the corrosion rate. While these parameters clearly play an important role in corrosion, any model based solely on these parameters is

0097-6156/86/0318-0172$06.25/0

likely to be incomplete because the effects of wet deposition and dry deposition of other air pollutants have been neglected.

Recent studies have shown that the deposition of a compound rather than the concentration should be used to assess the impact of SO_2 or other compounds on atmospheric corrosion (9,11-12). The dry deposition D_x of a reactive species x is defined by the following relationship:

$$D_x = \int_o^{t_e} F_x(t)dt \tag{1}$$

where
t_e = exposure time, and
$Fx(t)$ = the molar flux at time t, i.e.,

$$F_x(t) = [x(t)]v_d \tag{2}$$

where
$[x(t)]$ = the gas phase concentration at time t, and
vd = the dry deposition velocity.

One approach to developing a corrosion model for galvanized steel based on dry deposition is to assume that the corrosion C (in moles of Zn lost per unit area) can be represented as a linear combination of the corrosion induced in a clean air environment and that associated with air pollutant x, i.e.,

$$C = (A + \gamma[x]v_d)t_w \tag{3}$$

where
tw = the time of wetness,
A = a constant, and
γ = number of moles of Zn corroded per mole of x deposited.

If no antagonisms or synergisms occur, Equation 3 can be extended to include the effects of other compounds by converting the second term to a sum over all pollutants depositing onto the surface. Equation 3 assumes that deposition takes place only when the surface is covered by a film of moisture.

The inability to estimate dry deposition velocities and the lack of knowledge of the compounds that are likely to contribute to corrosion makes Equation 3 difficult to apply in evaluating corrosion data. The results of the experiments reported here will address both of these issues and will be used to develop the framework for a model for analysis of galvanized steel corrosion data obtained from field studies.

In the presence of a moisture film, but in the absence of pollutants other than CO_2, the effective operative electrochemical reactions taking place on a wet galvanized steel surface are believed to be (13)

Anode $Zn + 2H_2O \rightarrow Zn(OH)_2 + 2H^+ + 2e^-$

Cathode $O_2 + 2H_2O + 4e^- \rightarrow 4OH^-$.

The zinc hydroxide ($Zn(OH)_2$) formed will react with dissolved CO_2 to produce zinc carbonate ($ZnCO_3$), a compound that is slightly soluble in water, i.e., solubility=80 nmol/ml in water at 15°C (14). Buildup of the insoluble corrosion products such as $Zn(OH)_2$ and $ZnCO_3$ will create a protective layer that serves to inhibit further corrosion. The corrosion rate will then be small unless there is some means for either preventing the formation of the protective layer or destroying it once it has formed.

The details of the mechanism for SO_2-induced corrosion of galvanized steel have not been established; however, it is likely that the corrosion process is initiated by the reaction of Zn with the sulfurous (H_2SO_3) and sulfuric (H_2SO_4) acids generated in the dew. In particular, H_2SO_4 will react with either the base metal or the protective corrosion products, $ZnCO_3$ or $Zn(OH)_2$, producing soluble zinc sulfate ($ZnSO_4$) that has limited protective properties. Dissolution of the protective layer will stimulate further electrochemical corrosion which will tend to reform the insoluble layer. The extent of formation of the protective layer is determined by the SO_2 deposition to the surface. In clean air environments, a thick adhesive protective layer will form, whereas in polluted areas of high SO_2 concentrations only a very thin layer will be produced (15).

The production of soluble $ZnSO_4$ suggests that the precursor to the anion, gas phase SO_2, accelerates corrosion. *Based on this mechanism, it is likely that other air pollutants that readily adsorb on a surface, producing acidic compounds that can react with the corrosion products to form soluble Zn compounds, can accelerate the corrosion rate.*

The objective of this study was to determine whether air pollutants other than SO_2 accelerate the atmospheric corrosion rate of galvanized steel. Short-term laboratory experiments were conducted in which galvanized steel panels were exposed to the following mixtures in air: (1) NO_2, (2) irradiated propylene/nitrogen oxides (C_3H_6/NO_x), and (3) irradiated $C_3H_6/NO_x/SO_2$. The test panels were chilled below the dew point periodically to produce dew. Dew samples were collected and analyzed for Zn and their anion composition. To investigate the impact of acidic wet deposition, during some of the experiments, panels were removed periodically from the chamber and sprayed with either deionized water or dilute solutions of ammonium bisulfate (NH_4HSO_4; pH of 3.5). The results of these experiments, as well as the framework for a corrosion model for galvanized steel, are presented in this paper.

Experimental

The exposure system consists of two exposure chambers in parallel, coupled to an 11.3-m^3 aluminum and Teflon smog chamber (Figure 1). Light banks are located on two sides of the smog chamber and consist of 56 black lamps and 9 sun lamps. The smog chamber, operated as a continuous stirred tank reactor (CSTR), serves as a reservoir for the exposure chambers. Ambient air first passes through a clean air train, where pollutants and H_2O are removed, and then through a glass manifold where it is mixed with reactants to the desired concentration before entering the smog chamber. The flow rate through the smog chamber is 95 Lpm, producing a residence time of 112 min. Steam is injected into the chamber by the dew point control system that is programmed to maintain the air dew point at 15°C. The air temperature is not controlled.

A cross-sectional drawing of one of the exposure chambers is shown in Figure 2. The chambers are constructed of clean Teflon glued to a 13×13×152-cm welded aluminum frame. Glass piping is used to connect the exposure chambers to the smog chamber. Both exposure chambers have seven positions for mounting test panels. Each exposure position has a chiller back plate with supplied coolant so that the panel can be chilled below the air dew point to produce dew on the panels. Individual positions can be disconnected from the coolant manifold to remain dry. The coolant temperature enables dew to form in a sufficient quantity to drop into the collection trough below each panel. The collected dew then drains into a Nalgene bottle. The exposure chambers are also equipped with heating lamps. Blowers were mounted on the exposure chambers to maintain turbulent conditions (Reynolds number ~30,000 in the chambers), and the blower settings were selected to produce wind speeds of approximately 300 cm/s (~7 mi/h) in each chamber.

The galvanized steel panels were 8×13-cm Zn-coated (hot-dipped), 20-gauge steel plates. The average thickness of the coating was 20 µm, and the exposed surface area per panel was 84.7 cm^2. Before each exposure experiment, the panels were first cleaned by immersion into a 10% by weight solution of ammonium chloride (NH_4Cl) at a temperature of 60°-80°C for 2 min. The panels were rinsed in deionized water and then dried in methanol to remove the moisture.

During and/or at the completion of some of the exposure experiments, galvanized steel panels were sprayed with either deionized water or an NH_4HSO_4 (pH 3.5) solution. The deionized water quickly equilibrated with CO_2 in the laboratory air, producing a value of 5.6 for the pH. By removing the panels from the exposure chambers and mounting them into a Teflon rack, each panel could be sprayed with a specific volume of either deionized H_2O or an NH_4HSO_4 solution. The spray runoff was collected for chemical analysis. For a typical spray condition, ~10 s was the average residence time of a droplet on the surface, whereas the corresponding time for a dew droplet was ~1 h.

The galvanized steel panels were exposed to air masses that contained NO_2 and irradiated mixtures of C_3H_6/NO_X and $C_3H_6/NO_X/SO_2$. A brief description of the protocol used for each of the three experiments is presented.

The NO_2 exposure was a short-term experiment where the panels were exposed to a mixture of 649 ppb NO_2 in air. The total exposure time was 25 h. The microprocessor-controlled chiller system was used to generate two 7-h periods where the panels were covered with dew. The two wet periods were separated by a 5-h dry period. NO_X concentrations were monitored continuously. At the end of the experiment, the panels were sprayed with either 50 mL of deionized H_2O or NH_4HSO_4 solution (pH=3.5). The volume of collected dew was determined. The dew was then analyzed for NO_2^-, NO_3^-, $SO_3^=$, and $SO_4^=$ by ion chromatography and for Zn by atomic absorption spectroscopy. The rain rinse was analyzed in a similar way.

The C_3H_6/NO_X+hv experiment was a 14-day exposure which consisted of two 7-h dew periods per day separated by 5-h dry periods. Panels were weighed both prior to the experiment and at

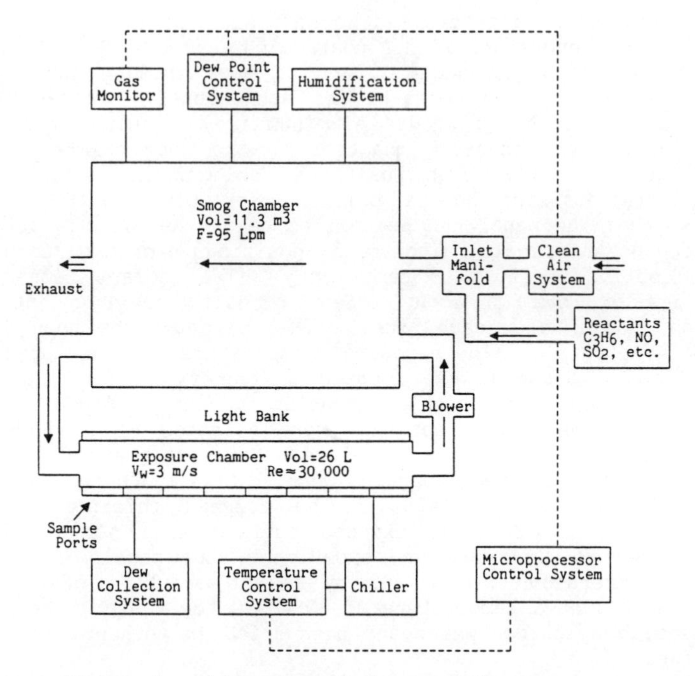

Figure 1. Schematic of Exposure System.

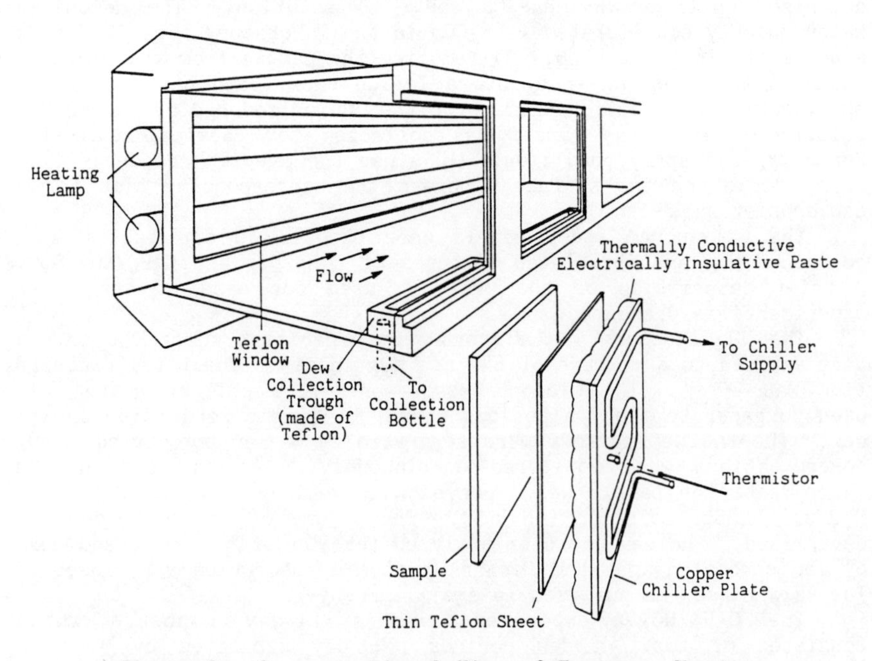

Figure 2. Cross-sectional View of Exposure Chamber.

the completion of the experiment after the corrosion products were removed using the cleaning procedure described above. A correction to the Zn weight loss was made to account for the loss of base metal during cleaning. Panels were exposed to (1) dew only, (2) dew plus acid rain, and (3) dew plus clean rain conditions. The panels exposed under conditions 2 and 3 were removed from the exposure chambers after four and nine days of exposure and sprayed with either 400 mL of deionized H_2O or NH_4HSO_4 (pH of 3.5) solutions. The panels were also sprayed at the end of the experiment. Dew samples were collected daily and analyzed. The NO_X and O_3 concentrations were monitored continuously and peroxyacetyl nitrate (PAN), nitric acid (HNO_3), and aldehyde concentrations were measured during both dry and wet periods.

The exposure conditions for the $C_3H_6/NO_X/SO_2$ + hv experiment were essentially the same as those used in the NO_2 experiment. The exposure period contained two 7-h wet periods separated by a 5-h dry period. All species measured in the C_3H_6/NO_X experiment were measured and, in addition, gas phase SO_2 concentrations were determined as well as $SO_3^=$ and $SO_4^=$ in the dew. No rain data were obtained for this experiment. To investigate the deposition of reaction products of an irradiated $HC/NO_X/SO_2$ mixture during dry periods, one panel was kept dry during the entire exposure.

NO_X concentrations were measured using a Bendix $NO/NO_2/NO_X$ analyzer. The analyzer was calibrated with a certified standard of NO in N_2 obtained from MG Scientific. Ozone was measured on a Bendix ozone analyzer calibrated using a Dasibi Environmental Corporation ultraviolet (UV) ozone generator. Formaldehyde (HCHO) and acetaldehyde (CH_3CHO) were measured by using the DNPH/high performance liquid chromatography technique (16). Calibrations were performed by using diluted solutions of twice-recrystalized hydrazone samples for each aldehyde. PAN was determined with an Analog Technology Corporation electron capture detector after separation on a packed column containing 10% Carbowax 400 on 80/100 Supelcoport operated at room temperature. The instrument was calibrated according to the procedure described by Lonneman *et al.* (17). SO_2 concentrations were determined using a Thermo Electron Corporation SO_2 monitor. The instrument was zeroed using clean air that had passed through brominated charcoal, reducing the SO_2 concentration to less than 0.5 ppb. An SO_2 permeation tube was used to calibrate the instrument. A CTE dew point hydrometer was used in the dew point control system and was calibrated with a saturated solution of sodium bromide (NaBr) (relative humidity = 58% at 20°C). Chamber HNO_3 concentrations were measured by drawing ~500 L of air through a 25-mm nylon filter (1-µm pore size), extraction of the filter with 10^{-5} M perchloric acid solution, and analysis of nitrate (NO_3^-) by ion chromatography.

The collected dew and rain rinse samples were analyzed for HCHO using the chromotropic acid method. In addition, the samples were analyzed for nitrite (NO_2^-), NO_3^-, sulfite ($SO_3^=$), and $SO_4^=$ using ion chromatography. The analysis was performed on a Dionex Auto Ion System 12 ion chromatograph with a 6-mm×250-mm fast run anion separator column. Calibrations curves for $SO_3^=$ and $SO_4^=$ were obtained using anhydrous sodium sulfate (Na_2SO_4) and sodium sulfite (Na_2SO_3) salts diluted with deionized H_2O. The $SO_3^=$ standards required an addition of 0.7% HCHO to the stock solution to deter oxidation of $SO_3^=$ to $SO_4^=$. NO_2^- and NO_3^- calibration curves were obtained by diluting the corresponding Na salts in deionized H_2O. The collected dew and rain rinse samples were analyzed for Zn using an Instrument Labs, Inc. atomic absorbance analyzer. The instrument was calibrated by measuring the absorbance of a series of American Chemical Society certified Zn standards.

Data Reduction

Analysis of Gas Phase and Dew Data. In principle, the amount of compound that deposits onto the panels during a wet period can be calculated two ways: (1) by using the difference in the gas phase concentrations during the dry and wet periods, or (2) by employing the concentration of the compound in the dew and the volume of dew. Deposition to a dry panel can also occur, but, as will be shown, for the majority of the compounds reported here, most of the deposition took place in the presence of dew. Therefore, emphasis will be placed on calculating the deposition during wet periods.

During wet periods, the gas phase concentrations decreased due to uptake by the panels. The time dependence of the gas phase concentration of species x during a period of dew formation can be approximated by the following equation:

$$\frac{d}{dt}[x(t)] = \frac{[x_d]}{\tau} - \left(k + \frac{1}{\tau}\right)[x(t)] \tag{4}$$

where

t = time parameter (t=0 corresponds to the onset of dew formation for the wet period under investigation),

$[x(t)]$ = gas phase concentration, and

$[x_d]$ = steady concentration of x during the preceding dry period.

The residence time τ is equal to V_s/Q, where Q is the flow rate through the smog chamber and V_s is the total volume of the system (smog chamber plus exposure chambers). The effective first-order rate constant for deposition to the panel is k and is equal to

$$\frac{v_d^{(1)}A}{V_s}, \; where\; v_d^{(1)}$$

is the dry deposition velocity and A is the total area of the wet panels. The solution to Equation 4 is as follows:

$$\frac{[x(t)]}{[x_d]} = \frac{1}{1 + k\tau} + \frac{k\tau}{1 + k\tau}\, exp\,[-(1 + k\tau)t/\tau] \tag{5}$$

Equation 5 can be further reduced if t is chosen such that the second term can be neglected. Under these circumstances, the panels remain wet long enough for a new steady state condition to be obtained; i.e., for $t/\tau \geq 2$,

$$[x(t)] \approx \frac{1}{1 + k\tau}[x_d] = [x_w] \tag{6}$$

Steady state concentrations were reached during the exposure experiments reported here since the time of wetness per cycle was 420 min and the residence time was 112 min. An expression for the dry deposition velocity $v_d^{(1)}$ is immediately obtained from Equation 6, i.e.,

$$v_d^{(1)} = \frac{Q}{A}\left(\frac{[x_d]}{[x_w]} - 1\right) \tag{7}$$

Use of Equation 7 is limited by the accuracy of the measurement of the decrease in the gas phase concentration and in practice can only be used to calculate dry deposition velocities of compounds that readily absorb onto the surface.

The amount of material deposited during a wet period is found by integrating the expression for the molecular flux over time, i.e.,

$$D_x^{(1)} = \int_o^{t_w} v_d^{(1)}[x(t)]dt = \frac{v_d^{(1)}[x_d]\tau}{1 + k\tau}\left\{\frac{t_w}{\tau} + \frac{k\tau}{1 + k\tau}(1 - exp[-(1 + k\tau)t_w/\tau])\right\} \tag{8}$$

where
t_w = the time of wetness per period.

The deposition can also be calculated from analysis of the dew if the fate of the dissolved gas is known. The deposition per wet period in terms of dew parameters is calculated by the following equation:

$$D_x^{(2)} = \frac{(x)_{dew}V_{dew}}{A_p} \tag{9}$$

where
(x) dew = the concentration of x in the dew,
V_{dew} = the amount of dew produced during a wet period, and
A_p = the area per test panel.

If the compound under investigation deposits readily onto the wet panel and can be associated with a particular dew ion, then both Equations 8 and 9 can be used to calculate the deposition. Equality of the two depositions establishes a mass balance for the system.

Approximate values for the dry deposition velocities can be obtained by equating the two expressions for the deposition (Equations 8 and 9) and solving for the deposition velocity. The result is

$$v_d^{(2)} = \frac{(x)_{dew} V_{dew}}{[x_w] A_p t_w} \quad (10)$$

where it has been assumed that the second term in Equation 8 can be neglected. (Calculations show that this approximation introduces at most a 10% error.) Equation 10 can be used to calculate the dry deposition velocity of compounds that readily adsorb onto a wet surface as well as weakly adsorbing species as long as the fate of the dissolved compound is known.

Analysis of Rain Rinse Data. The concentration of Zn in the rain rinse can be interpreted as a rain flux. The concentration of Zn in nmol/mL is equal to the Zn rain flux in nanomoles/cm^2-cm$_R$, i.e., the number of nanomoles of Zn dissolved into the rain per cm^2 of panel per cm of rain. The amount of Zn corroded by l_R cm of rain is

$$R_{zn} = [Zn_R] l_R \quad (11)$$

where

$[Zn_R]$ = the concentration of Zn in the rain rinse, and
R_{zn} = the Zn corrosion expressed as a deposition.

Results

NO_2 Exposure Experiment. The experimental conditions and results for the short-term NO_2 exposure experiment are given in Table I. The average concentration of NO_2 during the dry periods was 649 ppb which decreased to 627 ppb in the presence of dew, thus yielding (according to Equation 7) a value of 0.05 cm/s for the dry deposition velocity. Both NO_2^- and NO_3^- were detected in the dew with the NO_2^- concentration (203 nmol/ml) representing 95% of the NO_X^- concentration. Trace amounts of $SO_4^=$ were found in the dew. A value of 149 nmol/ml was found for the Zn dew concentration. Equation 10 yields a value of 0.06 cm/s for the NO_2 dry deposition velocity, in reasonable agreement with the value found from the decrease in gas phase concentrations.

A separate NO_2 exposure experiment was conducted to determine whether the amount of NO_2 deposited on the surface during a dry period could be ignored relative to that adsorbed in the presence of dew. It was found that saturation occurs on a dry galvanized steel surface and that the adsorbed NO_2 saturation surface density was approximately 2 nmol/cm^2. The saturation density is less than 5% of the amount of NO_2 deposited per 7 h of dew (41 nmol/cm^2), and therefore its contribution can be ignored.

Comparison of the NO_2 dry deposition velocity with that found for SO_2 under similar conditions (v_d (SO_2) = 0.8 cm/s; Table IV) indicates a substantial difference in the processes of uptake of the two gases on the test panels. The gas phase resistances for both compounds are expected to be essentially the same. Therefore, the difference between the dry deposition velocities is likely due to a difference in the aqueous reaction and diffusive mechanisms that take place on the metal surface covered by a thin film of moisture.

The disparity in velocities is consistent with the recent findings of Lee and Schwartz (18), who evaluated the rate of uptake of NO_2 by surface water. They found that the NO_2 deposition velocity was controlled by aqueous phase mass transport and/or reaction. In addition, Lee and Schwartz found that NO_2 reacts with liquid water according to the following reaction:

$$2\,NO_2\,(g) + H_2O\,(l) \rightarrow 2H^+ + NO_2^- + NO_3^-$$

This reaction cannot explain the results of the exposure experiment since the dominant NO_x^- compound is NO_2^-. This result may not be surprising since here the reactions took place on a reactive Zn surface.

Judeikis *et al.* recently investigated the uptake of NO_2 on a series of surfaces (19). They found that the identity of the dominant NO_x^- species in the reaction products depended on the chemical composition of the surfaces. Ten Brink *et al.* found that the presence of Zn particulates greatly enhanced the production rate of nitrous acid (HONO) in air containing NO_2 (20).

The chemical composition of the dew can be clarified by noting that

$$\phi = \frac{[Zn]}{0.5\,[NO_x^-]+[SO_4^=]+S} = 0.75 \qquad (12)$$

where S is the solubility of $ZnCO_3$. If $\phi = 1$, the chemical composition is consistent with a mixture of zinc nitrate ($Zn(NO_3)_2$), zinc nitrite ($Zn(NO_2)_2$), $ZnCO_3$, and $ZnSO_4$. The proposed minor compounds $Zn(NO_3)_2$ and $ZnSO_4$ are water soluble and can be produced by the reactions of the strong acids HNO_3 and H_2SO_4 with either the base metal or the relatively insoluble corrosion products $ZnCO_3$ and $Zn(OH)_2$. The solubility of $ZnCO_3$ is included to take into account the possibility that some of the Zn may have arisen from the dissolution of the corrosion products that is not associated with reactions with acidic components of the dew. This contribution is expected to be present even in the absence of air pollutants other than CO_2.

The presence of rather large amounts of NO_2^- suggests that $Zn(NO_2)_2$ is produced; however, no information on the proposed compound could be found in the literature. If produced, the formation mechanism is likely a reaction of the corrosion products with HONO. The fact that HONO is a weak acid could explain why a value less than 1 is found for ϕ, since the calculation assumes that all of the NO_2^- is associated with Zn. If the production of $Zn(NO_2)_2$ is ignored, a value of 1.5 is found for ϕ. Under these circumstances, there is excess Zn in the dew that cannot be accounted for.

The experimental results suggest that the deposition of NO_2 into dew accelerates the corrosion of galvanized steel; however, it is difficult to establish this definitively because the concentration of dissolved NO_x^- in the dew is small, thus its contribution to the production of soluble Zn corrosion products is similar in magnitude to that associated with the dissolution of the relatively insoluble corrosion products in non-acidic aqueous solutions.

The rain rinse data results are shown in Table I. The Zn concentration in the rain rinse with a pH of 3.5 was 13 times the value obtained from the analysis of the clean rain (pH of 5.6) rinse.

C_3H_6/NO_x + hv Exposure Experiment. The steady state reactant and product distribution in this experiment contained a complex mixture of oxidants whose concentrations were, in general, much larger than those that occur under ambient conditions but likely contains many of the air pollutants that are present during smog conditions (Table II).

The steady state NO_x, PAN, and O_3 concentrations were not very stable, and the cited dry concentration values are the average values for the entire experiment, i.e., $[x_d] \approx [x_w]$. The quantity NO_x-PAN was calculated to obtain an estimate of the NO_2 concentrations. It was assumed that the NO_x monitor responded to PAN as NO_x. The HNO_3 concentrations were so small (7 ppb) that it was not possible to observe a decrease in the concentration during periods of wetness. NO_2^-, NO_3^-, $SO_4^=$, HCHO, and Zn were detected in the dew.

Equation 10 is used to calculate the dry deposition velocity for NO_2. It is assumed that all of the NO_2^- in the dew came from the dissolution of NO_2. A value of 0.03 cm/s is found for the deposition velocity. To calculate the dry deposition velocity for HNO_3, it is assumed that the source of dew NO_3^- was the dry deposition of HNO_3. The calculated value, 2.0 cm/s, appears high relative to the velocities of other soluble gases such as HCHO (see Tables II and IV) and likely reflects that HNO_3 can be adsorbed onto a dry surface. Evidence to substantiate this point appears in the next section. It is also possible that some of the NO_3^- and/or NO_2^- arose from the deposition of small amounts of PAN; however, the ionic fate of dissolved PAN is not known and therefore more information is required for an assessment of its role in atmospheric corrosion.

The CH_3CHO deposition velocity is found using Equation 7. The value obtained is 0.35 cm/s. Deposition velocities for HCHO are found using Equations 7 and 10. Equation 7 yields a value of 0.93 cm/s whereas the dry deposition velocity based on the dew analysis is 0.35 cm/s. The discrepancy indicates either that there is an error in one of the measurements (gaseous HCHO or aqueous HCHO) or that the dissolved HCHO undergoes reaction. It is not apparent that any measurement errors were made, and, as will be shown, the discrepancy does not occur when analyzing the HCHO data for the $C_3H_6/NO_x/SO_2$ experiment. The difference in velocities can be explained if it is assumed that some of the dissolved HCHO is oxidized to formic acid (HCOOH), a species that is not detected by the chromotropic acid method. An estimate of the amount of HCOOH produced can be obtained by subtracting the amount of HCHO deposited, calculated by the dew analysis method (Equation 9) from the amount deposited using Equation 8. The results are as follows:

$$D_{HCHO} = 380\ \frac{nmol}{cm^2}\ \text{(Equation 8), and}$$

Table I. Results of NO_2 Exposure Experiment

Total Exposure Time: 25 h
Total Time of Wetness: 14 h (7-h periods)
Air Temperature: 26°C
Plate Temperature when Wet: 10°C
Rain Frequency: Once at the end of the experiment
Amount of Rain per Event: 0.59 cm

Gas Phase				
x	$[x_d]\,(ppb)$	$[x_w]\,(ppb)$	$v_d^{(1)}\left(\frac{cm}{s}\right)$	$v_d^{(2)}\left(\frac{cm}{s}\right)$
NO_2	649	627	0.05	0.06

Dew		
x	$[x]\left(\frac{nmol}{ml}\right)$	$D_x^{(2)}\left(\frac{n\,mol}{cm^2}\right)$
NO_2^-	203	39
NO_3^-	11	2
NO_X^-	214	41
$SO_4^=$	12	2
Zn	149	29

Rain		
pH	$[Zn]\left(\frac{nmol}{ml}\right)$	$R_{Zn}\left(\frac{n\,mol}{cm^2}\right)$
5.6	2	1.2
3.5	26	15

Table II. Results of C_3H_6/NO_X+hv Exposure Experiment

Total Exposure Time: 336 h
Total Time of Wetness: 196 h (7-h periods)
Air Temperature: 32°C
Plate Temperature when Wet: 10°C
Rain Frequency: After four and nine days exposure and at the end of the experiment
Amount of Rain per Event: 4.7 cm

Gas Phase

x	$[x_d]$ (ppb)	$[x_w]$ (ppb)	$v_d^{(1)}\left(\frac{cm}{s}\right)$	$v_d^{(2)}\left(\frac{cm}{s}\right)$
O_3	134			
PAN	57			
NO_X-PAN	359			0.03
HNO_3	7			2.0
HCHO	621	389	0.93	0.33
CH_3CHO	254	207	0.35	

Dew

x	$[x]\left(\frac{nmol}{ml}\right)$	$D_x^{(2)}\left(\frac{n\,mol}{cm^2}\right)$
NO_2^-	95	11
NO_3^-	118	15
NO_X^-	213	26
$SO_4^=$	6	1
HCHO	940	133
Zn	590	77

Rain

pH	$[Zn_R]\left(\frac{nmol}{ml}\right)$	$R_{Zn}\left(\frac{n\,mol}{cm^2}\right)$
5.6	4	17
3.5	27	127

$$D_{HCHO} = 133 \frac{nmol}{cm^2} \quad \text{(Equation 9)}$$

The difference in deposited HCHO converted into an equivalent HCOOH concentration gives a value of 1746 nmol/ml.

The chemical composition of the dew is analyzed in the same way as in the NO_2 experiment. The presence of NO_X^- in the dew suggests, as before, the formation of $Zn(NO_2)_2$, $ZnCO_3$, and $Zn(NO_3)_2$. Assuming that these, plus a small amount of $ZnSO_4$, are the only Zn compounds in the dew, produces the following result for ϕ:

$$\phi = \frac{[Zn]}{0.5\,[NO_x^-] + [SO_4^=] + S} = 3.0$$

The large value for ϕ indicates the presence of other soluble Zn compounds for which a prime candidate is zinc formate ($Zn(CHOO)_2$), a well known water-soluble compound that is formed when HCOOH reacts with Zn. If the estimated value for the $CHOO^-$ concentrations is included in the expression for ϕ, the following result is obtained:

$$\phi = \frac{[Zn]}{0.5\,[NO_x^-] + 0.5\,[CHOO] + [SO_4^=] + S} = 0.55$$

The introduction of $Zn(CHOO)_2$ narrows the difference between the calculated value for ϕ and the theoretical value of 1 for which the soluble Zn compounds have been totally accounted. However the introduction of $CHOO^-$ results in more anions than can be accounted for by the Zn concentration. A possible explanation is that HCOOH is a weak acid and that some of the acid remained undissociated. The presence of $CHOO^-$ in the dew is consistent with the findings of Knotkovà, who observed that HCHO accelerated the corrosion of a number of metals, including galvanized steel, and that the corrosion products contained $CHOO^-$ (21). In addition to HCHO oxidation in solution, the possibility exists that CH_3CHO coud be oxidized to acetic acid. However, no measurements were made during the experiments to evaluate this reaction.

Table III shows the rain rinse data for the experiment. The Zn dissolved per rain event increased by nearly a factor of 7 as the pH was decreased from 5.6 to 3.5, a result in qualitative agreement with that found in the NO_2 experiment. The Zn corrosion determined by weight loss measurements (D_{Zn}^{WL}) and the total amounts of Zn corrosion induced by dew and rain are also shown in Table III. The total Zn lost due to dew (D_{Zn}^{T}) is obtained by multiplying $D_{Zn}^{(2)}$ by 28, the number of dew cycles in the exposure experiment. Total Zn corrosion caused by rain (R_{Zn}^{T}) is found by multiplying the average Zn rain rinse concentration by the total rainfall (14.2 cm). The Zn corrosion determined by weight loss is consistent with the rain rinse data. The maximum Zn corrosion took place on those panels sprayed with rain having a pH of 3.5. For each exposure condition, the sum of the Zn found in the rain and dew is less than that found by weight loss. The difference is likely due to the presence of an insoluble protective layer that is not dissolved in the dew or rain but is dissolved by the cleaning process.

$C_3H_6/NO_x/SO_2$ + hν Exposure Experiment. Table IV contains the experimental conditions and results for the irradiated $C_3H_6/NO_x/SO_2$ mixture. The chemical composition was similar to the steady state mixture of the irradiated C_3H_6/NO_x system except for the presence of SO_2 and its oxidation products.

No decreases were found in the NO_x, O_3, or PAN concentrations during dew formation. The NO_2 and HNO_3 deposition velocities were calculated using the same method as was employed in the C_3H_6/NO_x experiment. The NO_2 dry deposition velocity was 0.02 cm/s, and the deposition velocity for HNO_3 was 2.0 cm/s, both in reasonable agreement with the previous results. Analysis of the dry panel, deionized H_2O rinse indicated that as much as 60% of the HNO_3 deposition took place during the dry periods. If this contribution is taken into account, Equation 11 yields a value of 0.7 cm/s for the HNO_3 dry deposition velocity.

The SO_2 dry deposition velocity based on the decrease in gas phase SO_2 concentration was 0.77 cm/s. The value obtained from the dew analysis was 0.82 cm/s. The approximate equality of the deposition velocities demonstrates a reasonable mass balance for the sulfur species. Good agreement is also found for the HCHO dry deposition velocities (0.45 cm/s vs. 0.47 cm/s). Analysis of the dry panels showed no evidence for substantial HCHO deposition to dry surfaces.

The dominant SO_x species in the dew was $SO_3^=$. This result differs from that found in single component SO_2 exposure experiments where the dominant species was $SO_4^=$ (12). It appears that a reaction took place which tied up the $SO_3^=$ or its precursor bisulfite (HSO_3^-) and prevented it from oxidizing to $SO_4^=$. The result that the $SO_3^=$ to aqueous HCHO ratio is 1.1 is consistent with the formation of a $HCHOHSO_3^-$ adduct, i.e.,

$$HCHO + HSO_3^- \rightarrow HCHOHSO_3^-,$$

the anion associated with hydroxymethane sulfonic acid. This species is detected as $SO_3^=$ by ion chromatography and as aqueous HCHO by the chromotropic acid method. Munger *et al.* have recently found this species present in the Los Angeles fog samples (22).

The dew molar ratios suggest that the dew consisted of a mixture of $Zn(NO_2)_2$, $Zn(NO_3)_2$, $ZnCO_3$, $ZnSO_4$, and $Zn(HCHOHSO_3)_2$. The corresponding value for φ is determined as follows:

$$\phi = \frac{[Zn]}{0.5\,[NO_x^-] + [SO_x^=] + 0.5\,[HCHOHSO_3^-] + S} = 1.0$$

where it has been assumed that the $HCHOHSO_3^-$ concentration is equal to the aqueous HCHO concentration, and the $SO_x^=$ concentration is equal to the sum of the $SO_4^=$ concentration and the difference between the aqueous HCHO concentration and the $SO_3^=$ concentration. This result shows that soluble Zn dew compounds can be completely accounted for by the above mentioned mixture. The presence of the $HCHOHSO_3^-$ ion indicates that the contributions of dissolved HCHO and SO_2 to corrosion are not additive. However, it is not clear to what

Table III. Comparison of Zn Corrosion Based on Weight Loss Determinations and That Obtained by Analysis of Dew and Rain Rinse Samples

Exposure Condition	$\left(\frac{n\,mol}{cm^2}\right)$		
	D^{WL}_{Zn}	D^{T}_{Zn}	R^{T}_{Zn}
pH=5.6 Rain + Dew	2,460	2,160	50
pH=3.5 Rain + Dew	3,450	2,160	380
Dew Only	2,880	2,160	0

Table IV. Results of $C_3H_6/NO_X/SO_2$+hv Exposure Experiment

Total Exposure Time: 25 h
Total Time of Wetness: 14 h (7-h periods)
Air Temperature: 32°C
Plate Temperature when Wet: 10°C
Rain Frequency: No rain events

Gas Phase

x	$[x_d]\,(ppb)$	$[x_w]\,(ppb)$	$v_d^{(1)}\left(\frac{cm}{s}\right)$	$v_d^{(2)}\left(\frac{cm}{s}\right)$
SO_2	1,190	798	0.77	0.82
O_3	240			
PAN	114			
NO_X-PAN	159			0.02
HNO_3	9			2.0
HCHO	1,550	1,195	0.47	0.45
CH_3CHO	792	715	0.17	

Dew

x	$[x]\left(\frac{nmol}{ml}\right)$	$D_x^{(2)}\left(\frac{n\,mol}{cm^2}\right)$
NO_2^-	35	4
NO_3^-	182	19
NO_X^-	217	23
$SO_3^=$	5,650	595
$SO_4^=$	826	91
$SO_X^=$	6,476	686
HCHO	5,100	560
Zn	4,174	441

extent the adduct would form under ambient conditions where other reaction pathways may be important.

Discussion

The results of the exposure experiments suggest that compounds other than SO_2 should be considered when designing and analyzing galvanized steel corrosion field data. These compounds include HNO_3, HCHO, and possibly NO_2. It was not possible to determine whether other photochemical oxidants such as CH_3CHO, PAN, H_2O_2, or O_3 accelerate corrosion.

SO_2, NO_2, and HCHO will deposit on a dry surface until saturation occurs. In the case of SO_2, previous work has shown that the saturation surface density is similar to that of a monolayer of adsorbed SO_2 (12). During wet periods, the dry deposition velocities of SO_2 and HCHO will be controlled essentially by the gas phase resistance of the atmosphere. However, the NO_2 deposition velocity is controlled by the surface resistance. The deposition of HNO_3 is more difficult to parameterize because, apparently, it is readily adsorbed on a dry surface. Each of the above mentioned compounds can form acids in dew that react to produce soluble Zn corrosion products. The corrosion induced per mole of compound deposited depends on the stoichiometric coefficient (γ_i) for the reaction. The coefficients for the compounds detected in this study and the experimentally determined average dry deposition velocities for the precursor gas phase species are given in Table V. It has been assumed that the acids produced in the dew are totally dissociated and react stoichiometrically with the layer compounds. The presence of the $HCHOHSO_3^-$ adduct in the dew produced in the irradiated $C_3H_6/NO_x/SO_2$ mixture indicates the possibility that antagonistic and/or synergistic effects will have to be considered when evaluating field data.

It has been shown that rain can induce Zn corrosion. Analysis of the rain rinse samples showed that the corrosion rate increased as the pH of the incident spray decreased. However, it is difficult to extrapolate the results to atmospheric conditions because the laboratory spray conditions were such that the residence time of a spray droplet was on the order of seconds. It is not clear whether there was adequate time for the reaction to take place.

The Zn corrosion products in the dew represent only the soluble corrosion products. Information on the insoluble corrosion products was obtained only in the C_3H_6/NO_x experiment in which weight loss determinations were made. It was found that the total Zn corrosion determined by the weight loss method was larger than that found in the dew and rinse. The difference is equal to the amount of insoluble corrosion product formed.

The results of the experiments suggest that the amount of soluble Zn corrosion product formed can be estimated if the deposition of the precursor gas phase species can be determined. However, the question of what controls the formation of the protective layer has not been discussed. In the next section, a model for the atmospheric corrosion of galvanized steel is formulated in which both the role of deposition and the parameters that control both the formation of soluble and insoluble products are addressed. For the remaining discussion, the term 'insoluble

Table V. Properties of Potential Corrosion Stimulations

Gas Phase	Anion	$v_d \left(\frac{cm}{s}\right)$	γ_i
SO_2	$SO_3^=$, $SO_4^=$	0.8	1.0
NO_2	NO_2^-	0.03	0.5
HNO_3	NO_3^-	2.0 (0.7)*	0.5
HCHO	$CHOO^-$	0.6	0.5
HCHO+SO_2	$HCHOHSO_3^-$	-	0.5

* The value in parentheses is based on the deposition obtained after subtracting the HNO_3 saturation surface density from the deposition obtained by dew analysis.

corrosion products' will be used to denote the compounds $ZnCO_3$ and $Zn(OH)_2$ that comprise the protective layer.

Formulation of Corrosion Model. The corrosion products formed on galvanized steel consist of insoluble compounds ($Zn(OH)_2$, $ZnCO_3$, etc.) and soluble compounds ($ZnSO_4$, $Zn(NO_3)_2$, etc.). First, the time evolution of the insoluble component will be addressed. The corrosion will be expressed in terms of change in surface thickness due to corrosion product formation.

As discussed previously, under clean air conditions, a protective layer will form on the surface inhibiting further corrosion. The time development of the protective insoluble corrosion product layer can be described by the following differential equation:

$$\frac{dT}{dt} = \frac{\alpha}{T} f \qquad (12)$$

where
- T = the thickness of the insoluble corrosion product layer,
- f = the fractional time of wetness, and
- α = a diffusion coefficient.

The parameter f is introduced to take into account the fact that corrosion only takes place when the surface is wet. The solution to Equation 12 is as follows:

$$T(t) = \sqrt{2 \alpha f t} \qquad (13)$$

where t is the total exposure time. T is a parabolic function of the exposure time controlled by the value of the diffusion coefficient α, whose value is not known, but is likely to be related to the diffusion coefficients that describe molecular transport through the solid protective layer.

The results of the experiments reported here are consistent with a model in which acidic components in the rain and dew react with the insoluble corrosion products. If it is assumed that all deposited compounds react stoichiometrically with the layer compound, Equation 12 can be re-expressed as Equation 14:

$$\frac{dT}{dt} = \frac{\alpha}{T} f - \beta \qquad (14)$$

where

$$\beta = A \left\{ f \sum_{i=1}^{N} \gamma_i v_{di} [x_i] + l(0.5[H^+] + S) \right\} \qquad (15)$$

and
- N = the number of gas phase species that deposits onto the surface and participates in corrosion;
- $[x_i]$ = the gas phase concentration of air pollutant i;
- v_{di} = the dry deposition velocity of air pollutant i;

γ_i = the stoichiometric coefficient for the reaction between the dissolved air pollutant i and the layer compound;
l = the amount of rain per unit time;
$[H^+]$ = the rain concentration of H^+;
S = solubility of $ZnCO_3$; and
A = a constant that converts the loss of material in moles per unit area to change in thickness.

The wet deposition component of β has been divided into a "clean" rain contribution and a term that represents the corrosion induced by the acidic components of the rain. The latter component is expressed in terms of H^+ concentration in the rain and assumes that it takes two H^+ ions to corrode one Zn atom. This mechanism is essentially identical to the one proposed for the production of soluble Zn corrosion products by deposition of acidifying air pollutants into the dew. The "clean" rain term contains the contribution due to the solubility of $ZnCO_3$ in H_2O. This type of term has not been included for dry deposition because, at least to a first approximation, the volumes of dew produced under ambient conditions, as opposed to that produced in the present laboratory, would be small. Therefore, the clean dew contribution can be ignored relative to that of the dew acids. More research is required to establish this assumption. In addition, it has been assumed that the two contributions to wet deposition-induced corrosion are additive. This may be an over-simplification of what actually occurs under ambient conditions.

The solution to Equation 14 is as follows:

$$T(t) = T_o\left\{1 - exp\left[-\frac{(T(t) + \beta t)}{T_o}\right]\right\}, \text{ where } T_o = \frac{\alpha f}{\beta} \tag{16}$$

and is the steady state thickness of the insoluble corrosion product layer. The total corrosion $C(t)$ is obtained by adding the contribution of the soluble corrosion products to $T(t)$, i.e.,

$$C(t) = T_o\left\{1 - exp\left[-\frac{(T(t) + \beta t)}{T_o}\right]\right\} + \beta t \tag{17}$$

Equation 16 is a transcendental equation that must be solved iteratively. The corrosion at exposure time t is the sum of the soluble and insoluble components; however, the insoluble contribution depends on the amount of soluble products produced via its dependence on βt.

The developed model differs significantly from the linear model represented by Equation 3. Analysis of galvanized steel corrosion field data shows that, in general, the corrosion is not a linear function of exposure time, particularly for short exposure times (5). The development of the steady state protective layer is nonlinear in time, and only after this layer is established will the corrosion appear to increase linearly. The thickness of the protective layer depends on the environmental condition. Under clean air conditions a thick protective layer will form; however, under highly polluted conditions (large β), T_o will be small and the corrosion will be dominated by the soluble contribution to Equation 17.

The proposed model represents a first step in the development of a corrosion model for galvanized steel. To validate the model, detailed testing of model predictions versus results of field exposure studies is required. Such an effort is presently being conducted in our laboratory.

Disclaimer

Although the research described in this article has been funded wholly or in part by the United States Environmental Protection Agency through contract 68-02-4033 to Northrop Services, Inc. - Environmental Sciences, it has not been subjected to the Agency's required peer and policy review and therefore does not necessarily reflect the views of the Agency, and no official endorsement should be inferred.

Literature Cited

1. Atteraas L.; Haagenrud, S. In "Atmospheric Corrosion"; Ailor, W.E., Ed.; J. Wiley and Sons: New York, 1982; pp. 873-891.
2. Legault, R.A. In "Atmospheric Corrosion"; Ailor, W.E., Ed.; J. Wiley and Sons: New York, 1982; pp. 607-614.
3. Mikhailvosky, V.N. In "Atmospheric Corrosion"; Ailor, W.E., Ed.; J. Wiley and Sons: New York, 1982; pp. 85-105.
4. Knotkova, D.; Barton, K.; Cerny, M. In "Atmospheric Corrosion"; Ailor, W.E., Ed.; New York: J. Wiley and Sons, 1982; pp. 991-1014.
5. Spence, J.W.; Haynie, F.H.; Edney, E.O.; Stiles, D.C. A Field Study for Determining the Effects of Dry and Wet Deposition on Materials. Presented at State of the Art Symposium on Degradation of Materials due to Acid Rain; Arlington, VA; 1985.
6. Flinn, D.R.; Cramer, S.D.; Carter, J.P.; Spence, J.W. Mat. Perform. Submitted.
7. Haynie, F.H.; Upham, J.B. Mat. Perform. 1970, 9:35.
8. Benarie, M. Metallic Corrosion as Functions of Atmospheric Pollutant Concentrations and Rain *p*H. BNL 35668. Upton, NY: Brookhaven National Laboratory.
9. Lipfert, F.W.; Benarie, M.; Dawn, M.L. Derivation of Metallic Corrosion Damage Functions for Use in Environmental Assessment. Draft Report. Brookhaven National Laboratory: Upton, NY; 1985.
10. Mansfield, F.B. Effects of Airborne Sulfur Pollutants on Materials. EPA-600/4-80-007. U.S. Environmental Protection Agency: Research Triangle Park, NC; 1980.
11. Haynie, F.H. • In "Durability of Building Materials and Components"; Sereda,P.J.; Litran, G.G., Eds.; American Society for Testing and Materials, 1980; pp. 157-175.
12. Edney, E.O.; Stiles, D.C.; Spence; J.W.; Haynie, F.H.; Wilson, W.E. Atmos. Environ. Submitted.
13. Barton, K. In "Protection Against Atmospheric Corrosion"; Wiley-Interscience, 1976; Chap. 3.
14. In "Handbook of Chemistry and Physics, 63rd Edition"; Weast, R.C., Ed; The Chemical Rubber Co.: Cleveland, OH, 1982.

15. Barton, K.; Beranek, E. Werkst. Dorros. 1959, 10:377.
16. Kuntz, R.; Lonneman, W.A.; Namie, G.R.; Hull, L.A. Anal. Lett. 1980, 13:409.
17. Lonneman, W.A.; Bufalini, J.J.; Namie, G.R. Environ. Sci. Technol. 1982, 16:655.
18. Lee, Y.N.; Schwartz; S.E. J. Geophys. Res. 1981, 86:11971.
19. Judeikis, H.S.; Siegel, S.; Stewart, T.B.; Hedgpeth, H.R.; Wren, A.G. In "Nitrogeneous Air Pollutants: Chemical and Biological Implications"; Grosjean, D., Ed.; Ann Arbor Science: Ann Arbor, MI, 1979; pp. 83-110.
20. Ten Brink, H.M.; Bontje, J.A.; Spoelstra, H.; Van DeVate, J.F. In "Atmospheric Corrosion"; Benarie, M., Ed.; Amsterdam: Elsevier, 1978; pp. 239-244.
21. Knotkova, D.; Vlckova, J. Werkst. Korros. 1970, 21:16.
22. Munger, J.W.; Jacob, D.J.; Hoffmann, M.R. J. of Atmos. Chem. 1984, 1:335.

RECEIVED January 13, 1986

11

A Field Experiment to Partition the Effects of Dry and Wet Deposition on Metallic Materials

John W. Spence[1], Fred H. Haynie[1], Edward O. Edney[2], and David C. Stiles[2]

[1]Atmospheric Sciences Research Laboratory, U.S. Environmental Protection Agency, Research Triangle Park, NC 27711

[2]Northrop Services, Inc., Research Triangle Park, NC 27709

One of the major research objectives of Task Group G, Effects on Materials and Cultural Resources, within the National Acid Precipitation Assessment Program is to derive material damage functions that relate the effects of dry and wet acid deposition on materials degradation. At an exposure site located at the Research Triangle Park, NC site, the Environmental Protection Agency has installed an automatic covering device to partition the effects of dry and wet acid deposition on the exposed materials. The device automatically covers the materials only during the rain event. Materials are exposed to both dry and wet deposition (uncovered exposure) and dry deposition (covered exposure). Corrosion data collected over a two year period for galvanized and weathering steels will be presented. The contributing effects of dry and wet deposition on the corrosion of these steels will be discussed.

At a material exposure site located at the Research Triangle Park, N.C., the Environmental Protection Agency began in October 1982 a study to partition the effects of dry and wet deposition on galvanized steel and Corten A weathering steel. In this study a mobile device covers a set of the metallic materials during each rain event. Another set of the metals is boldly exposed (uncovered) at the site. The results of the two-year exposure of these two metals is presented.

Experimental Procedures

Panels (10 x 15 cm) of 20 gauge galvanized steel and Corten A weathering steel were prepared. Each panel was identified by a code stenciled lightly on the ground exposed side of the panel. Prior to field exposure the galvanized panels were cleaned by immersion for 2 minutes in 10% by weight of ammonium chloride solution at a temperature of 60–80°C (1). The weathering steel panels were

0097–6156/86/0318–0194$06.00/0

cleaned by immersion in Clark's solution (1). All panels were rinsed in deionized water and then with methanol to remove moisture, and weighed to 0.1 mg.

The panels were exposed at 30° between ceramic insulators facing south at the site (2). Panels of each metal were exposed in triplicate for periods of 6, 12, and 24 months. The six-month panels were continually replaced with cleaned panels in order to study the corrosion during winter and summer exposures. Two sets of each metal panels were exposed at the site. An automatic device which is triggered by a sensor on a rain bucket collector covered one set of panels only during each rain event. This set of panels is exposed primarily to dry deposition. The remaining set of panels was exposed uncovered to wet and dry deposition. After each exposure period the panels were removed, then cleaned according to the procedure established prior to the exposure and reweighed to 0.1 mg.

Results

Weight-loss data for the galvanized and weathering steel for the covered and uncovered conditions are shown in Table 1 for each exposure period. The differences in weight-loss (Δloss) for covered and uncovered conditions are also shown in the table. After the initial six-month exposure period, the identity code of the weathering steel panels was not readable and new panels were prepared and exposed. The weight-loss of the unidentified panels was determined as the difference of the combined weight of the triplicate panels for each exposure period. This data is reported in Table 1 without computation of a standard deviation.

After two years of exposure the weight-loss of the galvanized steel for the uncovered exposure is nearly twice the weight-loss recorded for the covered condition. Whereas, for the weathering steel, the weight-loss for the covered exposure is greater than the uncovered exposure.

Discussion

During exposure, galvanized steel and weathering steel are expected to corrode by forming a protective surface film that retards corrosion. The rate of corrosion would then be diffusion controlled and depend upon the thickness of the film. Many of the corrosion products, particularly the carbonates, hydroxides, etc. within the film are soluble in acid solutions. In environments where increased acidity is present, dissolution of the protective film is a competing mechanism that accelerates the rate of corrosion of metals.

Previous exposure studies of these metals have shown that time-of-wetness and SO_2 concentrations contribute to the accelerated corrosion (3)(4). At the Research Triangle Park, NC exposure site, the ambient concentration of SO_2 is below the detectable limit (5 ppb) of air monitoring instrumentation and is not significantly different from zero. At these levels dry deposition of SO_2 is not expected to be an important factor in the corrosion of these metals. Time-of-wetness results from both the formation of dew and rain. With very low SO_2 levels, the dew is not expected to be very acidic.

Table I. Weight-Loss Data (mg/cm^2)* of Exposed Metals

Exposure Period	Galvanized Steel			Weathering Steel		
	Covered	Uncovered	ΔLoss	Covered	Uncovered	ΔLoss
6 Months						
10/82- 5/83	.33±.04	.52±.04	.19	20.6	23.4±1.3	2.8
5/83-11/83	.37±.06	.44±.05	.07	19.4±.05	18.3±1.3	-1.1
11/83- 5/84	.44±.11	.65±.14	.21	20.7±3.2	16.5±.3	-4.2
5/84-11/84	.27±.02	.54±.03	.27	19.8±0.1	18.6±.3	-1.2
12 Months						
10/82-11/83	.47±.01	.94±.06	.47	30.3	21.3	- .9
18 Months						
5/83-11/84	-	-	-	33.7±3.6	28.4±3.4	- .5
24 Months						
10/82-11/84	.82±.01	1.66±.09	.84	44.7	37.5	- .7

*Area of one side only.

The average pH of the rain at the site is about 4.5; however, specific rain events may have lower pH values.

The environmental rain data acquired at the site were compiled in Table 2 for each exposure period. Simple linear regression analysis was performed using each of the factors as the independent variable and the difference in corrosion (Δloss in Table 1) as the dependent variable. All of the regression coefficients were significant. The coefficients for the amount of rainfall were considered to be the most important, since delivery of $SO_4^=$, NO_3^- and H^+ (concentrations relatively constant) to the surface of the metals is dependent of rainfall.

In a clean environment, the corrosion film that forms on galvanized steel consist primarily of $ZnCO_3$. This process is consistent with the non-linear time function that has been observed for the corrosion of zinc (5). In pure water $ZnCO_3$ is reported to dissolve at 15°C to saturation at a rate of .001 gms per 100 grams or 1 mg per 100 cm^3. This would be equivalent to .01 $mg/cm^2/cm$ rain. Thus, a theoretical saturation coefficient for pure rain at 15°C would be .0052 mg per cm^2 per cm. The Research Triangle Park rain coefficient for galvanized steel is 0.00360 ± 0.00052 mg per cm^2 per cm. Although $ZnCO_3$ is more soluble in acidic solutions than in pure water, the amount of time rain is in contact with the surface of the galvanized steel panel is not expected to be long enough for the solution to become saturated. The magnitude of the regression coefficient is reasonable for a film dissolution mechanism. Thus, wet deposition (rainfall) is removing the protective film on the uncovered galvanized steel at the site.

In a recent study of weathering steel, Haynie reports a regression coefficient for rain of $-0.0798 \pm .0227$ μ/cm or $-0.061446 \pm .017479$ mg per cm^2 per cm (4). In this study the dissolution of the oxide film that forms on weathering steel is reduced by the amount of rainfall. The rain coefficient for weathering steel obtained from the Research Triangle Park exposure study is $-0.03442 \pm .01814$ mg per cm^2 per cm. The values of these two coefficients are not significantly different at the 95% confidence level using a t test. Thus, wet deposition (rainfall) has a beneficial effect on the corrosion of weathering steel, while having a detrimental effect on galvanized steel.

Long-term corrosion studies of weathering steel indicate that the corrosion time function should be nearly parabolic (6)(7). This function is associated with the formation of a passive protective film. It is apparent that two-year data are not sufficient to accurately predict long-term behavior.

Conclusion

The mobile covering device appears to provide a means to partition the effects of dry and wet deposition during field exposure of metallic materials. Wet deposition (rainfall) is having an effect on the weathering of galvanized and weathering steels at the Research Triangle Park, NC exposure site.

Table II. Environmental Rain Data

Exposure Period	Rainfall(cm)	SO_4(N moles/cm^2)	NO_3(N moles/cm^2)	H(N moles/cm^2)
6 Months	65.3	877.0	774.0	1436.0
	46.3	1334.0	1097.0	2943.0
	80.7	1014.0	879.0	2082.0
	50.2	1045.0	1000.0	2194.0
12 Months	111.0	2212.0	1872.0	4378.0
18 Months*	192.3	3226.0	2751.0	6460.0
24 Months	242.4	4271.0	3752.0	8655.0

*For weathering steel only.

Literature Cited

1. American Society for Testing and Materials, 1970 ASTM Standards, Part 31, p. 928-931.
2. American Society for Testing and Materials, 1980a ASTM Standards, Part 10, p. 985-992.
3. Haynie, F.H., Spence, J.W., and Upham, J.B. Effects of Gaseous Pollutants on Materials - A Chamber Study, EPA-600/3-76-015, February 1976.
4. Haynie, F.H. Environmental Factors Affecting the Corrosion of Weathering Steel. To be published in Proceedings of ACS Meeting, Degradation of Materials Due to Acid Rain, Arlington, Va., June 1985.
5. Guttman, H. In: Metal Corrosion in the Atmosphere, ASTM STP 435, 1968, p. 223-239.
6. Haynie, F.H., Spence, J.W., and Upham, J.B. In: Atmospheric Factors Affecting the Corrosion of Engineering Metals, ASTM STP 646, 1978, p. 30-47.
7. Haynie, F.H. and Upham, J.B. Effects of Atmospheric Pollutants on Corrosion Behavior of Steels, Materials Protection and Performance, 10:18-21, November 1971.

RECEIVED January 2, 1986

12

Effects of Acid Deposition on Poultice-Induced Automotive Corrosion

R. C. Turcotte, T. C. Comeau[1], and Robert Baboian

Texas Instruments Incorporated, 34 Forest Street, Attleboro, MA 02703

The chemistry of wet poultice deposits (road dirt and debris) on automobiles is affected by factors such as road salts and the composition of wet and dry deposition. The rate of metallic corrosion where poultice deposits accummulate on automobiles is highly dependent on this chemistry. The composition of aggressive species in poultice deposits was determined by sampling in Montreal, Detroit, Boston, and Dallas. A correlation between the corrosivity and the poultice chemistry was made by mounting corrosion test coupons on vehicles in each city and by simulating poultice effects in the laboratory. Results show that corrosion damage increases in a complex manner in areas of heavy road salt use and acid deposition.

Factors affecting the corrosion of auto-body steel and trim are in two major catagories, design and environmental. Design factors include the choice of material of construction, galvanic coupling of dissimilar metals, the creation of lips or cups that may hold aggressive environments and other entrapment areas. Environmental factors include all the various chemical species to which the automobile is subjected; major contributors include road salts such as sodium chloride and calcium chloride, sea salt aerosols, and precipitation, especially if it has been acidified. Figure 1, which graphically depicts the automotive corrosion milestones that have accompanied road salt usage in the United States (1-2), and Figure 2, which superimposes automobile corrosion rates and average rain pH in the United States (1), support these claims.

Poultice deposits are accumulations of road dirt, vegetation, tar, and other debris which collects in crevices around trim materials, in wheel wells, and in other entrapment areas (3). Since the basic design in those locations allows poultice deposits to become wetted without being washed away, poultices remain for a considerable period of time and can cause severe corrosion leading

[1]Current address: University of Rhode Island, Kingston, RI 02881

0097-6156/86/0318-0200$06.00/0

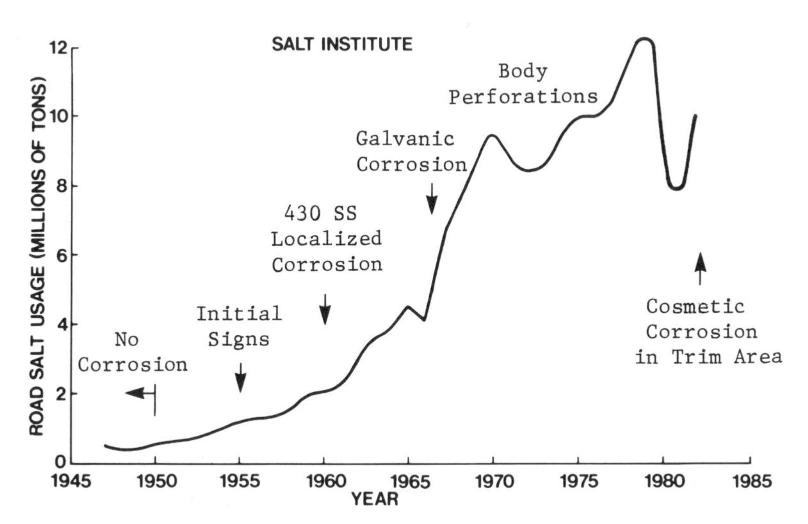

Figure 1. Automobile corrosion milestones as a function of road salt usage.

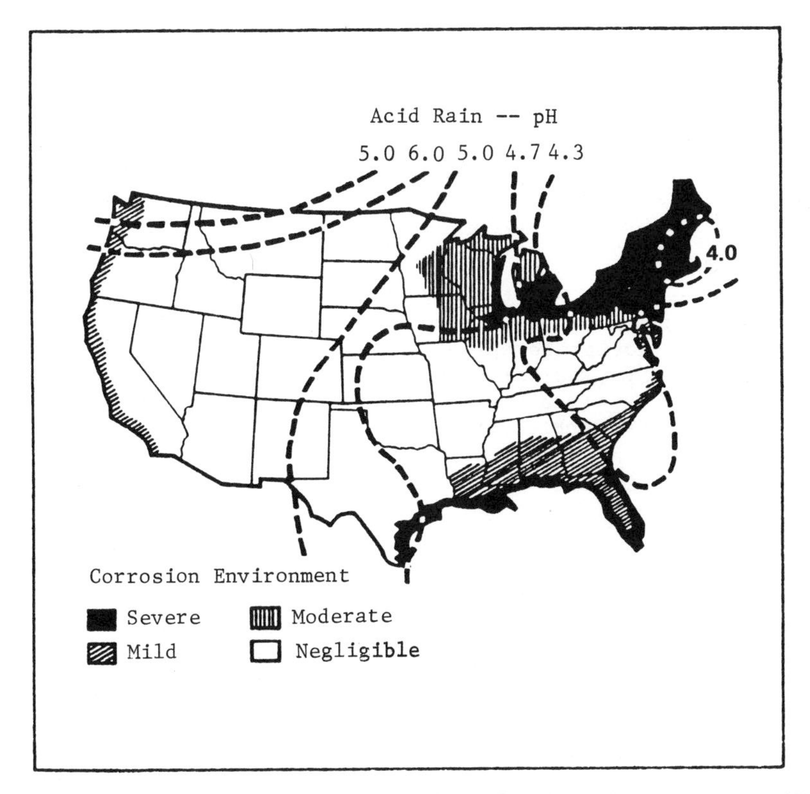

Figure 2. Geographic distribution of acid rain and automobile corrosion in the United States.

even to perforation (4). This paper examines the chemistry of aggressive species in poultice samples from Boston, Dallas, Detroit, and Montreal. A correlation between corrosivity and poultice chemistry is presented.

Theory

When organic fuels are burned, carbon dioxide and water vapor are released along with various amounts of sulfur dioxide and nitrogen oxides. The sulfur and nitrogen oxides in the atmosphere are then further oxidized with the assistance of ultraviolet solar radiation; when these gases are scrubbed from the air by precipitation, a dilute solution of sulfuric acid and nitric acid forms. Carbon dioxide itself hydrolyzes to carbonic acid and is important in the marine carbonate buffer system; however, it is a weak organic acid and atmospheric concentrations typically lower the pH of distilled water only to about 5.7 (5-6).

Metal oxides are typically more soluble in acidic media. Hence, low pH solutions should adversely affect the formation of protective films on metal surfaces.

Chlorides, on the other hand, are known to induce corrosion through a catalytic mechanism. Chloride ions are small and mobile and therefore affect the protective films on metals causing local breakdown and pitting. Once pitting has initiated, migration of chloride ion into the pit causes a lowering of pH within the restriction by the formation of HCl hydrolysis products. This occurs in both pits and crevices and is shown in Figure 3 (7). Additionally, calcium chloride is deliquescent, absorbing moisture from the air and maintaining a wet surface above a critical humidity. It is known that calcium chloride is highly corrosive to automobiles, at least partly because of this deliquescence (8).

The mechanism of poultice corrosion is shown diagramatically in Figure 4. Corrosion initially occurs uniformly over the whole metal surface. However, as the process continues, the oxygen reduction cathodic reaction may become restricted to a band near the surface where oxygen is readily available. Corrosion of metal then takes place preferentially slightly below the band. The pH rises in the cathodic reaction area due to an increase in hydroxyl ion concentration while the anodic area decreases in pH due to the hydrolysis of metal chloride reaction products.

Wet Poultice Chemistry

Figure 2 shows that the northeastern United States suffers especially severe automotive corrosion. Annual automobile corrosion surveys conducted by Texas Instruments Inc. have indicated that Montreal, Canada is also highly corrosive. Therefore Boston, Detroit, and Montreal were chosen as ideal sampling sites that should have the greatest amounts of corrosive species; for comparison, Dallas, Texas was chosen as typical of what should be a mildly corrosive area.

Representative samples (100 grams) of poultices were collected from cars in each city (Montreal 31, Detroit 24, Boston 15, Dallas 4). Soluble species were then extracted into distilled water by means of a ball mill, which ran 8 hours on and 16 hours off for 3

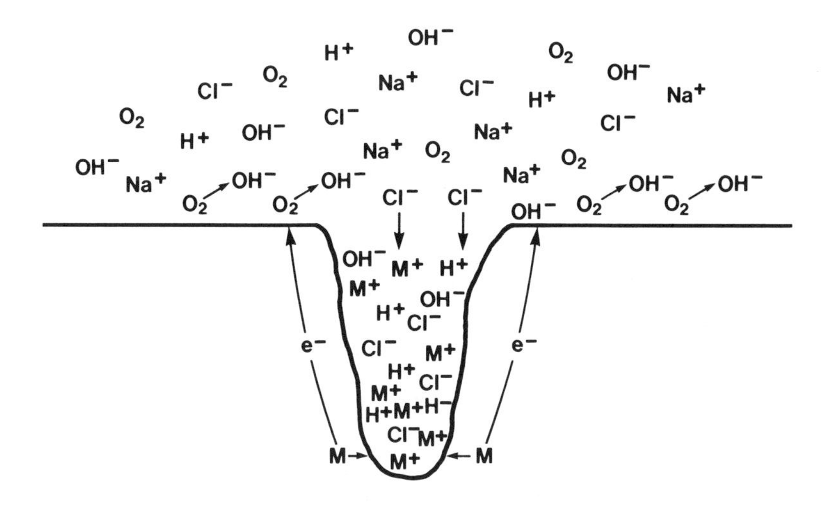

Figure 3. Autocatalytic pitting mechanism.

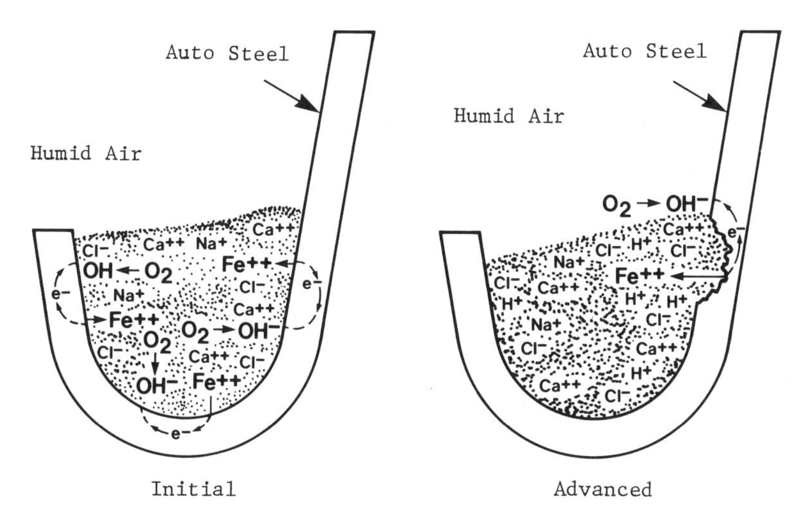

Figure 4. Mechanism of poultice corrosion of auto steel showing initial and advanced stages.

days; the solution was then filtered (0.45 μm) before analysis (9). Qualitative analyses were done using DC-Plasma emission spectroscopy and a series of anionic specific methods. The primary chemical constituents were then quantitatively determined by methods listed in Table I (9-10). EPA water quality standards were used as a check on the methods used.

Table I. Methods of Analysis

Species		Technique Employed
CATIONS	Na^+ Ca^{2+} K^+ Mg^{2+}	ATOMIC ABSORPTION SPECTROSCOPY OR ATOMIC EMISSION SPECTROSCOPY
ANIONS	Cl^-	ION SELECTIVE ELECTRODE
	NO_3^-	ION SELECTIVE ELECTRODE
	SO_4^{2-}	UV-VISIBLE SPECTROSCOPY
	PO_4^{3-}	UV-VISIBLE SPECTROSCOPY
	HCO_3^-	TITRIMETRY
	CO_3^{2-}	TITRIMETRY

Figure 5 compares the ions that might be attributed to road salt use and acidic deposition in the northern cities. These include sodium, calcium, chloride, and sulfate; they were major constituents. While the northern cities have expectedly high levels of these ionic species, the levels in Dallas were unexpected. Further observation pointed out important differences in the chemistry of the environments as described below.

Comparisons of the pertinent ionic species in the poultices are shown in Figure 6. In Boston, Detroit, and Montreal, there is an approximate balance between equivalents of chlorides and sulfates versus equivalents of calcium and sodium; in Dallas, there is an overabundance of cations, indicating that unidentified anionic species exist.

The poultices from different cities also have varying characteristic soluble species. Montreal poultices are typically very high in sulfate and calcium ions with significant levels of chloride and sodium. Detroit poultices are highest in chloride and calcium ions with significant levels of sulfate and sodium ions. Boston, while possessing less of these soluble species, contains mainly sulfate and calcium ions with a fair proportion of chloride and sodium ions. Dallas has high cation levels, especially calcium, but chloride and sulfate levels were similar to those of Boston.

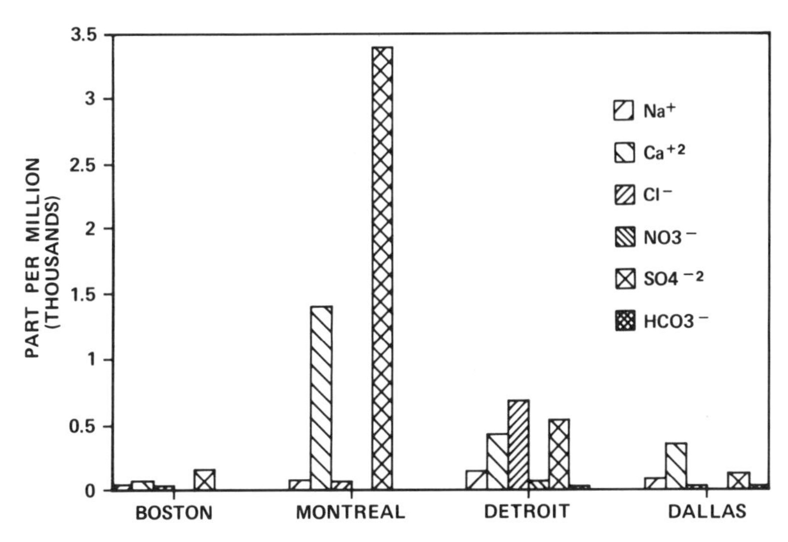

Figure 5. Concentration of road salt and acid deposition ions in poultices.

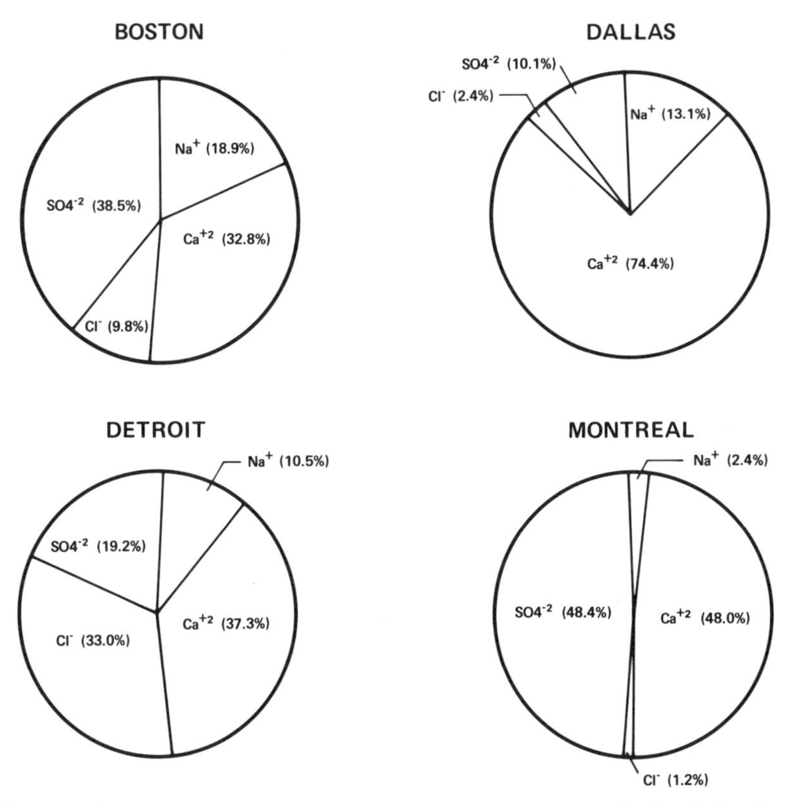

Figure 6. Proportions of poultice ions on an equivalent basis.

Automotive Corrosion

In the winter of 1984-1985, test samples on Plexiglass plates were exposed in each of the test cities. These included samples of bare weighed auto-body steel, weighed galvanized steel, and painted auto-body steel with stainless steel trim. Where painted panels were involved, cut edges were coated; in the stainless steel trimmed panels, the ends of the stainless steel bolts used for electrical continuity were coated as well. Nylon bolts were used to attach the other test specimens to the Plexiglass plate which was mounted on the front bumper of an automobile.

Bare steel from all sites except Dallas generally had the same visual appearance, with Figure 7 being typical. Dallas steel plates showed very little rusting. Upon retrieval all the steel plates were cleaned using Clarke's solution (ASTM G1) to determine the weight loss. The calculated mils per year penetration rates are listed in Table II. The cities may be listed in decreasing order of aggressiveness as:

Montreal $>$ Boston $>$ Detroit $\gg$ Dallas

with differences in aggressiveness being slight in the Northeast in this test.

Table II. Atmospheric Corrosion Rates of Bare Automotive Steel Plates During the Winter of '84 - '85

City	Corrosion Rate (mpy)	Average Corrosion Rate (mpy)
	1.83	
Boston	2.24	2.25
	2.68	
Dallas	0.0491	
	0.114	0.114
	0.143	
	0.148	
Detroit	1.41	
	1.75	1.78
	2.19	
Montreal	1.99	
	2.28	2.55
	2.29	
	3.64	

Galvanized steel plates showed some visual differences between cities, as seen in Figure 8. Some differences were also observed within the same city (e.g. Figure 9). These plates were cleaned using an ammonia wash followed by a chromic acid and silver nitrate dip (ASTM G1). Mils per year penetration rates are shown in Table III. Decreasing corrosivity follows the order:

Montreal > Boston > Detroit > Dallas

Galvanizing not only protects the steel, but actually corrodes more slowly.

Table III. Atmospheric Corrosion Rates of Galvanizing on Automotive Steel During the Winter of '84 - '85

City	Corrosion Rate (mpy)	Average Corrosion Rate (mpy)
Boston	0.0590 0.0932 0.193	0.115
Dallas	0.0152 0.0185 0.0202 0.0319	0.0214
Detroit	0.0364 0.0560 0.0626	0.0517
Montreal	0.198 0.238 0.284 0.319	0.260

Galvanic couples of stainless steel trim with painted auto-body steel produced very dramatic differences in cosmetic corrosion between cities, as seen in Figure 10. Numerical corrosion ratings are listed in Table IV. The decreasing order of aggressiveness is:

Montreal > Detroit > Boston > Dallas

In addition, rusting of the stainless steel trim material was observed in the northern cities, especially in Montreal.

To reproduce poultice corrosion in the lab, samples were immersed in washed sand saturated with a solution of interest, as described previously (8). When bare steel samples were tested in

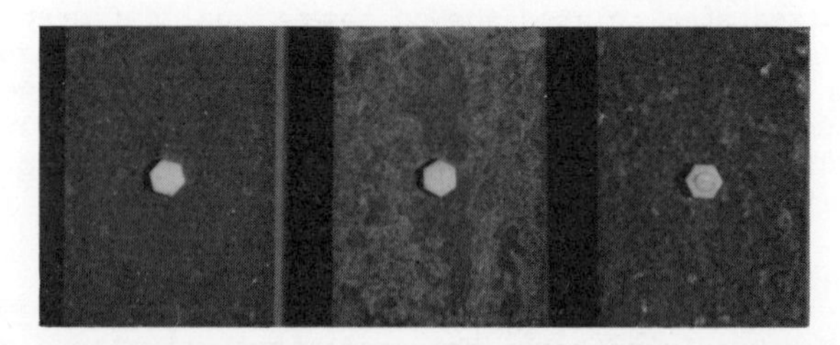

Figure 7. Typical appearance of bare auto-body steel following one season exposure in Detroit.

Figure 8. Visual comparison of typical galvanized steel auto-body plates, one season exposure. Right to left: Boston, Dallas, Detroit, Montreal.

Figure 9. Examples of the range of appearance for galvanized auto-body steel following one season exposure in Detroit.

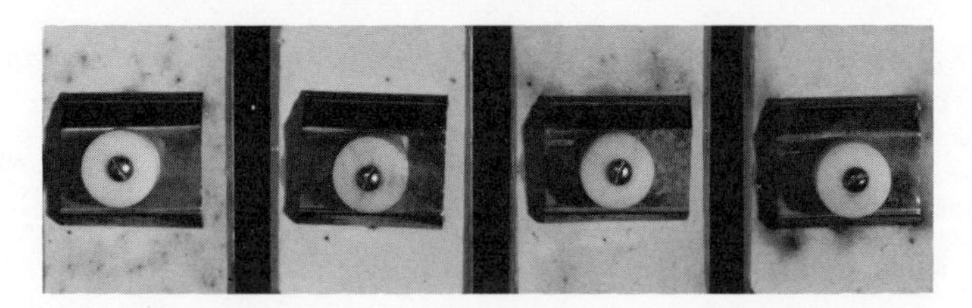

Figure 10. Visual comparison of stainless steel/painted auto-body steel galvanic couples, one season exposure. Right to left: Boston, Dallas, Detroit, Montreal.

Table IV. Rating of SS Trim/Painted Auto Body Steel Galvanic Corrosion

City	Corrosion Severity Rating	Average Corrosion Severity Rating
Boston	1.9 2.1 2.3	2.1
Dallas	1.1 1.4 1.4 1.9	1.4
Detroit	2.2 3.3 3.6	3.0
Montreal	3.6 4.1 4.1	3.9

Note: Increasing corrosion is 0 to 5 with 5 the most severe.

acid rain (distilled water adjusted to pH 4 with a 3:7 nitric acid/sulfuric acid mixture) and 1% calcium chloride, the data in Table V resulted. That level of calcium chloride is more aggressive than acid rain; however, if the two are combined, there is initially a synergistic effect that greatly increases corrosion. With repeated wetting and drying, the synergistic effect is lost. The reason for this phenomenon is that alternate wetting and drying of poultices can affect solution chemistry.

Table V. Corrosion Data for Auto-Body Steel in Poultice Test

	Acid Rain	$CaCl_2$	Acid Rain and $CaCl_2$
Avg. Penetration 2 weeks; no drying (mpy)	1.3	12.5	26.3
Avg. Penetration 12 weeks; drying and wetting (mpy)	19.3	87.4	82.8

A solution of sulfuric and nitric acids, with and without calcium chloride addition, was subjected to cyclic drying. After three wet/dry cycles, the pH of both had risen, with the $CaCl_2$ solution rising more. Analysis of the original and of the cycled solutions is shown in Table VI. Note that the nitric acid evaporated away and that chloride was depleted to very low values. This is due to evaporation and the reaction:

$$CaCl_2 + H_2SO_4 \longrightarrow CaSO_4\downarrow + 2HCl\uparrow$$

in which HCl is driven off upon drying.

Table VI. Effect of Repeated Wetting and Drying on Acid Rain Chemistry

Sample		pH	SO_4^{-2} (mg/1)	NO_3^- (mg/1)	Ca^{+2} (mg/1)	Cl^- (mg/1)
H_2SO_4 HNO_3 $CaCl_2$	no drying	0.96	590	291	350	940
H_2SO_4 HNO_3 $CaCl_2$	3 wet-dry cycles	1.57	350	6.8	390	<10
H_2SO_4 HNO_3	3 wet-dry cycles	1.39	650	37	-	-

Discussion

The main source of sulfate in automobile poultices from aggressive northern sites is acid deposition. Road salts introduce both sodium chloride and calcium chloride. However, because of wet/dry cycles, the chemistry of the poultices is complex, involving the formation of calcium sulfate and the depletion of nitrate and chloride ions with a reduction in acidity. Thus corrosion tests based on the analysis of solubles within a poultice at any one time may not reproduce field results; the history of the poultice is important.

The high levels of calcium sulfate in Montreal are indicative of the heavy use of road salts, especially calcium chloride, and of the acid deposition that affects that area. Field corrosion tests showed how aggressive this environment can be even when the solution can quickly run off and evaporate; lab tests indicate that after dying out and reacting with calcium chloride, the acid deposition may no longer be important to the corrosion process unless it is renewed.

The calcium chloride levels in Detroit also indicate a high road salt use. However, the acid deposition is not as severe as in Montreal, as indicated by the poultice chemistry.

Boston shows a lower salt usage than Montreal or Detroit with sufficient acid deposition to consume most of the chloride. The ratio of calcium to sodium also indicates probably less $CaCl_2$ being used in the road salt. Given the aggressiveness of chloride salts and especially $CaCl_2$, one would expect Boston to be the least aggressive of the three northern cities. The fact that it is more aggessive than Detroit in two of the field tests may indicate the influence of increased acid deposition activity in the Boston area.

In Dallas, $CaCl_2$ is used for dust control purposes. The absolute levels of chloride and sulfates are similar to those found in Boston. However, environmental factors such as precipitation and humidity are different and therefore Dallas is less aggressive. In addition, Dallas analyses do not appear to account for all ionic species.

Regardless of the environment, tests indicate that automobile manufacturers can significantly reduce corrosion in acid depostion areas by using galvanized steel. Even in the very severe Montreal environment, the corrosion of galvanizing is roughly an order of magnitude less than that of bare steel. Corrosion performance can also be improved by careful consideration of galvanic effects in the trim areas.

Summary and Conclusions

Representative automobile poultice samples from Boston, Dallas, Detroit, and Montreal show a common chemistry in the northern states involving road salts and acid deposition. There were also significant amounts of chloride from dust control operations in Dallas. However, the different sites varied in the levels of ionic species and also in their proportions.

As in past surveys, field tests during the winter of '84-'85 showed that areas of high road salt use and serious acid deposition are very corrosive. Total time of wetness is also important. This research indicates a decreasing order of aggressiveness on bare or galvanized steel to be:

$$\text{Montreal} > \text{Boston} > \text{Detroit} > \text{Dallas}$$

Simple laboratory tests may not be adequate to reproduce field results. The reason for this was demonstrated in the lab with an artificial solution containing road salt and acid rain constituents. Allowing such a solution to evaporate drives off the volatile nitric acid component and hydrochloric acid formed when calcium sulfate precipitates. Hence the history of a poultice is important in determining the aggressiveness of a particular environment.

Literature Cited

1. Baboian, R. "Automotive Corrosion by Deicing Salts"; National Association of Corrosion Engineers: 1981.
2. Baboian, R. "Causes and Effects of Corrosion Relating to Exterior Trim on Automobiles;" (831835) in Proceedings of the 2nd Automotive Corrosion Prevention Conference; Society of Automotive Engineers, Inc.: Dec. 1983.
3. Rowe, L. C.; and Chance, R. L. "Fundamentals of Corrosion Testing"; in AUTOMOTIVE CORROSION BY DEICING SALTS; R. Baboian, Ed.; N.A.C.E.: 1981.
4. Baboian, R. "Chemistry of the Automotive Environment"; Society of Automotive Engineers; Reprint 780908: 1978.
5. Riley, J. P.; and Chester, R. "Introduction to Marine Chemistry"; Academic Press: N.Y., N.Y., 1971.
6. Pierce, W. C.; Haenisch, E. L.; Sawyer, D. T. "Quantitative Analysis"; 4th ed.; John Wiley & Sons Inc.: New York, N.Y., 1966.
7. Fontana, M.; Greene, N. D. "Corrosion Engineering"; McGraw-Hill Book Co., N.Y., 1967.
8. Baboian, R.; and Turcotte, R. C. "Development of Poultice Corrosion Tests for Automobiles"; National Assoc. of Corrosion Eng.: Paper Number 383, presented at N.A.C.E. Corrosion/85, March, 1985.
9. "Soil Survey: Laboratory Methods and Procedures for Collecting Soil Samples"; Soil Conservation Service, U.S. Dept. of Agriculture.
10. Franson, M. A. H., Managing ed. "Standard Methods for the Analyses of Water and Wastewater"; 15th edition; American Pub. Health Assoc.: 1981.

RECEIVED January 2, 1986

13

Effect of Acid Rain on Exterior Anodized Aluminum Automotive Trim

Gardner Haynes and Robert Baboian

Texas Instruments Incorporated, 34 Forest Street, Attleboro, MA 02703

Initial attempts to use aluminum for automotive trim were unsuccessful due to the corrosion behavior of the metal. It is therefore anodized for automotive trim applications to provide a protective oxide surface which acts as a barrier coating for corrosion protection.[1,2] Aluminum and its alloys are susceptible to pitting and crevice corrosion in chloride containing environments. The corrosion resistance of anodized aluminum is therefore highly dependent on the quality of the anodized surface and the absence of scratches and other damage sites.

The anodizing process consists of: 1) trim forming, 2) cleaning, 3) anodic electrolysis usually in sulfuric acid and, 4) sealing in an aqueous bath. Each of these steps are important, for example, the anodized oxide layer must be controlled to provide an optimum thickness. Sealing the anodized layer is also very important and must be complete. Corrosion problems associated with anodized aluminum trim include etching of the oxide film (known as "blush and bloom") and pitting of the aluminum.

Pitting of aluminum occurs at defects in the anodized coating due to poor process control or mechanical forces in service such as stone pecking.[3,4] It is characterized by a central pit which is surrounded with a white halo of corrosion product. Pitting is more prevelant in high chloride containing environments.

Blush and bloom are caused by the etching of the anodized oxide which changes the surface reflectivity of the trim material. Problems of a milky white appearance (blush and bloom) on the surface of anodized aluminum trim have frequently been reported. However, this effect has been more pronounced in recent years in the Northeast USA and Canada. Although in previous years the problem has been related to quality variations, recent data indicates that acid precipitation enhances this effect.

This investigation was conducted to determine the relative

0097-6156/86/0318-0213$06.00/0

effects of acid precipitation and road salt usage on corrosion of anodized aluminum trim. A second objective was to develop an accelerated test which could duplicate the mechanisms of corrosion and separate the effects of acid precipitation from those of road salt usage.

Experimental

Anodized aluminum test panels (4 inch by 6 inch) were cut from automotive rocker panels purchased at a local dealership. Before testing cut edges were masked with a potting compound to eliminate edge effects.

Test solutions were prepared from reagent grade chemicals and distilled water. Appropriate salt concentrations were obtained using a mixture of equal weights of NaCl and $CaCl_2$. The pH of the test solutions was adjusted with a mixture of three parts H_2SO_4 and one part HNO_3. The resulting matrix of test solutions had a range of total salt concentration of 0 to 20 percent and a pH range of 0.2 to 7. Triplicate test panels were tested for each condition within the matrix. The MIST test consisted of a spray to the horizontal trim test panel every two hours for 3 cycles and subsequent drying for 18 hours. The test panels were evaluated for degree of blush and bloom and pitting daily.

Results and Discussion

Results in Table I show that MIST test road salt spray alone did not produce blush and bloom although pitting was observed at the higher salt concentrations. Acid precipitation had a very pronounced affect at or below pH=4. This pH range is easily obtained on automobiles due to a drop in pH of the solution as a result of evaporation as well as the fact that the pH of precipitation has been less than 4 on many occasions. The appearance of the test panels was similar to that observed on vehicles after one or two years in Montreal or after several years in milder environments such as Detroit or Boston.

The combination of acid precipitation with road salt spray produced the worst effects on anodized aluminum. As the MIST test pH became more acidic and the amount of salt increased the time to achieve an equivalent milky white appearance was reduced significantly (Table I). At more neutral high chloride concentrations severe pitting occurred while at more acid high chloride concentrations blush and bloom predominated. The latter environment is similar to that existing in the Northeast USA and Canada and, therefore, the results can explain the problems of blush and bloom in these areas. Corrosion surveys by automotive companies and trim producers in these areas have shown that blush and bloom and pitting have become increasingly more severe over the last ten years.[5] These problems have led to a shift away from anodized aluminum as an automotive trim material in recent years.

Table I. Blush and Bloom of Anodized Aluminum During Mist Test (Days to Equivalent Milky Appearance)

pH-Adjusted with 3:1 H_2SO_4-HNO_3 (by Volume)	0	1	2	5	20
7.0	No Effect	Pitting			
4.0	Pitting and Blush and Bloom				
2.0	12	8	10	12	8
1.0	6	4	2	1	1
0.5	6	3	2	1	1
0.2	6	3	2	1	1

Percent Salt (1:1 NaCl-$CaCl_2$ by Weight)

Conclusions

An accelerated MIST test procedure has been developed which duplicates the mechanisms of corrosion of anodized aluminum trim in automotive environments. Blush and bloom of anodized aluminum automotive trim is more severe in environments with acid precipitation and this effect can be duplicated in an acidified MIST test procedure. Pitting of anodized aluminum is more prevelant in automotive environments with high chloride concentrations and this effect can be duplicated in a neutral chloride MIST test procedure. A change in the mechanism of corrosion of anodized aluminum trim from pitting to blush and bloom in chloride containing environments occurs in the pH range of 2 to 4. These results indicate that blush and bloom of anodized aluminum will become more severe as the acidity of precipitation increases. Thus more expensive trim materials such as bimetal are being used by the automotive industry.

Literature Cited

1. H.A. Kahler and R.E. Harvie, Corrosion Resistance of Trim Materials, Paper presented at SAE Meeting, 1963.
2. G.F. Bush, "Corrosion of Exterior Automotive Trim." Paper 650A presented at SAE Meeting, Detroit, January, 1963.
3. G.F. Bush, Metal Progress, Vol 90, p 56, 1966.
4. P.W. Bolmer, "Bright Aluminum Trim-Alloy and Process Developments," Paper 17 presented at NACE Meeting, March, 1970.
5. R. Baboian, Proceedings of the 2nd Automotive Corrosion Prevention Conference, pp 223-227, Dec. 1983.

RECEIVED January 2, 1986

14

Effects of Acid Rain on Indoor Zinc and Aluminum Surfaces

J. D. Sinclair[1] and C. J. Weschler[2]

[1]AT&T Bell Laboratories, Murray Hill, NJ 07974
[2]Bell Communications Research, Lincroft, NJ 07738

Acidic precursors of acid rain can readily penetrate buildings. Regions of the United States where "acid rain" has been identified as a problem will most likely have elevated concentrations of such precursors that may influence indoor surfaces. If the available information on outdoor, indoor, and surface concentrations of ammonium, sulfate, nitrate, calcium, and other species is considered, an estimate of the loading above the background can be made for all areas where outdoor concentrations are known. The relative degree of hazard for the different environments can be estimated from the developed methodology and, based on previous experience with zinc and aluminum surfaces, provides a framework for qualitative rankings of comparable environments. For metals and other materials for which surface acidity is often strongly buffered by the corresponding metal oxide layer, the anion loading may contribute more to enhanced degradation than the acidity. This behavior is especially true for top side horizontal surfaces, where coarse alkaline particles will tend to neutralize acidity contributed by anthropogenically derived substances.

The intent of this paper is to present a methodology for estimating, from available information on concentrations and deposition velocities, the potential effects of anthropogenically derived acidic substances on indoor surfaces. Surface accumulation rates are derived that are applicable to all types of indoor surfaces. The discussion of the possible effects of the accumulated substances will concentrate on zinc and aluminum surfaces because data exists on the behavior of these metals in indoor environments (1). Aluminum forms a passivating oxide which protects against corrosion in most environments, while zinc is expected to corrode at a roughly linear rate over its lifetime.

The term 'acid rain' is commonly used to describe phenomena associated with both wet and dry deposition. The effects of man-made airborne substances on indoor surfaces are expected to be different from and less dramatic than those on outdoor surfaces. In the outdoor environment, both the species contained in the rain or fog droplets and the added influence of substances previously deposited on the surfaces by dry deposition will impact the surfaces. For very brief rain events, the surface concentrations of many species are likely to be substantial, due to the combined effects of accumulated dry deposition and the high concentrations of species that have been commonly associated with the early stages of a rain event. The available data indicate that roughly 70 percent of the total deposition is dry deposition. Hansen and Hidy (2), in their assessment of existing rain chemistry data, discuss the influence of geographical factors on pH and state that in western arid areas of the United States, soil-derived substances elevate the pH in the early stages of a rain event, while in the northeastern

0097–6156/86/0318–0216$06.00/0

part of the country, anthropogenically derived acidic substances result in rain of lower pH. For long rain events, much of the soluble substances on top side horizontal and vertical surfaces may be eliminated by the continuous washing of the surfaces. Dry deposition will also affect these surfaces between rain or fog episodes, the degree depending not only on the concentrations of species but also on the amount of moisture in the environment and its variability. If the critical relative humidity (CRH) of substances on these surfaces is exceeded or the temperature of these surfaces is sufficiently low that the dew point is reached, the effects of acidic substances are likely to be more dramatic than in dry climates.

For the indoor environment, the effects of anthropogenically derived substances on surfaces will usually be limited to those resulting from dry deposition of acidic substances that are able to pass through filtration systems, to penetrate through leakage pathways or to be carried indoors by humans or various articles that have been outdoors. If the standard indoor environment is taken to be one with (1) closed windows, (2) only occasional opening of doors to the outdoor environment, (3) an air handling system that filters recirculated air, and (4) leakage pathways that remove nearly all particles > 1 μm diameter, then the effect of a rain episode by itself will likely be limited to an increased rate of degradation associated with a higher ambient relative humidity (RH) and concomitant higher concentration of surface moisture rather than a substantially increased influx of acidic substances. The incoming air may well be somewhat cleaner during a lengthy rain event due to the cleansing effect of the rain on the outdoor pollutant concentrations through impact with particles and solubilization of gases.

Very little acidic material in the form of acid rain or fog droplets is expected to enter the indoor environment. The size distribution ranges from 0.5-50 μm diameter for fog droplets and from 200-5000 μm diameter for rain droplets (3). Most of the < 1 μm diameter fog droplets that pass through standard filtration systems are likely to react with air duct walls before entering the indoor environment. The indoor surfaces are taken to be those surfaces that are not associated with the air handling system or leakage pathways.

Methodology For Estimating Accumulation Rates

For the purposes of this discussion, it is reasonable to assume that the outdoor environment is the source of most of the anthropogenically derived substances (4) that are present in the indoor environment. The accumulation rates of species on indoor surfaces are related to the outdoor concentrations of these substances through the relationships among the indoor and outdoor concentrations and the indoor deposition velocities of these species. A substantial amount of data is available on outdoor concentrations (4-13). Simultaneous measurements of outdoor and indoor concentrations are less numerous. Very few measurements of indoor deposition velocities have been made. Estimated ratios of outdoor to indoor concentrations will be used that are based on field data, where available, or best judgments. From the limited experimental measurements, taking into account the relative variations in outdoor deposition velocities as a function of particle size, indoor deposition velocities will be estimated. Using these approximate indoor/outdoor ratios and deposition velocities, the indoor surface accumulation rates for substances contained in airborne particles can then be estimated from prevailing outdoor concentrations.

Estimating the deposition velocities of gaseous species is considerably more complex than estimating those for substances in particles, in part due to the uncertainties in the sticking and reaction probabilities. Such estimates have not been made but the potential effects of some of the typical gases can be surmised from available data on surface accumulation rates, e.g. sulfate accumulation on indoor zinc and aluminum surfaces is predominantly a result of particulate sulfate deposition rather than a corrosion reaction involving sulfur dioxide (1).

Concentrations and Deposition Velocities of Acidic Substances

As has been documented in a number of continuing and completed studies, the concentrations of the precursors of acid rain and the concentrations of the acidic substances in wet and dry deposition are quite variable from region to region. The anthropogenically derived particles associated with acid rain are expected to have aerodynamic diameters < 2.5 μm. Such particles may include sulfate, nitrate, chloride, and ammonium. However, natural coarse particles > 2.5 μm diameter may also play a role in the effects of anthropogenically derived substances on surfaces. Such particles are typically rich in calcium, are somewhat alkaline, and may neutralize acidity in the man made substances. Concentrations of relevant solid and gas phase species that are likely to be encountered in urban and rural environments are given in Table I (4–13). Background values that would be expected in the absence of anthropogenic

Table I. Typical Tropospheric Concentrations of Species Associated with Acid Rain (μg/m^3)

Species	Urban	Rural	Background
Sulfate[a]	20[12]	4[6]	0.6[5,11]
Nitrate[a]	4[6]	2[6]	0.2[5,11]
Chloride[a]	0.3[7]	0.02[7]	0.02[11,4]
Ammonium[a]	5[6]	2[6]	0.2[11]
Calcium[b]	2[25]	1[25]	
Sulfur dioxide	130[8]	3[8]	0.3[9]
Nitrogen oxides	12[8]	0.1[8]	0.05[9]
Hydrogen chloride	2[7]	0.2[7]	
Organic acids (formic)	130[9]	9[9]	
Hydrogen peroxide	30[8]	1[8]	1[10]
Ozone	140[8]	60[8]	
Ammonia (NH_3)	20[8]	1[8]	0.1[9]

[a] Fine particle fraction.
[b] Coarse particle fraction.

influences are also listed. While the values given are representative of the various environments, deviations from these values will only affect the magnitude of the impact on surface degradation and not the methodology used to examine the impact. The concentrations of solid substances are intended to include the solid materials in the particles, as well as adsorbed materials or substances contained in an aqueous shell. Thus, fine ammonium acid sulfate particles and adsorbed sulfuric acid associated with these and other types of particles contribute to the total sulfate concentration.

Table II (4,14,15) reports indoor/outdoor (I/O) ratios for the type of building described in the Introduction. These values should at least provide an indication of ratios to be expected. For the species associated with fine particles, two different I/O ratios are listed. The first is for structures with standard filtration systems and the second is for buildings with high efficiency filtration systems. To a first approximation, such a distinction does not apply to the gas phase species. From these ratios, outdoor data then becomes useful in understanding indoor surface degradation processes. Indoor concentrations based on typical outdoor concentrations are given in Table III.

In order to obtain surface accumulation rates from the indoor concentrations, indoor deposition velocities are needed. These are expected to be considerably lower than outdoor deposition velocities, primarily because of reduced turbulence. Data from the authors (4) and other sources (16) suggest that indoor deposition velocities for substances associated with particles are approximately a factor of 100 lower than outdoor values; this factor has been used to estimate values where experimental data are not available. Values for substances in airborne particles are summarized in Table IV. As discussed above, data are not included for gaseous species.

Table II. Typical Indoor/Outdoor Ratios for Office Buildings

<table>
<tr><th rowspan="2">Species</th><th colspan="2">I/O Ratio</th></tr>
<tr><th>Standard Filters</th><th>High Efficiency Filters</th></tr>
<tr><td>Fine Particles
(Sulfate)
(Nitrate)
(Chloride)
(Ammonium)</td><td>0.3[4]</td><td>0.07[4]</td></tr>
<tr><td>Coarse Particles
(Calcium)</td><td>0.05[4]</td><td>0.02[4]</td></tr>
<tr><td>Sulfur dioxide</td><td colspan="2">0.6[14,15]</td></tr>
<tr><td>Nitrogen oxides</td><td colspan="2">0.6[14,15]</td></tr>
<tr><td>Organic acids</td><td colspan="2">0.6[14]</td></tr>
<tr><td>Ozone</td><td colspan="2">0.25[14]</td></tr>
<tr><td>Hydrogen peroxide</td><td colspan="2">0.25[14]</td></tr>
<tr><td>Hydrogen chloride</td><td colspan="2">0.1[15]</td></tr>
<tr><td>Ammonia</td><td>1.5[14]</td><td></td></tr>
</table>

Table III. Estimated Indoor Concentrations of Species Associated with Acid Rain ($\mu g/m^3$)

<table>
<tr><th rowspan="2">Species</th><th colspan="2">Urban</th><th colspan="2">Rural</th><th colspan="2">Background</th></tr>
<tr><th>STD[a]</th><th>HE[b]</th><th>STD</th><th>HE</th><th>STD</th><th>HE</th></tr>
<tr><td>Sulfate</td><td>6</td><td>1.4</td><td>1.2</td><td>0.3</td><td>0.2</td><td>0.04</td></tr>
<tr><td>Nitrate</td><td>1.2</td><td>0.3</td><td>0.6</td><td>0.1</td><td>0.06</td><td>0.01</td></tr>
<tr><td>Chloride</td><td>0.08</td><td>0.02</td><td>0.006</td><td>0.001</td><td>0.006</td><td>0.001</td></tr>
<tr><td>Ammonium (NH_4^+)</td><td>1.5</td><td>0.4</td><td>0.6</td><td>0.1</td><td>0.06</td><td>0.01</td></tr>
<tr><td>Calcium</td><td>0.1</td><td>0.04</td><td>0.05</td><td>0.02</td><td></td><td></td></tr>
<tr><td>Sulfur dioxide</td><td colspan="2">78</td><td colspan="2">1.8</td><td colspan="2">0.2</td></tr>
<tr><td>Nitrogen oxides</td><td colspan="2">7.2</td><td colspan="2">0.06</td><td colspan="2">0.06</td></tr>
<tr><td>Hydrochloric acid</td><td colspan="2">0.6</td><td colspan="2">0.02</td><td colspan="2"></td></tr>
<tr><td>Organic acids</td><td colspan="2">78</td><td colspan="2">5.4</td><td colspan="2"></td></tr>
<tr><td>Hydrogen peroxide</td><td colspan="2">7.5</td><td colspan="2">0.25</td><td colspan="2">0.25</td></tr>
<tr><td>Ozone</td><td colspan="2">35</td><td colspan="2">15</td><td colspan="2"></td></tr>
<tr><td>Ammonia (NH_3)</td><td colspan="2">30</td><td colspan="2">1.5</td><td colspan="2">0.15</td></tr>
</table>

[a] STD signifies standard filters.
[b] HE signifies high efficiency filters.

Table IV. Estimated Indoor Deposition Velocities (cm/sec)

Species	Velocity
Sulfate[a]	0.004[4,16]
Nitrate[a]	0.004[4,16]
Chloride[a]	0.004[4,16]
Ammonium[a]	0.004[4,16]
Calcium[b]	0.6[4,16]

[a] Fine particle fraction
[b] Coarse particle fraction.

Surface Accumulation Rates

To the best of our knowledge, the only measurements of surface accumulation rates of specific species on indoor surfaces have been conducted by us and our coworkers (1,4). These data can be represented in terms of contributions from fine and coarse particles using a procedure described elsewhere (1). Average accumulation rates on zinc and aluminum surfaces derived from studies at eleven urban and three rural areas are given in Table V. Also given in Table V

Table V. Surface Accumulation Rates ($\mu g/cm^2$ yr)

Species	Urban			Rural			Background	
	Measured	Calculated		Measured	Calculated		Calculated	
		STD[a]	HE[b]		STD	HE	STD	HE
Sulfate	0.17[c]	0.76	0.18	0.07[c]	0.15	0.04	0.03	0.005
Nitrate	0.15[c]	0.15	0.04	0.03[c]	0.08	0.01	0.008	0.001
Chloride	0.02[c]	0.01	0.003	0.01[c]	0.0008	0.0001	0.0008	0.0001
Ammonium	d	0.19	0.04	d	0.08	0.01	0.008	0.001
Calcium	0.27[e]	1.9	0.77	0.52[e]	0.95	0.38		

[a] STD signifies standard filtration
[b] HE signifies high efficiency filtration
[c] Accumulation due to fine particles
[d] Measurements show that ammonium does not accumulate appreciably on these surfaces after a brief initial exposure.
[e] Accumulation due to coarse particles

are calculated accumulation rates based on the indoor concentrations of Table III and the deposition velocities of Table IV. Considering that the experimental accumulation rates are based on a small amount of data and the calculated values result from a series of estimated values, the agreement is good.

This procedure for deriving indoor surface accumulation rates from outdoor concentrations provides a means for estimating the degree to which anthropogenically derived substances interact with indoor surfaces in different geographic regions. By comparing the typical urban and rural accumulations rates with the background rates, the contributions due to anthropogenic influences can be calculated. These estimates are given in Table VI.

Table VI. Surface Accumulation Rates Due To Anthropogenic Influences ($\mu g/cm^2$ yr)

Species	Urban		Rural	
	STD[a]	HE[b]	STD	HE
Sulfate	0.73	0.17	0.12	0.03
Nitrate	0.14	0.04	0.07	0.01
Chloride	0.01	0.003	-	-
Ammonium	0.18	0.04	0.07	0.01

[a] STD signifies standard filtration
[b] HE signifies high efficiency filtration

Expected Interactions with Typical Indoor Surfaces

As anthropogenically derived substances accumulate on indoor surfaces, reactions with the surfaces and among the substances on the surfaces can occur. The acidity of these surfaces will be a result of a complex interplay among many factors, including the amounts of both natural and anthropogenic acidic and non-acidic substances, and the natural acidity and buffering capacity of the surfaces. The concentrations of natural coarse particles will also affect the acidity and, for this reason, a distinction should be made between horizontal and vertical surfaces.

For zinc surfaces, the rate of accumulation of substances is expected to affect the rate of corrosion of the metal. For aluminum in most environments, the passive surface oxide will protect the metal from further significant corrosion. Any interactions that occur on the surface will be among the accumulated substances and perhaps with the outermost layer of metal oxide. Interestingly, a few materials perform better in industrial atmospheres than in clean environments. Suzuki, *et al.* (17), have discussed the fact that weathering steel utilizes atmospheric pollutants, particularly sulfur dioxide gas, to form protective layers.

Prior to considering the effects of man-made airborne substances on indoor surfaces, it is instructive to consider studies on corrosion of metals in outdoor environments. Most of the atmospheric corrosion investigations have used weight gain (or loss if the surfaces are cleaned before weighing) to measure the degree of corrosion. These results have been compared with measurements of specific pollutants (18). For metals that corrode, comparisons between the corrosion rate and sulfur dioxide concentration have shown that a CRH typically in the range of 70 - 80 percent is required for rapid corrosion to occur (18,19), even in the presence of high concentrations of sulfur dioxide. The CRH in this context can be considered as the minimum RH at which the thickness of the adsorbed water film is sufficient for the water to behave like bulk water. The effect of adsorbed moisture on corrosion processes has also been related to time-of-wetness measurements (18). While Mansfeld (20) has shown that the reproducibility of such measurements is subject to variables that are difficult to control, the concept of time-of-wetness is useful for understanding atmospheric corrosion of metals. Rice and Phipps (21) have shown that, for pure metal surfaces, less than a monolayer of water is adsorbed at < 20 percent RH. At 60 percent RH several monolayers can be adsorbed. When the RH is sufficiently high for a moisture film to exist on metal surfaces or on contaminants on metal surfaces, corrosive pollutant gases can dissolve in the moisture film and will then be readily available to react with the metal. Further, the creation of active electrochemical cells between anodic and cathodic areas on the metal surface, possibly caused by particle deposition or other contamination, requires the presence of moisture.

The aqueous corrosion literature (22) provides a further basis for understanding some aspects of the surface degradation of metals. Examination of the Pourbaix diagrams (23) for the zinc-water and aluminum-water systems shows that these metal surfaces should be insensitive to small pH changes. The aqueous corrosion rates increase rapidly above pH 12 and below pH 4 for zinc and above pH 10 and below pH 1 for aluminum. The rates change very little in the intermediate ranges. We have measured the relative acidity of these surfaces after extended indoor exposures and find that they are mildly acidic, with aluminum surfaces being somewhat more acidic than zinc surfaces (1). Deposition of acidic substances would have to cause a substantial decrease in pH for the increased acidity to significantly alter the corrosion rate. For other metals, unless the small change occurs over a critical pH which is defined by a phase change in the Pourbaix diagram for the metal-water system, the effects on these metals are also likely to be small. Even if all the acidity that accumulated on these surfaces over 10 years remained active and contributed to a pH decline, the contribution of accumulated H_3O^+ ions to urban indoor surfaces (estimated by assuming that most of the deposited acidity arrives on the surface as ammonium acid sulfate) is not likely to exceed 0.08 $\mu g/cm^2$. The oxides on zinc and aluminum are good buffers and will tend to maintain the pH at a slightly acidic level. The thickness of oxide that would be required to dissolve in order to neutralize the acidity from ten years of accumulated ammonium acid sulfate is < 100 Å. Further, some of the deposited acidity is likely to be neutralized by co-deposited calcium salts. On top-side horizontal surfaces, if all the calcium that deposits is assumed to behave like calcium carbonate, there would be more than sufficient basicity in the calcium salts to neutralize the ammonium acid sulfate. The acidity of dry deposition, then, is not likely to have a significant impact on zinc or aluminum degradation.

The RH in most indoor environments is usually not above 70 percent and, thus, the CRH of most common metals is seldom exceeded. The time-of-wetness will be quite small. The corrosion rate is likely to be comparable to the outdoor rate (at similar contaminant levels) when the surfaces are dry. Such rates are insignificant compared to the wet rates for most metals (18). In many cases, the anions associated with deposited substances may play the dominant role in surface processes (24). The concentrations of sulfate, nitrate, and chloride, which accumulate on these surfaces, are likely to increase continuously. After 10 years exposure, total anion concentrations of five to ten $\mu g/cm^2$ can be expected in urban environments. These anions, especially chloride, are well known to dramatically affect the corrosion rates of many metals in aqueous solutions. This acceleration is often a result of solubilization of the surface metal oxide through complexation of the metal by the anions. Chloride, in particular, can dramatically lower the RH above which a moisture film is present on the surface, since chloride salts often have low CRHs (e.g., zinc chloride - <10 percent; calcium chloride - 30 percent; and aluminum chloride - 40 percent). The combination of the low CRHs of chloride salts and the well documented ability of dissolved chloride to break down metal oxide passivation set chloride apart from the other common anions in ability to corrode indoor metal surfaces. Some nitrate salts also have moderately low CRHs (e.g., zinc nitrate - 38 percent; calcium nitrate - 49 percent; aluminum nitrate - 60 percent).

The corrosion rate of most metal surfaces during dry conditions in outdoor environments has historically been considered to be insignificant and thus of little concern. Since the indoor environment is normally dry, corrosion rates indoors have also been considered to be insignificant for most metals relative to outdoor rates. While the accumulation rates can provide a framework in which degradation of metal surfaces may be considered, it is not possible to predict, directly, the corrosion rates. Nevertheless, the degree of degradation hazard associated with the various environments should follow readily from the calculated accumulation rates. Previously, the authors used measured accumulation rates of airborne substances at telephone company electronic switching equipment locations in cities across the United States to estimate the degree of hazard to the electronic equipment (1). The decreasing degree of hazard was as follows: New York City > Philadelphia > Albany > Cleveland, Houston > Colorado, San Francisco > Chicago, Seattle > Minneapolis, Omaha, North Jersey > Indianapolis > Lubbock > Wichita, Los Angeles, Orlando. The same ranking should be appropriate when considering degradation associated with other types of indoor materials.

Consideration of the chemical and physical properties of these accumulated substances permits some further conclusions. The volatility of surface reaction products is likely to be important, as are the ambient concentrations of ammonia, carbon dioxide, and other gases that could influence the surface acidity. Indeed, the ratio of ammonium to sulfate that has been found on zinc and aluminum surfaces after extended exposures is much lower than typical ratios for ammonium to sulfate in airborne particles (4). Clearly, ammonium is able to volatilize from these surfaces through a hydrolysis reaction or, after redistribution of anions and cations on the surface such as

$$(NH_4)HSO_4 + NaCl \rightarrow NH_4Cl + NaHSO_4$$

through a reaction such as

$$NH_4Cl \rightarrow NH_3 + HCl$$

We have also noted that chloride accumulation on indoor zinc or aluminum surfaces can occur at dramatically different rates (1). At locations where large differences are observed, hydrogen chloride has been implicated to be the source of much of the chloride accumulation. This difference in behavior was attributed, in part, to the slightly higher acidity and higher ammonium content of the aluminum surfaces, both of which will inhibit the reaction of hydrogen chloride with aluminum and will enhance the volatilization of hydrogen chloride.

Conclusions

The intent of this paper has been to provide a methodology by which the interaction of the environment with indoor surfaces can be related to prevailing outdoor concentrations of airborne substances. The complexity of the combined effects of these airborne substances frustrates making quantitative predictions of the degree of degradation. The relative degree of hazard for the different environments can be estimated from the methodology and, based on previous experience with zinc and aluminum surfaces in known environments, provides a framework for qualitative rankings of comparable environments.

Literature Cited

1. Sinclair, J. D., Psota-Kelty, L. A., *Proceedings of the International Congress on Metallic Corrosion*, 1984, Vol. 2, pp. 296-303.
2. Hansen, D. A., Hidy, G. M., *Atmos. Environ.*, 1982, 16, 2107-2126.
3. Pruppacher, H. R., Klett, J. D., "Microphysics of Clouds and Precipitation," D. Reidel Publishing Co., Holland, 1978.
4. Sinclair, J. D., Psota-Kelty, L. A., and Weschler, C. J., *Atmos. Environ.*, 1985, 19, 315-323.
5. Prospero, J. M., Savoie, D. L., Nees, R. T., Duce, R. A., Uematsu, M., and Merrill, J. M., *SEAREX Newsletter*, 1982, 5(2), 19-21.
6. Kleinman, M. T., Tomczyk, C., Leaderer, B. P. and Tanner, R. L., *Ann. NY Acad. Sci.*, 1979, 322, 115-123.
7. Rahn, K. A., Borys, R. D., Butler, E. L. and Duce, R. A., *Ann. NY Acad. Sci.*, 1979, 322, 143-151. Matusca, P., Schwarz, B. and Bachman, K., *Atmos. Environ.*, 1984, 18, 1667-1675.
8. Graedel, T. E. and Weschler, C. J., *Revs. Geophys. and Space Phys.*, 1981, 19, 505-539.
9. Graedel, T. E., "Chemical Compounds in the Atmosphere," Academic Press, New York, 1978.
10. Logan, J. A., Prather, M. J., Wofsy, S. C., and McElroy, M. B., *Geophs. Res.*, 1981, 86, 7210-7254.
11. Huebert, B. J. and Lazrus, A. L., *J. Geophys. Res.*, 1980, 85, 7337-7344.
12. Tanner, R. L., *Ann. NY Acad. Sci.*, 1980, 338, 39-49.
13. Total chloride (fine and coarse mode) background concentration was 0.12 $\mu g/m^3$; fine mode backgroun concentration was 0.12 $\mu g/m^3$ (see Reference 9).
14. Yocum, J. E., *J. Air Polln. Contr. Assoc.*, 1982, 32, 500-520.
15. Rice, D. W., Cappell, R. J., Kinsolving, W. and Laskowski, J. J., *J. Electrochem. Soc.*, 1980, 127, 891-901.
16. Pruppacher, H. R., Semonin, R. G., Slinn, W. G. N., Proceedings of the Fourth International Conference on Precipitation Scavenging, Dry Deposition, and Resuspension, 1982, Vol. 2.
17. Suzuki, I., Hisamatsu, Y., Masuko, N., *J. Electrochem. Soc.*, 1980, 127, 2210-2215.
18. Dean, S. W. and Rhea, E. C., Atmospheric Corrosion of Metals, ASTM Special Technical Publication 767, May, 1980.
19. Vernon, W. H. J., *Trans. Faraday Soc.*, 1935, 31, 1668.
20. Ailor, W. H., *Atmospheric Corrosion, The Corrosion Monograh Series*, 1982, p. 139.
21. Rice, D. W., Phipps, P. B. P., Tremoureux, R., *J. Electrochem. Soc.*, 1980, 127, 563-568.
22. *Proceedings of the International Congress Metallic Corrosion*, 1984, Volume 1, pp. 79-143, 274-332; ibid, Volume 2 pp. 99-179; ibid, Volume 3, pp. 394-467; ibid, Volume 4, pp. 506-550.
23. Pourbaix, M., "Atlas of Electrochemical Equilibria in Aqueous Solutions," translated from French by J. A. Franklin, National Association of Corrosion Engineers, Houston, Texas.
24. *Proceedings of the International Congress on Metallic Corrosion*, 1984, Volume 1, pp. 208-240; ibid, Volume 2, pp. 296-316; ibid, Volume 3, pp. 192-228.
25. Rahn, K. A., "The Chemical Composition of the Atmospheric Aeorosol," Technical Report, Graduate School of Oceanography, University of Rhode Island, Kingston, Rhode Island, 1976.

RECEIVED February 3, 1986

MASONRY DETERIORATION

15

Limestone and Marble Dissolution by Acid Rain: An Onsite Weathering Experiment

Michael M. Reddy[1], Susan I. Sherwood[2], and B. R. Doe[3]

[1]U.S. Geological Survey, Lakewood, CO 80225
[2]Preservation Assistance Division, United States Park Service, Washington, DC 20006
[3]U.S. Geological Survey, 12201 Sunrise Valley Drive, Reston, VA 22092

In this paper we describe an experimental research program, conducted in conjunction with the National Acidic Precipitation Assessment Program (NAPAP), to quantify acid-rain damage to commercial and cultural carbonate-rock resources. We are conducting a carbonate-rock onsite weathering experiment, using the change in rainfall-runoff composition to quantify the interaction of acid rain with a carbonate-rock surface. Initial results of this experiment show that carbonate-rock dissolution and associated surface recession increase with increasing acid deposition to the rock surface. The interaction of acid rain with carbonate rock has a stoichiometry, consistent with the reaction:

$$CaCO_3 \text{ (solid)} + H^+ \rightleftharpoons Ca^{2+} + HCO_3^-$$

A statistically significant linear relation has been found between carbonate-rock surface-recession rate and hydrogen ion loading to the rock surface.

Weathering, and subsequent deterioration of rock exposed to the environment, are slow processes. Weathering rates are influenced by temperature, the presence of moisture, the presence of organic acids, and the carbon dioxide partial pressure in the solution in contact with the rock surface. In contemporary industrial societies, natural weathering processes are accelerated by elevated pollutant concentrations. Acidic pollutants in particular, in both air and rainfall, are recognized as serious hazards to carbonate rock used in commerce and cultural resources (1).

In general, acids present in wet and dry deposition, and gaseous sulfur dioxide (in the presence of surface moisture) are important agents causing deterioration of limestone and marble (1). Damage may be the result of direct acid reaction with the rock surface, or it may proceed through formation of secondary mineral phases, such as gypsum. Gypsum will form on carbonate-rock surfaces, following deposition of gaseous or dissolved sulfur species. Capillary transport of surface sulfate into rock pores may cause rock-volume increases, accelerating rock breakup. Little quantitative information is available to assess the effect of nonsulfur-containing pollutants in carbonate-rock damage (1); however, the presence of organic acids on the stone surface, for example, could accelerate rock damage markedly.

Mechanistic details of carbonate-rock deterioration in polluted environments are needed to protect and refurbish structures and monuments at risk to acid rain and air pollution. For example, if diffusion of reaction products away from the rock surface is limiting the rate of dissolution, then modeling rock deterioration, using simple chemical models, will be difficult. In the case of a diffusion-limited reaction, hydrodynamic factors, such as turbulent flow, may influence the dissolution process. Humidity fluctuations also may alter the solute concentrations in the moisture film at the rock surface, and they may result in the dissolution and recrystallization, not only of the carbonates, but also of the secondary minerals, such as calcium sulfate. Wetting and drying cycles may lead to measureable changes in fluid composition at the stone surface.

In this paper, we describe an onsite weathering experiment designed to identify acid-rain increased dissolution of carbonate rock. This experiment is based on the measurement of the change in rainfall-runoff composition from the interaction of a rock surface with incident acid rain (2). The experiment involves conducting long-term exposures of two commercially and culturally important calcium carbonate dimension stones (i.e., Indiana Limestone (commercial name for Salem Limestone) and Vermont Marble (commercial name for Shelburne Marble)) (3-5). This technique appears to give a direct measurement of the chemical dissolution of carbonate rock from the combined reactions of wet and dry deposition. Preliminary results from the initial months of onsite operation are presented to illustrate the technique.

Materials and Methods

The onsite experiments described in this paper are being conducted at four locations in the Eastern United States: Newcomb, New York; Chester, New Jersey; Washington, D.C.; and Research Triangle Park, North Carolina (Figure 1) (3-5). Each site has four rock-exposure racks fabricated of polypropylene. Racks contain two stone slabs each (Figure 2). Exposure racks have been designed to facilitate collection of runoff from the rock surface with a minimum of contamination.

Figure 1. Rock exposure-site locations in the Eastern United States.

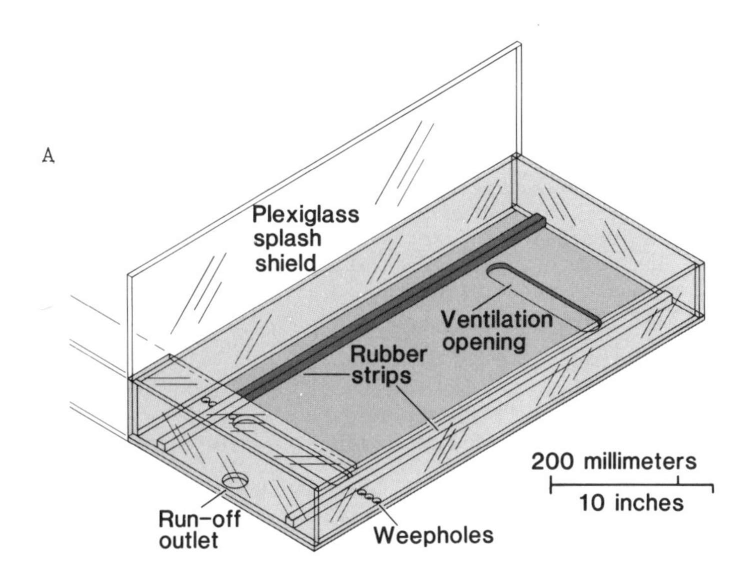

Figure 2. A. An isometric projection of a single research rock-exposure rack. Racks are mounted in pairs with four sets per location. B. A typical completed research rock-exposure site with research stones in place.

Approximately one liter of rainfall-runoff solution was collected from each of three experimental racks following rainfall events. Limestone and marble reference rocks were in two of the racks; a third rack without a rock was used as a control (hereinafter referred to as a blank sample). Runoff volume, specific conductance, and pH measurements were determined at the exposure site as soon as possible after collection, typically within a few hours. Samples then were filtered through a 0.45-micrometer pore size membrane filter, placed in a clean polyethlyene bottle, and sent to the U.S. Geological Survey laboratory in Denver, Colorado, for chemical analysis. Sample preparation, analytical procedures, and laboratory quality-control protocols are described by Skougstad et al. (6).

Each exposure site has other materials-damage experiments, in addition to those discussed in this report (3). To supplement all damage studies, an array of meteorological, climatological, air-pollution, and rainfall-chemistry data also will be available for each site. Research-rock-procurement information recently has been published (7-8).

Results

Results obtained at Research Triangle Park, North Carolina during the last half of 1985 are summarized in Tables I and II. As a first approximation, because data on daily sulfur dioxide concentrations and times-of-wetness were not available, antecedent sulfur dioxide loading to the stone surface was estimated by multiplying the average atmospheric sulfur dioxide concentration by the number of days between rainstorms.

Analytical results for sulfate ions are for the net change in rainfall composition, after the rainfall flows over the surface of the exposed stone (Tables I and II). The net concentration, resulting from the interaction of rain with the stone surface, is calculated by subtracting the blank-sample concentration value from the value measured for the stone runoff. For very small rain amounts (typically 2 to 4 millimeters), blank volumes were too small to allow complete chemical analysis. Sulfate ion net concentration, in excess of that possible from sulfate loading, was calculated to identify the presence of soluble sulfate-containing minerals on the rock surface. Loss of rock mass, measured by the calcium concentration in the runoff, is expressed as surface recession, in micrometers of rock surface lost per rainstorm. The net calcium concentration in the runoff solution is converted to an equivalent volume of rock removed; this result divided by the surface area of the exposed rock, yields a measurement of surface recession (2).

Discussion

Relation between environmental variables and surface recession. Carbonate-rock surface recession and the environmental variables measured in this program (i.e., rain amount, rain pH, hydrogen ion loading to the rock surface, and antecedent sulfur dioxide) exhibited a range of values at the North Carolina site during the last half of 1984. To examine qualitatively the relation between

Table I. Rainfall quantity and pH, antecedent sulfur dioxide, net sulfate, hydrogen ion loading, and surface recession for a marble reference rock at Research Triangle Park, North Carolina, June to October 1984

U.S. Geological Survey Sample no.	Rainfall (millimeters)	pH	Antecedent sulfur dioxide (parts per billion times days)	Net sulfate ion in runoff (milligrams per liter)	Hydrogen ion loading (milliequivalents per meter squared)	Surface recession (micrometers)
1	17	4.53	0	1.00	0.50	0.13
2	16	3.98	12	.74	1.63	.12
4	41	4.43	16	.67	1.51	.21
5	17	4.19	12	.90	1.07	.13
6	5	4.38	16	2.63	.21	.04
7	3	4.04	12	7.60	.23	.02
8	32	4.46	12	.00	1.10	.19
10	29	4.27	4	1.00	1.57	.17
11	46	5.55	4	-.15	.13	.17
12	7	4.06	4	2.35	.58	.03
13	4	4.91	16	5.50	.05	.02
14	6	4.71	8	1.60	.11	.05
16	55	4.49	16	1.10	1.77	.35
21	21	4.30	40	.30	1.07	.18
26	2	4.67	20	14.00	.03	.02
28	10	4.51	4	2.30	.31	.07
30	51	4.35	4	-.80	2.27	.30
31	2	3.83	4	11.00	.23	.02

Table II. Rainfall quantity and pH, antecedent sulfur dioxide, net sulfate, hydrogen ion loading, and surface recession for a limestone reference rock at Research Triangle Park, North Carolina, June to October 1984

U.S. Geological Survey Sample no.	Rainfall (millimeters)	pH	Antecedent sulfur dioxide (parts per billion times days)	Net sulfate ion in runoff (milligrams per liter)	Hydrogen ion loading (milliequivalents per meter squared)	Surface recession (micrometers)
1	17	4.53	C	0.35	0.50	0.13
2	16	3.98	12	-.52	1.63	.13
4	39	4.43	16	.95	1.51	.27
5	17	4.19	12	3.60	1.07	.21
6	5	4.38	16	1.83	.22	.02
8	32	4.46	12	.00	1.10	.26
10	29	4.27	4	2.30	1.57	.19
12	7	4.06	4	1.15	.58	.05
14	6	4.71	8	.70	.11	.06
16	55	4.49	16	6.40	1.77	.63
21	21	4.30	40	1.40	1.07	.16
28	10	4.51	4	7.30	.31	.04
30	51	4.35	4	5.20	2.27	.57

these variables selected data for each stone type have been grouped into two classes: large rain amount and small rain amount (Table III). This comparison illustrates the influence of rain amount and hydrogen ion loading on the observed surface recession for both limestone and marble. An increase in recession occurs with both increasing rain amount and increasing hydrogen ion loading. The increase in surface recession with increasing rain amount is more pronounced than the increase in surface recession with increasing hydrogen ion loading. Surface-recession values for both stone types are similar at small rain amounts. At large rain amounts, the limestone-surface recession, for the same event, is somewhat greater than the marble-surface recession.

In addition to the influence of acid loading on rock-surface recession, sulfur containing compounds that accumulate on a rock surface may lead to damage. Data in Table IV illustrate the influence of the major processes associated with sulfur redistribution during carbonate-rock weathering. Marked differences occur between the two stone types: limestone and marble. For marble, excess sulfate is highest in the small-rain amount group; as the rain amount increases, the excess sulfate values decrease. Small rain amounts appear to remove surface-accumulated sulfur species with a minimum of dilution; this lack of dilution leads to large, excess sulfur concentrations in the rock-surface runoff.

The influence of antecedent sulfur dioxide exposure on excess-runoff sulfate concentration is less apparent than the effect of rain amount for marble. At small rain amounts, larger antecedent sulfur dioxide values are associated with large excess-sulfate values. This observation is consistent with surface-sulfur accumulation by gaseous deposition; however, sample 12 (Table IV) is an exception to this trend. For large rain amounts, excess sulfate values are low or negative, and they decrease with decreasing antecedent sulfur dioxide exposure. The negative values for excess sulfate at large rain amounts and small-antecedent sulfur dioxide exposure suggest that interaction with a marble surface removes sulfate from the incident rain.

Sulfur deposition and subsequent remobilization by incident rain for a limestone surface differs from that exhibited by a marble surface. This difference may be the result of larger porosity values of limestone compared to marble. Larger limestone porosity adsorbs a larger portion of incident rain than marble. Thus, fewer low-rain events occur for limestone. Where small rain-amount data are available for both limestone and marble, excess sulfate values consistently are lower for limestone runoff than for marble runoff. This observation suggests that limestone surface-accumulated sulfate is less mobile than marble surface-accumulated sulfate.

The situation differs for large rain amounts. For large rain amounts, excess sulfate values for limestone exceed excess sulfate values for small rain amounts. Large-excess sulfate values for a large-rain-amount for limestone are associated with twice the surface recession of a marble sample for the same event. Some of this surface loss may be the result of the removal of a calcium sulfate mineral.

Table III. Incident rainfall, rain pH, hydrogen ion loading, and surface recession for limestone and marble at Research Triangle Park, North Carolina, June to October 1984

U.S. Geological Survey Sample no.	Rainfall (millimeters)	pH	Hydrogen ion loading (milliequivalents per meter squared)	Surface recession (micrometers)
Limestone				
6	5	4.38	0.22	0.02
14	6	4.71	.11	.06
12	7	4.06	.58	.05
16	55	4.49	1.77	.63
30	51	4.35	2.27	.57
Marble				
26	2	4.67	.03	.02
31	2	3.83	.23	.02
7	3	4.04	.23	.02
13	4	4.91	.05	.02
6	5	4.38	.21	.04
14	6	4.71	.11	.05
12	7	4.06	.58	.03
16	55	4.49	1.77	.35
30	51	4.35	2.27	.30
11	46	5.55	.13	.17

Dissolution by Acid Rain. The dissolution of limestone and marble by acid rain may be written as:

$$H^+ + CaCO_3\ (\text{solid}) \rightleftharpoons Ca^{2+} + HCO_3^- \qquad (1)$$

where one mole of hydrogen ion reacts with excess calcium carbonate to produce one mole of calcium ion and one mole of bicarbonate ion in solution. In addition to this reaction, other reactions of hydrogen ion (and/or) other hydrogen ion donating species) need to be considered (9). Each reaction may influence the observed stoichiometry of the calcium carbonate dissolution.

Carbonate-rock acid-rain reaction stiochiometry is best examined using a mathematical model of the ionic species present in a dilute carbonate solution. Appropriate models have been recently developed and are used with an electronic digital computer. Although a discussion of the application of mineral-reaction modeling is outside the scope of this paper, several recent publications explain its use in the study of carbonate-mineral reactions in aqueous solution (9-11). A reaction model, such as PHREEQE (11), the model used in this study, incorporates changes in gas-solution and solution-solid equilibria, as well as ionic equilibria in solution, in arriving at a final equilibrium-solution composition (4).

Table IV. Incident rainfall, antecedent sulfur dioxide, net sulfate removed from the stone surface and surface recession for limestone and marble at Research Triangle Park, North Carolina, June to October 1984

U.S. Geological Survey Sample no.	Rain- fall (milli- meters)	Antecedent sulfur dioxide (parts per billion times days)	Net sulfate (milligrams per liter)	Surface recession (micro- meters)
Limestone				
6	5	16	1.83	0.02
14	6	8	.70	.06
12	7	4	1.15	.05
16	55	16	6.40	.63
30	51	4	5.20	.57
Marble				
26	2	40	14.00	.02
31	2	4	11.00	.02
7	3	16	7.60	.02
13	4	4	5.50	.02
6	5	12	2.63	.04
14	6	8	1.60	.05
12	7	4	2.35	.03
16	55	8	1.10	.35
30	51	4	-.80	.30
11	46	4	-.15	.17

In Figure 3, the stoichiometric difference between the calcium ion concentration in the rock runoff and the hydrogen ion concentration in the incident rain is plotted versus the rain pH. Lines are drawn for the two calculated stoichiometric differences for both open and closed systems. The representative points from the onsite experiment fall within these two limiting cases. These results illustrate that the onsite stoichiometry is dependent on the initial pH of the incident rainfall and on whether the reacting solution is open or closed to exchange of carbon dioxide in the atmosphere.

Acid-Rain-Induced Calcium Carbonate Surface Recession. Calcium carbonate surface recession may be correlated with rainfall characteristics, using standard statistical techniques. The relation may be written as:

$$\text{Surface recession} = A + (B \times (\text{Independent variable})) \qquad (2)$$

where A and B are constants determined from a least-squares regression analysis; and the independent variable may be rainfall amount, rainfall pH, rainfall hydrogen ion concentration, or rainfall hydrogen ion loading. For example, in equation 2, with

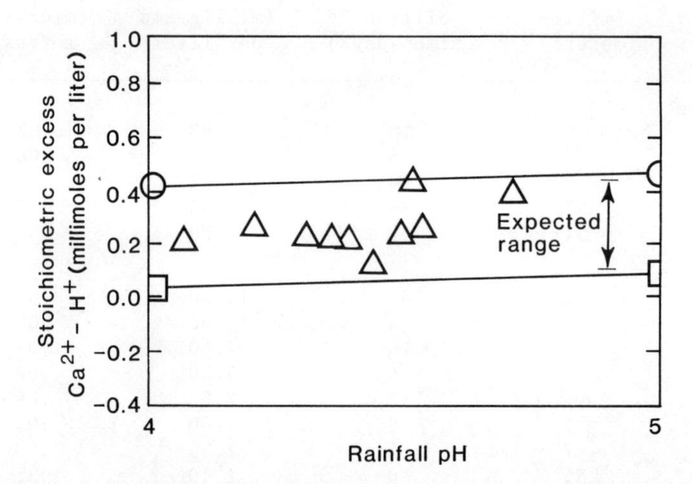

Figure 3. A plot of calcium minus hydrogen concentration difference (open triangles) in units of millimoles per liter versus incident rainfall pH. Calculated points for solutions open (open circles) and closed (open squares) to exchange with atmospheric carbon dioxide, also shown, define the anticipated range of values.

hydrogen ion loading as the independent variable, the regression coefficient, "B," reflects damage from acid deposition, while the constant term, "A," reflects the average of all other effects.

Marble-surface recession at the North Carolina site was correlated with rainfall amount. The correlation coefficent was significant at greater than the 99-percent confidence level. Marble-surface recession also was correlated with hydrogen ion deposition to the rock. The linear relation between surface recession and hydrogen ion deposition was found to be:

$$\text{Recession} = 0.03 + (0.12 \times (\text{hydrogen ion deposition})) \quad (3)$$

where recession is in units of micrometers per rainfall event; and hydrogen ion deposition is in units of millimoles per meter square. Units of the regression coefficient (0.12) are micrometers/millimole of hydrogen ion deposited per square meter to the rock surface by rain. This relation is significant at greater than the 99-percent confidence level. No statistically-significant relation between surface recession and antecedent sulfur dioxide exposure occurs at the study site in Research Triangle Park, North Carolina.

Sulfate net concentration in rock runoff is significantly larger than zero for marble (significant at the 90-percent confidence level), and significantly larger than zero for limestone (significant at the 99-percent confidence level). That is, rainfall runoff from both reference rocks has sulfate concentrations significantly above sulfate concentration values in ambient rainfall. This excess shows that dry deposition and(or) precipitation reactions must occur, resulting in the formation of sulfate or dry deposition of sulfate on the rock surface.

Conclusions

This report describes early measurements of the deterioration of limestone and marble building stones by acidic components of air pollution and acid rain. This onsite technique allows measurement of carbonate-rock deterioration under a wide range of ambient conditions. Limestone and marble deterioration have been related quantitatively to acid deposition to the rock surface. The procedures described here appear to have general applicability for developing reliable rock-damage relations for environmental contaminants.

Literature Cited

1. "Acid rain: A review of the phenomenon in the EEC and Europe" Environmental Resources Limited, Unipub, New York, 1983.
2. Reddy, M. M.; Werner, M. "Chemical analysis of samples from a rainfall-runoff experiment using limestone and marble at Research Triangle Park, North Carolina," U.S. Geological Survey Open-File Report 85-630.
3. Sherwood, S. I.; Doe, B. EOS 1984, 65, 1210.

4. Reddy, M. M.; Sherwood, S.; Doe, B. Proc. of Research and Design '85, Los Angeles, March 14-18, 1985, The American Institute of Architects Foundation, Washington, D.C., pp. 383-388.
5. Reddy, M. M.; Sherwood, S.; Doe, B. Proc. of the Vth International Congress on Deterioration and Conservation of Stone, 1985, Presses Polytechniques Romandes, Lausanne, Switzerland, pp. 517-526.
6. Skougstad, M. W.; Fishman, M. J.; Friedman, L. C.; Erdmann, D. E.; and Duncan, S. S. "Methods for determination of inorganic substances in water and fluvial sediments: Techniques of Water-Resources Investigations of the United States Geological Survey," Book 5, Chap. A1, p. 626, 1979.
7. Ross, Malcolm; Knab, Larry "Selection, procurement and description of Salem Limestone samples used to study effects of acid rain," U.S. Department of Commerce, NBSIR 84-2905, p. 22, 1984.
8. Ross, M. "Description, selection, and procurement of Shelburne Marble samples used to study effects of acid rain," U.S. Geological Survey Open-File Report 85-594.
9. Reddy, M. M. Sci. Geol. 1983, 71, 109-117.
10. Nordstrom, D. K., et al., in Chemical Modeling in Aqueous Systems, Jenne, E. A., Ed., ACS SYMPOSIUM SERIES, No. 93, Washington, D.C., 1979, pp. 857-892.
11. Parkhurst, D. L.; Thorstenson, D. C.; and Plummer, L. N. "PHREEQE - A computer program for geochemical calculations," U.S. Geological Survey Water-Resources Investigations 80-96, pp. 210. Available as National Technical Information Services Report, PB 81-167 801, Springfield, VA, 22161, 1980.

RECEIVED January 21, 1986

16

Effects of Acid Deposition on Portland Cement Concrete

R. P. Webster and L. E. Kukacka

Department of Applied Science, Process Sciences Division, Brookhaven National Laboratory, Upton, NY 11973

Presented are the results of a program, sponsored by the Environmental Protection Agency, conducted to determine the state-of-the-art knowledge pertaining to the effects of acid deposition on the properties of portland cement concrete structures (PCC). Information was collected from a computerized literature survey, interviews, and replies to mail and telephone inquiries addressed to cement and concrete researchers and to governmental agencies and private firms active in the maintenance and restoration of concrete structures. In general, the study revealed very little qualitative or quantitative information on the effects of acid deposition on PCC structures. The rate of deterioration of reinforced PCC structures in polluted areas, however, appears to be increasing, and available information makes it readily apparent that acids and acid waters significantly affect the durability of concrete, and that SO_2, NO_x, and HCl accelerate the corrosion of reinforcing steel. On the basis of this evidence, it was recommended that an experimental test program, consisting of both laboratory and field tests, be developed and implemented to quantitatively measure the effects of acid deposition on PCC structures.

Over the last 25 years, an increasing amount of evidence has been gathered indicating that acid deposition, resulting from the emission of oxides of sulfur and nitrogen into the atmosphere, has a significantly adverse effect upon man's environment including the acidification of lakes, rivers and groundwaters, acidification and demineralization of soils, reduction of forest productivity, damage to crops, and deterioration of building materials (1-9). Although much of this evidence pertains to the effects of acid deposition

0097-6156/86/0318-0239$06.00/0

on the natural environment, i.e. lakes, rivers, forests and crops, it is becoming increasingly clear that acid deposition has a significant impact also on the durability of building materials such as metals, masonry, building stone, and concrete (10-24). Since concrete is the most widely used construction material in the United States, any comprehensive assessment of the economic impact of acid deposition on materials should include consideration of PCC and steel reinforced PCC. To this end, Brookhaven National Laboratory (BNL) has carried out a program as part of the National Acid Precipitation Assessment Program (NAPAP) Task Force Project G3-1.05, sponsored by the Environmental Protection Agency/Atmospheric Sciences Research Laboratory (EPA/ASRL), entitled "Effects of Acid Deposition on the Properties of Reinforced Portland Cement Concrete Structures."

The specific objectives of the BNL program were (a) to determine the state-of-the-art-knowledge pertaining to the effects of acid deposition on the properties of PCC, and (b) if the results indicated a need for quantitative data, to develop recommendations for an experimental test program to be submitted for Task Group G approval and implementation.

State-of-the-Art Review

General Summary. Information for the state-of-the-art review was obtained from a computerized literature survey, interviews, and replies to mail and telephone inquiries addressed to cement and concrete researchers and to governmental agencies and private firms active in the maintenance and restoration of concrete structures.

In general, the computerized literature survey indicated that an abundance of literature on acid precipitation is available, but most of it deals with the chemistry of acid precipitation and its effects on the natural environment. Literature dealing with the effects of acid deposition on buildings and building materials does exist, however, very little of it pertains to cement or concrete. The information that was found regarding the effects of acid deposition on buildings and building materials indicates that the increasing acidity of precipitation enhances normal weathering and corrosion processes. In addition, private communications indicated that a rapidly increasing number of reinforced concrete structures in cities are showing deterioration which the respondents attributed to SO_2, NO_x, and HCl.

The literature also indicates that while it is felt, by some, that acid precipitation does have an adverse effect on the performance of building materials, no work apparently is being done to investigate these effects. Ashton and Sereda (24) report that: "Monitoring of the concentration of pollutants in the atmosphere and in rain is done by various government agencies in many countries, but little effort is directed to the study of their effect on building materials, particularly that of acidic components in rainwater."

None of the individuals and organizations responding to the mail and telephone inquiries were aware of any documented information dealing specifically with the effects of acid deposition on PCC structures, or of any research that had been or was being done in this area. Comments on the need for such research work were varied: some respondents thought it was needed because the large volume of concrete structures in the United States could present a potentially large problem; others thought the need for such research was open to question because they considered other mechanisms of deterioration to be more important.

The study revealed very little qualitative or quantitative information on the effects of acid deposition on PCC structures. The rate of deterioration of reinforced PCC structures in polluted areas, however, appears to be increasing, and available information makes it readily apparent that acids and acid waters significantly affect the durability of concrete, and that SO_2, NO_x, and HCl accelerate the corrosion of reinforcing steel.

Resistance of Concrete to Chemical Attack

Because the literature on the effects of acid deposition on PCC is limited, the large amount of literature dealing with the corrosive effects of acids, acid waters, and sulfates on concrete was reviewed in an attempt to estimate the effects of acid deposition on PCC (25-48).

Effects of Acids. Since concrete, chemically, is a basic material, having a pH of about 13, it is subject to attack by acids. Woods (32) reports that concrete is not very resistant to strong solutions of sulfuric, sulfurous, hydrochloric, nitric, hydrobromic, or hydrofluoric acids, and is destroyed by prolonged contact with any of these, though not necessarily at the same rate. Weaker solutions (<1%) attack concrete at a slower rate, but in some cases the severity of the attack can be very significant. Woods further states that for all practical purposes, an acidity of pH 5.5 to 6 may be considered the limit of tolerance of high quality concrete in contact with any of these acids, although the pH value is not invariably a good criterion of the aggressiveness of acids. The chemical composition of the acid is at least as important as pH in influencing the rate at which concrete is attacked.

Galloway (1) reports that the relative contribution of H_2SO_4, HNO_3, and HCl to the acidity of precipitation is difficult to determine because the acids are not present as such in solution but rather as dissociated ions. However, using the absolute concentration of SO_4, NO_3 and Cl it is possible to determine their relative contribution. Likens and Bormann (2) and Glass et al (3) report that precipitation data for the northeastern United States indicate that 60 to 70% of the acidity in acid precipitation is due to sulfuric acid, 30 to 40% to nitric acid, and ~5% to hydrochloric acid. For this reason, the following disucssion of the effects of acids on concrete will focus on effects of sulfuric acid.

In general, acid solutions attack concrete in any combination of four ways: (a) by dissolving both hydrated and unhydrated cement compounds, (b) by dissolving calcareous aggregates present in the mix, (c) through physical stresses induced by the deposition of soluble sulfate and nitrate salts and the subsequent formation of new solid phases within the pore structure, and (d) by salt-induced corrosion of the reinforcing steel.

The first two forms of attack involve the same mechanism: the leaching away of water-soluble salts formed by reaction of the acid with the calcium compounds in the cement paste and aggregate. When calcareous materials are attacked by sulfuric acid, the sulfate radicals in the acid react with the calcium carbonate ($CaCO_3$) to produce calcium sulfate ($CaSO_4$), or gypsum. Since gypsum is much more soluble in water than calcium carbonate, it is readily washed away. This process eventually results in the complete destruction and removal of any calcareous material exposed to attack. This type of deterioration has been documented to be one of the major mechanisms of the damage occurring to many of the ancient statues, monuments, and buildings made with calcareous building stone in and near industrialized areas of Europe (4-6,10-13,17-19).

Acid attacks cement paste the same way: it reacts with the calcium compounds in the paste, such as calcium hydroxide and calcium carbonate, producing soluble salts that are easily leached away. Gradually, the acid also attacks the hydrated minerals in the cement paste, again producing soluble salts. The leaching process results in the gradual loss of cement paste from the surface of the concrete and eventual exposure of the aggregate. In addition, Tremper (27) reports that the leaching process is not limited to the surface of the concrete, but also extends into the concrete. He states that as calcium carbonate is removed from the surface, calcium hydroxide (lime) diffuses from the interior to the surface and is precipitated as calcium carbonate. When calcium hydroxide is thus precipitated, the water held in the pores of the concrete becomes unsaturated and more calcium hydroxide is taken into solution. There is thus a continuous travel of calcium hydroxide from the interior to the surface, resulting in a general loss of lime throughout the body of the concrete. In its early stages, this form of attack is characterized by a slight etching of the surface, and, in later stages by severe pitting and scaling, followed by a gradual decrease in strength.

The third form of attack is a secondary effect of the first two forms. The reaction of acids with the various calcium compounds present in the cement paste or aggregate leaves a residue of soluble salts, which accumulates on or just beneath the surface. The salts at the surface are leached away by rainwater, the salts accumulated beneath the surface can crystallize with the absorption of water, which increases their volume. This results in the development of enormous stresses within the pores of the cement paste or aggregate, which can eventually lead to blistering and spalling of the surface. In addition, some of the sulfates formed, such as gypsum, react with the hydrated tricalcium aluminate in the cement paste to produce ettringite ($3CaO \cdot Al_2O_3 \cdot 3CaSO_4 \cdot 31H_2O$), which also occupies a large volume and, thus, can also cause cracking.

The accumulation of salts beneath the surface can also lead to the formation of crusts in protected areas that are shielded from washing by rainwater. These impermeable crusts can hold water and salts within their pore structure, causing the concrete or stone to spall off in layers rather than gradually eroding.

The accumulation of salts within the concrete pore structure can also lead to the corrosion of reinforcing steel, the fourth form of deterioration identified above. This corrosion is accompanied by an increase in the volume of the steel, which eventually causes the concrete to crack and spall. In discussing the atmospheric corrosion of concrete reinforcements, Skoulikidis (21) notes: "The increase of atmospheric pollution intensifies the corrosion tendency of the reinforcements in the atmosphere. The cracking of the concrete was observed more frequently with an increase of the atmospheric pollution (SO_2, CO_2, NH_3, NO_x, etc.) and the acceleration of the corrosion by the formation of a more conductive environment, that also chemically attacks the metals."

In addition to the forms of attack already discussed, cracking and spalling of concrete due to acid-induced corrosion can also lead to and accelerate other forms of attack having other causes, most notably freeze-thaw deterioration. Prudil (30) found that concrete which normally withstood attack due to freeze-thaw cycling was subject to attack after exposure to acid solutions.

Effects of Carbon Dioxide. Concrete is known to be affected by the take-up of CO_2 from ambient air, i.e. carbonation (32,34-37). Woods (32) states: "The reaction between atmospheric carbon dioxide and dense hardened concrete is very slow, and even after a considerable number of years, may affect only a thin layer nearest the exposed surfaces. A principal product of the reaction is calcium carbonate, the presence of which may enhance the early resistance of concrete to attack by some chemicals in solution, such as sulfates. In practice, however, any beneficial effect that may exist appears to be of relatively small moment." The harmful effect of carbonation arises when the carbonated layer created on the surface of reinforced concrete over the years reaches the steel reinforcement. The alkaline protective layer is then considerably less alkaline, and the steel bars may start to rust.

Carbon dioxide will also react with water to form carbonic acid. There are, however, conflicting data regarding the rate at which carbonic acid will attack concrete.

Bertacchi (28) reported, for standard portland cement concrete, a weight loss of ~3% and compressive and flexural strength reductions of ~90% after a 7-yr exposure to distilled water into which CO_2 was continuously bubbled and which was replaced periodically as the dissolved lime content increased so that its pH varied from ~4 to 5.5. Tremper (27) reported reductions in compressive strength of 5 to 22% for portland cement concrete subjected to carbonic acid solutions of pH 6.9 to 6.1, respectively, for 8 mo.

On the other hand, Greschuchna (38) reports that a carbonic acid solution saturated at 760 Torr (14.7 psi) and 25°C (77°F) has a pH of 3.7, and that for pH >3 the corrosion rate should be hardly greater than that due to leaching by pure water, i.e. the acid effect becomes negligible. There is also evidence, however, that carbonic acid attack is enhanced by the presence of sulfates (39).

Tremper (27) developed a damage function to describe the deterioration observed in his work. The damage function, which is expressed as

$$\log L = K \cdot \log T$$

where L = the percentage of the original lime lost from the concrete

T = the time in days for which the concrete has been exposed

K = a constant which varies with the pH of the solution to which the concrete is exposed and the surface to volume ratio of the concrete,

predicts the rate of deterioration due to acid attack based upon the loss of lime from concrete. Based upon the results of a series of laboratory tests, Tremper concluded that for purposes of computation the mechanical failure of average portland cement concrete will occur when 50% of the original lime content has been removed. Values for K are developed based upon the results of the laboratory tests, however, Tremper had to make several assumptions during the development of these values, thereby limiting the accuracy of the damage function.

Friede (43) published a series of equations for calculating the depth of the corroded zone for concrete specimens exposed to carbonic acid attack as a function of the physical and mechanical properties of corroded and noncorroded specimens, i.e. density, volume, mass, modulus of elasticity and strength. The equations, however, do not take into consideration such factors as the composition of the concrete or the degree of attack to which it is subjected. In addition, the wide variation in test results, obtained from laboratory tests performed to varify the theoretically derived equations, limit their applicability.

Effects of Sulfur Dioxide. Sulfur dioxide, when dry, has little or no effect on dry PCC. It does, however, combine with water to form sulfurous acid (H_2SO_3), which gradually reacts with oxygen to form sulfuric acid (H_2SO_4), both of which will attack concrete (32). Most of the damage to materials from SO_2 is attributed to highly reactive sulfuric acid formed either in the atmosphere or on the surface of materials. Damage to limestone products, concrete, and marble has been observed in those areas experiencing relatively high levels of SO_2 over a prolonged period (23).

As previously discussed, sulfuric acid attacks concrete (a) by converting calcium carbonate to gypsum, which is subsequently leached away, and (b) by reacting with calcium compounds to form salts which crystallize, producing enormous stresses within the pores of the cement paste which eventually lead to spalling and cracking. The latter form of attack is commonly known as sulfate attack. Both mechanisms of deterioration were identified by Hansen et al (31) in their work regarding the corrosion of concrete due to sulfuric acid attack, in which they conclude that exposure to sulfuric acid can progress from a straight corrosive attack to a combination of corrosion and sulfate attack.

The effects of sulfate attack on portland cement concrete have been well documented (25,26,32,33,40-42,44-48). Kuenning (26) describes the mechanism of sulfate attack as follows: "The destructive action of sulfates on concrete is primarily the result of their reaction with either C_3A or the C_3A hydration products to form the high-sulfate form of calcium sulfoaluminate (ettringite). The crystalline reaction product is of larger volume than the original aluminate constituent, and expansion results. The concrete or mortar increases in strength at first, because of the increase in solid matter, even though it is changing chemically. As the process continues the concrete or mortar expands, cracks, becomes progressively weaker, and finally disintegrates."

Jambor (42) has published a damage function describing sulfate attack in terms of the percent of SO_3 bound in hardened cement paste, which he reports as being the prime cause of sulfate corrosion. The damage function (DC) which is expressed as

$$DC = (0.11S^{0.45})(0.143t^{0.33})(0.204e^{0.145C_3A})$$

takes into consideration the concentration of the acting sulfate solution (S), the period of time of its action (t), and the tricalcium aluminate (C_3A) content of the portland cement used. The damage function was developed on the basis of experimental test results which demonstrated the effects, with time, of sodium sulfate solutions, with varying SO_4 concentrations, on the dynamic modulus of elasticity, compressive and flexural strength, volmetric and mass changes, and changes in the bound SO_3 content of portland cement mortar specimens.

Jambor has, apparently, been able to relate the changes observed in the physical and mechanical properties of the specimens to changes in the bound SO_3 content. These data, however, were not given in the paper. The data presented relate bound SO_3 content to sulfate concentration, C_3A content, and time of testing. It is these data upon which the damage function was based. As Jambor points out, the damage function is limited in that it does not take into account temperature effects, influence of the cement content in the mortars and concrete, total porosity of the composite material, as well as the influence of the cross-section size of the structure. It does, however, serve to give a first approximation.

Even though the mechanisms are not fully understood, it has been fairly well established that SO_2 accelerates the corrosion of carbon steel (14-16,49-51). This results in the creation of a layer of rust, i.e. iron oxides, on the surface of the steel, which occupies more than twice the volume of the iron from which it was produced. In addition, Haynie and Upham (15) report that iron oxides catalyze the oxidation of SO_2 to SO_3 as well as react with SO_2 to form sulfates. Both conditions, i.e. the expansion of reinforcing steel due to corrosion and sulfate attack, have been shown to cause the deterioration of PCC.

Effects of Nitrogen Oxides. Very little information is presently available in the literature regarding the effects of nitrogen oxides (NO_x) on PCC. Gauri (19) reports that nitrogen dioxide produced primarily during combustion processes by the oxidation of atmospheric nitrogen is the main cause of the acidity of precipitation in the Los Angeles Basin where NO_3^- is more than twice as concentrated as SO_4^{-2}. NO_3^- is also present at a lesser extent in the northeastern United States. But due to higher solubility, its lodgement time in the atmosphere is much shorter. It is perhaps due to this that nitrates have not yet been identified in stone and concrete structures.

Nitrogen oxides will react with water or, as ammonia (NH_3), with oxygen to form nitrous (HNO_2) and nitric (HNO_3) acid. Biczok (25) reports that although nitric acid is not as strong as sulfuric acid, it is more harmful to concrete on brief exposure as it transforms the $Ca(OH)_2$ of concrete into highly soluble calcium nitrate. Nitric acid is destructive enough to bring about extensive deterioration even in highly diluted solutions.

Conclusions and Recommendations

The results of the literature survey, private discussions, and the responses to mail and telephone inquiries have indicated that very little qualitative or quantitative information is available dealing specifically with the effects of acid deposition on PCC structures, but there is a considerable amount of information available indicating that acids and acid waters have a significant effect on the durability of concrete. This effect may not be sudden or dramatic, but it is a cause for concern.

It has been well documented that high levels of pollutants (SO_2, NO_x, etc.) have greatly accelerated the deterioration of many of the ancient statues, monuments, and buildings made using calcareous building stone in and near the industrialized areas of Europe. Evidence is now beginning to indicate that rapidly increasing numbers of reinforced concrete structures are also showing increased rates of deterioration which are attributed by some to be due to exposure to high levels of SO_2, NO_x, and HCl.

On the basis of this evidence, it is recommended that an experimental test program, consisting of both laboratory and field tests, be developed and implemented to quantitatively measure the effects of acid deposition on both the asthetic and structural properties of PCC structures. It is, however, recommended that a preliminary series of controlled, accelerated laboratory tests be carried out before a full-scale field evaluation program is instituted. The objectives of the accelerated laboratory test program should be to identify the magnitude of the problem and to attempt to differentiate between the effects of wet deposition, dry deposition, and normal weathering. The preliminary test program should concentrate on surface chemistry effects and penetration rates of SO_4, NO_x, and Cl^- as deposited from wet and dry deposition. The tests can be run using small portland cement mortar specimens, formulated to simulate the quality of cement pastes normally encountered in concrete construction. Parameters to be studied in the test program should include:

wet deposition - various simulated acid rain mixtures to differentiate between the effects of SO_2, NO_x, and the normal background components of rain.

- simulated acid rain mixtures of varying pH values.

dry deposition - various levels of SO_2 and NO_x.
- various relative humidities.

normal weathering - freeze-thaw cycles.

Test methods used in the evaluation should include; (a) tests to evaluate changes in the physical and mechanical properties of the specimens, (b) chemical analyses to determine the depth and rate of penetration of the aggressive solutions, and (c) tests to monitor and control the treatment solutions and to analyze them for materials being leached from the test specimens.

If the results of the preliminary test program indicate that a problem does exist, full-scale field tests should be initiated, along with continued inter-related laboratory-scale bench tests and environmental chamber studies. As with the preliminary test program, the field tests should be designed to differentiate between the effects of wet deposition, dry deposition, and normal weathering on both the asthetic and structural properties of portland cement concrete. Exposure parameters studied in the preliminary program should also be studied in the field test program. In addition, several different qualities of portland cement concrete, i.e., structural quality and decorative, should be incorporated into the program. If possible, more than one field test site should be used to obtain environmental exposure conditions representative of various areas of the United States. The field test program should also include surveys of existing PCC structures in areas where air quality, rain chemistry and meteorological data are readily available for analyses in order to obtain a comparison between the effects observed in existing structures with those observed in the test specimens.

Literature Cited

1. Galloway, J. N. Acid Rain, Proceedings of a Symposium Sponsored by ASCE. April 2-6, 1979, 1-20.
2. Likens, G. E., and F. H. Bormann. Science. 184(4142), 1974, 1176-9.
3. Glass, N. R., et al. Environmental International. 4, 1980, 443-52.
4. Fisher, T. Prog. Arch. 7, 1983, 99-105.
5. Stanwood, L. The Construction Specifier. Nov. 1983, 74-9.
6. Gauri, K. L. Proc. of a Session Held at the National Convention, ASCE. Boston, MA, April 2, 1979, 70-91.
7. Cowling, E. B. Environ. Sci. Technol. 16(2), 1982, 110A-123A.
8. Martin, H. C. Materials Performance. 21(1), Jan. 1982, 36-9.
9. Likens, G. E. Chem. and Engr. News. 54(48), Nov. 22, 1976, 29-44.
10. Dornberg, J. Pan Am Clipper. July 1982, 18-22 and 77-9.
11. Fassina, V. Atmospheric Environment. 12, 1978, 2205-11.
12. Gauri, K. L. and G. C. Holden, Jr. Environ. Sci. and Technol. 15(4), April 1981, 386-90.

13. Longinelli, A. and M. Bartelloni. Water, Air, and Soil Pollution. (10) 1978, 335-41.
14. Kucera, V. Ambio. 5(5,6) 1976, 243-8.
15. Haynie, F. H. and J. B. Upham. Materials Protection and Performance. 10(11) 1971, 18-21.
16. Ericsson, R. and T. Sydberger. Werkstoffe und Korrosion. 31, 1980, 455-63.
17. Schreiber, H. Air Pollution Effects on Materials. Report for Panel 3 ("Environmental Impact") of the NATO/CCMS Pilot Study on Air Pollution Control Strategies and Impact Modelling, 1982.
18. Von Ward, P., ed. Impact of Air Pollutants on Materials. A Report for Panel 3 ("Environmental Impact") of the NATO/ CCMS Pilot Study on Air Pollution Control Strategies and Impact Modelling, 1982.
19. Gauri, K. L., Polluted Rain (Environmental Science Research). 17, 1980, 125-45.
20. Dijkstra, G. Effects of Air Pollution on the Materials in the Netherlands. Report for Panel 3 ("Environmental Impact") of the NATO/CCMS Pilot Study on Air Pollution Control Strategies and Impact Modelling, 1982.
21. Skoulikidis, T. N. Atmospheric Corrosion of Concrete Reinforcements and Their Protection. Date unknown, 807-25.
22. Skoulikidis, T. N. Proc. Acid Deposition, A Challenge for Europe. Preliminary Edition, Sept. 1983, 193-226.
23. Gillette, D. G. J. Air Pollution Control Assoc. 25(12), Dec. 1975, 1238-43.
24. Ashton, H. E., and P. J. Sereda. Durability of Building Materials. 1, 1982, 49-66.
25. Biczok, I. Concrete Corrosion, Concrete Protection. Akademia Kiado, Budapest, 1972.
26. Kuenning, W. H. Highway Research Record. (113), 1966, 43-87.
27. Tremper, B. ACI Journal. 28(1), Sept. 1931, 1-32.
28. Bertacchi, P. RILEM Symposium, Durability of Concrete-1969, Final Report, Part II. Academia Prague, 1970, C159-C168.
29. Mlodecki, J. RILEM Symposium, Durability of Concrete-1969, Preliminary Report, Part II. Academia Prague, 1970, C221-C240.
30. Prudil, S. RILEM Symposium, Durability of Concrete-1969, Preliminary Report, Part I. Academia Prague, 1970, A59-A68.
31. Hansen, W. C., et al. ASTM Bulletin. 231, July 1958, 85-8.
32. Woods, H. Durability of Concrete Construction. ACI Monograph No. 4, 1968.
33. Czernin, W. Cement Chemistry and Physics For Engineers. Chemical Publishing Co., New York, NY, 1962.
34. Sentler, L. Proc. Third International Conference on the Durability of Building Materials and Components. Espoo, Finland, Vol. 3. Aug. 12-15, 1984, 569-80.
35. Verbeck, G. IN: ASTM STP-205. 1958, 17-36.
36. Leber, I. and F. A. Blakey. Journ. Amer. Conc. Institute. 53-16, Sept. 1956, 295-308.
37. Smolczyk, H. G. RILEM Symposium on Durability of Concrete-1969, Preliminary Report, Part II. Academia Prague, 1970, D59-D75.

38. Greschuchna, R. RILEM Symposium, Durability of Concrete-1969, Final Report, Part II. Academia Prague, 1970, C189-C191.
39. Idorn, G. M. Ph.D. Thesis, Technical University of Denmark, 1967.
40. Brown, P. W. Cement and Concrete Research. 11, 1981, 719-27.
41. Heller, L., and M. Ben-Yair. J. App. Chem. 14, Jan. 1964, 20-30.
42. Jambor, J. IN: "Durability of Building Materials and Components"; ASTM STP-691, 1980, 301-12.
43. Friede, H. IN: "Durability of Building Materials and Components"; ASTM STP-691, 1980, 355-63.
44. Reading, T. J. IN: "Durability of Concrete"; ACI SP-47, 1975, 343-66.
45. Mehta, P. K. and M. Polivka. IN: "Durability of Concrete"; ACI SP-47, 1975, 367-79.
46. Ludwig, V. IN: "Durability of Building Materials and Components"; ASTM STP-691, 1980, 269-81.
47. Klieger, P. IN: "Durability of Building Materials and Components"; ASTM STP-691, 1980, 282-300.
48 Mehta, P. K. IN: "Durability of Building Materials and Components"; ASTM STP-691, 336-45.
49. Haynie, F. H. and J. B. Upham. IN: "Corrosion in Natural Environments"; ASTM STP-558, 1974, 33-51.
50. Mikhailovskii, Y. N. and A. P. San'ko. Protection of Metals. 15(4), 1979, 342-5.
51. Matsushima, I. and T. Ueno. Corrosion Science. 11(3), 1979, 129-40.

RECEIVED January 2, 1986

17

Deterioration of Brick Masonry Caused by Acid Rain

A. E. Charola[1] and L. Lazzarini[2]

[1]Metropolitan Museum of Art, New York, NY 10028
[2]Laboratorio Scientifico della Misericordia, 30121 Venice, Italy

Several different mechanisms are operant in the deterioration of brick masonry through the action of acid rain. The bricks are susceptible to acid rain through the selective dissolution of their glassy phase. The mortar is affected mainly by the reaction of its calcareous components. The soluble salts resulting from these reactions, in solution with rain water or condensed moisture, will migrate through the porous matrix of the masonry. In the places where the water evaporates the salts will be deposited.
Repeated dissolution and recrystallization of these salts leads to the mechanical disruption of the masonry structure. Since the salts will concentrate in the more porous material, either the brick or the mortar will be more seriously affected, depending on their relative porosity.

Do we build a house to last forever?
The Epic of Gilgamesh (1)
(originated about third millenium BC)

Deterioration of brick masonry is a problem that has worried man ever since the first brick wall was constructed. It is interesting to remember that already by the first century A.D. bricks were considered more resistant to deterioration than marble., which had an estimated useful lifetime of about 80 years (2). Today, the problem is still in study, if only due to the increased number of masonry structures.

The deterioration of brick masonry is a complex problem, in which two variable ingredients, brick and mortar, make up the whole. As each component can have a large variation in composition and structure, when both are combined in a wall, the number of variations that result is given by all the possible combinations of the two components.

Both materials are more or less susceptible to attack by acid rain, but the overall deterioration is not a simple chemical attack

0097-6156/86/0318-0250$06.00/0

on the individual materials. Other phenomena, such as the crystallization of the salts formed by this attack, can compound the effect, and in some instances contribute significantly to the deterioration of brick masonry.

Brick

Bricks are ceramic bodies manufactured from clays, molded and fired. The nature and quality of the bricks will depend on all the above mentioned parametes: the clays used in the manufacture, the molding technique used, the firing temperature and the residence time in the firing kiln or time at temperature (3). Bricks produced at higher temperatures and longer times at temperature will have a more vitreous matrix. Also, their hardness and strength will increase and the water absorption capacity will decrease as the pore size distribution shifts towards smaller pores. This last feature is important as the deterioration of bricks is usually related to the amount of water that can be absorbed by them. The presence of a large number of smaller pores (<1 μm) increases the susceptibility of bricks to freeze-thaw and salt crystallization damage (3,4).

The chemical durability of brick will depend on the chemical stability of its components viz a viz the aggressive environment. The glassy matrix in a brick, --like glass itself--, is the component most susceptible to chemical dissolution by neutral and by acid or alkaline solutions. The result of the attack by water is the extraction of the alkali ions from the surface of the glass thus forming an alkali-depleted layer. The rate of dissolution of the glass appears to be a diffusion controlled reaction --through the leached layer-- for short times and at low temperatures (5,6). At longer times, and especially at higher values of pH (pH >9), a second surface controlled reaction that dissolves the leached layer can occur simultaneously (7).

All silica glasses are particularly susceptible to dissolution by solutions above pH 9 due to the nucleophilic attack of the hydroxyl ion on the silicon-oxygen bond (7). This is relevant in our particular discussion on bricks, as these are set in masonry with mortar, from which lime may be leached out such that solutions reaching the neighbouring pores could well reach pH 11 or 12.

In the case of dilute acid solutions (pH >2), the alkalis and basic oxides are dissolved preferentially, but the amount of silica dissolved is less than that removed by water alone (7,8,9). This can be explained by the fact that the acid neutralizes the leached alkalis, so that the pH does not rise to higher values where silica dissolution becomes important. The preferential leaching produces an alkali-depleted layer that can be twice as thick as the one obtained by neutral water (10). Because of the replacement of the alkali ions by the smaller hydrogen ions, stresses will be induced in this layer which can cause it to crack. Further shrinkage can also occur if this hydrated silica layer loses water (10,11).

This type of deterioration was observed in the glassy matrix of a brick exposed to concentrated sulfuric acid where the leached layer crumbled away exposing the more acid resistant minerals (12). Sulfuric acid solutions formed from atmospheric constituents were estimated to increase tenfold the rate of attack on bricks (13). Gaseous SO_2 in a water vapour saturated atmosphere was found to deteriorate bricks by increasing their porosity, mainly in respect to the larger pores (14).

Nitric and hydrochloric acid were found to be even more aggressive to glass than sulfuric acid (15). It could therefore be expected that nitrogen oxides would be more aggressive than sulfur oxides with respect to brick deterioration.

Another important factor in the dissolution of glass is the ratio of surface area exposed to volume of leaching solution. The higher this ratio the higher the amount of silicate dissolved though the amount of alkali ion extracted does not appear to be quite so affected (7,10). Considering the particular case of bricks exposed to acid rain, the dissolution of the glassy matrix will occur in the films of water that wet the exposed surface and the inner walls of the pores of the brick. In either case the SA/V ratios will be very high thus accelerating dissolution. Because of the constant renewal of the wetting solution the dissolution will be even more important on the exposed surface film (16). It should also be pointed out that the dissolution effect of water run-off may be more important than the pH of the solution. This has been shown to be the case even for acid susceptible calcareous materials. The run-off effect was more important than the actual pH of the solution, for pHs above 3 (17).

The appearance of the surface of 16th century bricks in Venice, Italy, exposed to rain water run-off is shown in Figure 1. Surface erosion, as can be seen, is fairly appreciable, and especially noticeable is the uneveness in the erosion due to difference in vitrification. These bricks, and even older ones (12th century) have been studied because they have resisted rain water run-off for seven centuries and high air pollution for the past half century, the latter due to the proximity of Venice to the refinery at Porto Marghera.

SEM examination of these brick samples show in detail the result of the prolonged exposure to rain-water run-off (Figure 2). The eroded surface is apparent even at low magnification. Nodules of highly vitrified material can be seen as well as the "channels" carved around them by the running water. Figure 3 shows the detail at higher magnification of the residual platy minerals which were exposed as the glassy phase eroded away.

Figures 2 and 3 are of local Venetian bricks fabricated in the 12th century. These bricks are for historical reasons called "altinelle" and have a fairly uniform size of approximately 17 cm x 7 cm x 5 cm (18). Table I gives a typical chemical analysis for the matrix of the brick and for the more resistant, vitrified nodules.

Table I. Chemical Analysis of 12th century Venetian "altinella" brick

	SiO_2 %w/w	Al_2O_3 %w/w	Fe_2O_3 %w/w	MgO %w/w	CaO %w/w	Na_2O %w/w	K_2O %w/w	TiO_2 %w/w
Matrix	87.10	6.94	1.31	0.13	0.24	0.28	0.74	0.56
Nodules	71.43	7.14	15.42	0.13	0.36	0.20	0.75	0.46

Figure 1. Surface of a 16th century Venetian brick eroded by rain water run-off. A highly-vitrified nodule resists deterioration better than the less vitrified matrix.

Figure 2. SEM photomicrograph of an eroded surface of a 12th century brick, "altinella", from Venice. The quartz grains are exposed as the more susceptible matrix deteriorates faster.

Figure 3. SEM photomicrograph showing in detail the matrix of the "altinella" brick. Platy agglomerates start to be exposed as the glassy matrix is dissolved away by the acid rain.

As can be seen, the nodules have a higher Fe_2O_3 content which has the double function of increasing the fusibility of the clay, giving rise to a less porous matrix, and producing a chemically more resistant glassy phase.

It is most interesting that these bricks, even though they were fired at relatively low temperatures (850-900°C) and are not highly vitrified, are more durable than any modern bricks (18). This can be explained by the fact that these bricks are very homogeneous throughout and lacking the highly vitrified outer skin which characterizes modern bricks. Their pore size distribution is centered around larger pore sizes (1-2 μm) thus avoiding the critical smaller range (<1 μm). Therefore, even though a certain amount of material is lost due to rain and acid rain attack, the bricks is still resistant. In the case of many modern bricks, once the outer, hardened skin has been attacked, by acid rain or any other phenomenon, the soft interior of the brick will deteriorate at an extremely fast rate.

Mortar

The nature of mortar has changed considerably over time. The primitive clay based mortars gave way to the lime-sand formulation of the Romans with additions of plaster, crushed brick, and/or volcanic earth (pozzolan). The rediscovery of natural cements occurred in the eighteenth century and finally Portland cement was developed. Mortars in use changed accordingly to include the new products. In each case the type, size and amount of charge added in the mix has a large influence on properties such as the bulk density and porosity of the final product.

The effect of acid rain on mortars will depend on the particular mortar in consideration. The most susceptible mortars will be the lime-sand ones. The carbonated lime will be particularly attacked due to the small crystal size of the formed calcite (19,20). The resulting calcium sulfate can crystallize as gypsum [$CaSO_4.2H_2O$] inducing mechanical stresses into the matrix of the mortar.

Figure 4 shows the platy gypsum crystals that formed in a crack of a lime mortar. The continuing growth of these crystals will cause the crack to develop even further and thus mechanically disrupt the matrix of the mortar.

Portland cement concrete is attacked by acid rain which softens and disintegrates its surface (21). The presence of highly reactive free calcium hydroxide, portlandite, formed during the hydration of the hydraulic components in Portland cement will make it especially susceptible to acid attack. Even though this compound will tend to carbonate, --and still be available for reaction with acid rain--, century old mortars have been found to contain significant amounts of it. The amount present was higher in the more hydraulic mortars (22). The lower porosity of mortars with a higher proportion of Portland cement could account for that. The resulting sulfates produced by the reaction can further degrade concrete by reacting with calcium aluminate [C_3A] to produce ettringite [$3CaO.Al_2O_3.3CaSO_4.32H_2O$] which induces large internal stresses during its crystallization.

Portland cement concrete is even susceptible to pure rain water which can hydrolize the alkali silicates and aluminosilicates present in it and give rise to the formation of trona [$Na_2CO_3.NaHCO_3.2H_2O$] (23). Furthermore, water can leach out of the cement any soluble salts that are present in it, thus weakening its structure.

Discussion

It can be seen that even though both materials used in the construction of masonry can be affected by acid rain, the degree of susceptibility to chemical dissolution is usually fairly low. The main damage to brick masonry is produced by the reaction products of that dissolution, i.e., soluble salts that can crystallize in it.

The origin of salts like gypsum, trona and ettringite were mentioned. Also, not to be forgotten, is the possibility that the bricks themselves contain Na_2SO_4 produced during the firing if the kilns use sulfur rich fuels and if the temperature was not sufficient to decompose this salt. Such bricks have an inherent vice: Na_2SO_4, which in the presence of water will be dissolved and as the water evaporates can recrystallize either as the anhydrous salt, thenardite, or as the decahydrate, mirabilite, depending on the temperature and relative humidity. It furthermore can recrystallize and/or change the degree of hydration as the environmental conditions change. The stresses involved in these processes are sufficient to eventually destroy the brick completely. This salt can also be found in marine environments as a reaction product of air pollutants and saline aerosols.

Another frequently occurring soluble salt is magnesium sulfate, usually crystallized as the heptahydrate, epsomite, produced by acid rain attack on mortars with magnesian lime. Salts used to prevent the freezing of mortar during construction, or as de-icing agents around the structure, are usually the source of chlorides that crystallize as NaCl [halite] or KCl [sylvite]. Ground water is an important source of soluble salts; the distribution that these salts can have in masonry has been studied extensively (24).

The deterioration that occurs in masonry due to soluble salt crystallization depends on where the evaporation of the salt-containing solution occurs. The crystallization of the salt in a steady state system will depend on the relative rates of the two phenomena involved: rate of capillary migration of the solution and rate of evaporation of the water (12, 25). Where the equilibrium of the two rates is established will depend on the relative porosities of the brick and the mortar. If the mortar is very porous, most of the evaporation will take place there, the salt will concentrate in the mortar and eventually destroy it completely. An example of this type of deterioration is shown in Figure 5. If such a wall were then repointed with a very strong, impermeable Portland cement mortar, the end result, if the sources of soluble salts are not isolated, is shown in Figure 6: the mortar subsists but the bricks deteriorate. The fact that salt induced deterioration will occur in the most porous material has been proved experimentally (26). If both materials have similar porosities, the decay will occur gradually and homogeneously over the whole masonry unit and will take a much longer time.

The porosity of the bricks has an important function during the setting of the mortar between them. If the brick is fairly porous,

Figure 4. SEM photomicrograph showing gypsum crystals in a crack within a lime mortar. The growth of these crystals induces mechanical stresses which progressively enlarge the crack and break up the matrix of the mortar. The extremely fine-grained calcite (formed during the setting by carbonation of the lime mortar) is especially susceptible to acid rain dissolution.

Figure 5. Mortar deterioration due to salt crystallization in masonry. The mortar is more porous than the brick allowing for fast evaporation and concentration of the soluble salts.

Figure 6. Brick deterioration due to salt crystallization in masonry. In this case the bricks are more porous than the mortar and the salts will accumulate in them leading to their deterioration. Considerable variation in the degree of deterioration is evident.

it will take water from the setting mortar thus inhibiting the hydrolysis of the hydraulic components in the adjacent area to the brick-mortar interface. As a consequence the amount of portlandite produced in that area will be reduced. The bond between the brick and mortar will be based on the crystallization of ettringite that forms as the water that moves into the brick concentrates the soluble gypsum at the interface where it reacts with the calcium aluminate. If the brick is not very porous, more water will be available in the mortar and the result is that both ettringite and portlandite will form at the interface. Larger pore sizes in brick favour brick-mortar adhesion because of the lower suction produced by the large diameter capillaries and the better adherence of the ettringite layer as it can crystallize in the pores of the brick (27). The development of a good bond between brick and mortar is not only important from a structural point of view but will also determine the walls' resistance to water penetration.

Conclusions

Brick masonry, even though susceptible to acid rain attack, owes its deterioration mainly to the crystallization of the soluble salts produced in that reaction or from other, more important, sources of soluble salts: the inherent vice in bricks and ground water salts. The capability of a given masonry structure to resist deterioration will be directly related to its resistance to water penetration, which is the main single agent responsible for the decay process by salt crystallization.

The design and construction mode of a given wall is highly important. The quality of the bricks, the mortar to be used in setting them, and their relative porosities will determine the durability of the wall. The actual setting of the bricks in mortar is also an important factor contributing to an effective bond between bricks and mortar especially with regards to the resulting masonry's resistance to water penetration.

In summary, the main cause for masonry deterioration is the presence of water in it. A dry wall will not effloresce, the salts contained in it will not be able to recrystallize and other deterioration processes such as freeze-thaw will also be avoided.

Literature Cited

1. "The Epic of Gilgamesh"; Radice, B., Ed.; Penguin Books: Reading, 1983; pp. 106-7.
2. Vitruvius, "De Architectura"; Book II, Chapter 8.
3. Robinson, G. C. Ceramic Bulletin 1984, 63, 295-300
4. Torraca, G. "Porous Building Materials"; ICCROM: Rome, 1985; pp. 30-2.
5. Rana, M. A.; Douglas, R. W. Physics and Chemistry of Glasses 1961, 2, 179-95, 196-205.
6. Das, C. R.; Douglas, R. W. Physics and Chemistry of Glasses 1967, 8, 178-84.
7. Paul, A. "Chemistry of Glasses"; Chapman and Hall: London, 1982; pp. 108-47
8. Holland, L. "The Properties of Glass Surfaces"; J. Wiley & Sons: New York, 1964; pp. 133-4.

9. Bacon, F. R. Glass Ind. 1968, 49, 438-9, 442-6.
10. Clark, D. E.; Pantano, C. G.; Hench, L.L. "Corrosion of Glass"; Books for Industry: New York, 1979; pp. 22-39.
11. Newton, R. G. Glass Technology 1985, 26, 21-38
12. Lewin, S. Z.; Charola, A. E. Proc. Conf. Il Mattone di Venezia, 1979, pp. 189-214.
13. Robinson, G. C. In "Conservation of Historic Stone Buildings and Monuments"; National Academy Press: Washington, D.C., 1982; pp. 145-62
14. Baronio, G.; Binda, L.; Contro, R.; Scirocco, F. Proc. XV ANDIL Congress, 1980, pp. 81-93.
15. El-Shamy, T. M.; Morsi, S. E. J. Non-Cryst. Solids 1975, 19, 241-50.
16. Clark, D.E.; Pantano, C. G.; Hench, L. L. "Corrosion of Glass" Books for Industry: New York, 1979; pp. 40-54, 64-7.
17. Guidobaldi, F. In "The Conservation of Stone II"; Rossi-Manaresi, R., Ed.; Centro per la Conservazione delle Sculture all'Aperto: Bologna, 1981; pp. 483-97
18. Fazio, G.; Hreglich, S.; Lazzarini, L.; Piredda, U.; Verita, M. Proc. Conf. Il Mattone di Venezia, 1982, pp. 227-91.
19. Thomson, G.; White, R. Studies in Conservation, 1974, 19, 190-1.
20. Charola, A. E.; Koestler, R. J. Wiener Ber. Naturwiss. Kunst, (in press)
21. Ashurst. J. "Mortars, Plasters and Renders in Conservation"; Ecclesiastical Architects' and Surveyors' Assoc.: London, 1983; pp. 34-5.
22. Charola, A. E.; Dupas, M.; Sheryll, R. P.; Freund, G. G. Symp. Scientific Methodologies Applied to Works of Art, 1984, (in press)
23. Charola, A. E.; Lewin, S. Z. Scanning Electron Microscopy, 1979, I, 378-86.
24. Arnold, A. Proc. IV Int. Congr. Deter. and Preserv. Stone Objects, 1982, pp. 11-28.
25. Lewin, S. Z. In "Conservation of Historic Stone Buildings and Monuments"; National Academy Press: Washington, D.C., 1982; pp. 120-44.
26. Binda, L.; Baronio, G. Proc. 7th Int. Brick Masonry Conf., 1985, pp. 605-16.
27. Goodwin, J. F.; West, H. W. H. Proc. British Ceram. Soc. 1982, 30, 23-37.

RECEIVED January 2, 1986

Cultural Resource Monitoring: Concurrent Aerometric and Materials Deterioration Studies at Mesa Verde National Park

D. A. Dolske[1] and W. T. Petuskey[2]

[1]State Water Survey Division, Illinois Department of Energy and Natural Resources, 2204 Griffith Drive, Champaign, IL 61820
[2]Department of Chemistry, Arizona State University, Tempe, AZ 85287

The effects of acid deposition on the sandstone masonry of Anasazi cliff-dwelling ruins at Mesa Verde National Park, Colorado, are being investigated. Potential correlations are sought between surface erosion rates and continuously monitored aerometric parameters. A mechanism of deterioration by acceleration of natural erosion processes is hypothesized. Gaseous and particulate pollutant concentrations and meteorological variables are measured at Spruce Tree House ruin. Two test walls, closely resembling Anasazi structures, have been constructed of sandstone specimens typical of masonry units found at Mesa Verde. Deterioration of the stone occurs by a combination of chemical and mechanical processes which weaken the material which bonds the quartz grains. The surface recession rate is periodically recorded using photogrammetry, and microstructural and microchemical alterations in the bonding phase are observed using optical and electron probing techniques. Backscattered-imaging electron microscopy is sensitive to the chemistry of constituent phases, with spatial resolution equivalent to secondary electron imaging. The monitoring scheme is discussed in the context of ongoing research.

The four corners region of the United States contains some of this country's treasured cultural resources in the form of dwelling ruins of prehistoric Anasazi civilization. The ruins have been exposed for a thousand years to a natural environment which was largely unaffected by human activity. However, in very recent years rapidly increasing visitation of cultural sites in the area by tourists, construction and operation of large man-made installations such as power plants which burn fossil fuel, and rapid regional population growth have changed this situation. Local air chemistry and climatic conditions may be sufficiently altered as to affect the mode of deterioration of the masonry of the ruins. A system has been

0097-6156/86/0318-0259$06.00/0

developed for monitoring the effects of the deposition of acidifying substances on certain building materials found in these cultural resources. A pilot test system has been erected near Spruce Tree House ruin at Mesa Verde National Park, Colorado. The objectives of the project are to devise and implement systems for simultaneous monitoring of aerometric and materials-deterioration parameters. The target material is the sandstone masonry of the ruins. The system will provide data allowing correlations to be studied concerning the influence of local air chemistry on the rate of deterioration.

Two specific hypotheses form the basis for the pilot monitoring system design:

(1) Correlations between the rate of deterioration of the sandstone masonry and fluctuations in the concentrations of acidic components delivered by the local atmosphere can be detected.

(2) Inputs of acidic components accelerate deterioration by natural processes by increasing the rate of dissolution of the bonding phases in the sandstone microstructure.

Methods

Simultaneous materials-deterioration and aerometric measurements are made at one location. Correlations of deterioration parameters with varying atmospheric physical and chemical parameters will be examined using multivariate statistical methods when sufficient data are accumulated. Deterioration parameters being measured include rates of surface erosion, surface chemical changes, and microstructural alterations. Atmospheric monitoring includes particulate sulfate, nitrate, and sulfur dioxide concentrations, as well as temperature, relative humidity, and parameters of wind and water erosion.

Materials Monitoring

The objectives of the materials monitoring portion of this project are to (1) provide a means for periodically assessing any changes in the chemistry of the sandstone masonry used in Anasazi Indian ruins, and (2) provide a means of assessing the rate of their deterioration. Towards this end, a monitoring system has been devised based on the photography of stone surfaces and the chemical analyses of specimens taken at regular intervals. The information obtained is to be compared to timewise changes in the air chemistry and meteorological conditions to determine the existence of any statistically significant correlations.

The sandstone masonry used by the Anasazi is generally thought to originate from nearby cliffs from which huge portions spalled off due to the expansion-cracking forces associated with salts which percolate from behind the surface. These were fashioned, dressed, and used in the dwellings for which these ancient people are known.

Monitoring consists of periodic examination of two test walls constructed of actual ruins stones. These were selected in consultation with Park personnel and were taken from rubble piles left by crews who reconstructed and stabilized the existing ruins. Generally, the stones have lost their context with respect to any existing ruins and therefore are of small archaeological value. The selection criterion was that each stone have an exposed and well-weathered surface which was reasonably flat for photographic inspection.

A series of five stones were selected for each test wall. One wall is located on a rock ledge fully exposed to the elements and is near the aerometric station. Twenty-five meters away, the second site is located at the extreme down-canyon side of Spruce Tree House Ruins which is under the protection of rock overhang. At each site is an array of sandstone specimens cored out of a single sandstone block. Four cores each are mounted in fused quartz holders which themselves are held in an acrylic rack. A single holder with four cores is removed each month and subjected to a variety of chemical analyses.

In devising a scheme for monitoring the rate of deterioration of the sandstone, the assumption was made that any effect that atmospheric pollutants may have on the stones' integrity will appear at or near the surface. It is not a common occurrance that masonry is saturated throughout by water. Consequently, an adequate measure of erosion is the rate of recession of the surface. Here, the bonding material is dissolved or disrupted by repeated exposures to moisture which may contain chemicals which accelerate the process. The rate of recession is characterized by the rate at which the loosened quartz grains are subsequently removed.

The rate of recession is measured by periodically examining selected sandstone surfaces. Photographs are made at magnifications of 8X and 32X. Sapphire and polycrystalline alumina rods, about 1-mm diameter, are embedded in the sandstone surfaces to provide positional references. By comparing photographs taken at monthly intervals, it is possible to detemine when each surface grain disappears from the surface. This allows an average recession rate to be calculated. Under some circumstances, agglomerates of several grains will dislodge from the surface at once. Since it is difficult to determine the precise number of grains involved, stereophotogrammetry is used to determine the depth of the voids (or peaks if existent prior to removal). Currently, it appears that monthly intervals between photographs are sufficient to resolve such events in a timewise manner.

Chemical changes in the sandstone are monitored by a combination of schemes. One of the most important tools is the use of backscattered electron imaging microscopy (BSEI) of prepared cross sections of sandstone specimens. Unlike scanning electron microscopy (SEM), which produces a topographical image of the surface of materials, BSEI produces an image of flat-polished surfaces which is sensitive to the composition of the material. Examination of a sandstone specimen can reveal several different phases which are distinguished by their image brightness. Basically, the brightness is a function of the efficiency at which the primary electrons are scattered by each compound. This is a function of the density of electrons in the structure and therefore depends on the average atomic number. Compounds consisting of high levels of heavy elements are efficient scatterers and appear lighter in tone than compounds of lighter elements which appear to be darker. This sensitivity to the average atomic number, or Z-contrast, will make it possible to detect phase transformations that might occur. For instance, it would be possible to detect the solid state transformation of calcium carbonate to a sulfate phase in the presence of high sulfate concentrations.

It is also important to assess what atmospheric contaminants are deposited on the surface.Consequently, two examination schemes have been adopted for this purpose. In general, the contaminants of greatest concern are those which are water soluble. Therefore, sandstone cores are pulverized and dispersed in distilled water. After centrifugation and filtering, the decanted liquid is analyzed for sulphates, nitrates and chlorides using ion chromatography.

Finally, the surfaces of the cores are examined both optically and by secondary electron microscopy to determine the extent of microstructural changes that are occuring due to atmospheric exposure. Energy dispersive analysis of x-rays (EDAX) also serves to detect atmospheric particles which have deposited onto the core surface. All this information can then be used to at least qualitatively identify deterioration processes that may be occuring.

Aerometric Monitoring

The aerometric monitoring system was designed to provide adequate temporal resolution for the observation of features in the atmospheric signal that may drive the hypothesized mechanisms of accelerated sandstone deterioration. Meteorological parameters being measured at Spruce Tree House include temperature, relative humidity, horizontal and vertical wind velocity, wind direction, incident solar radiation, and rainfall rate. Atmospheric chemistry is monitored in several ways. The chemical composition of precipitation is obtained from the Mesa Verde National Atmospheric Deposition Program regional rain sampling station, which is 1.6 km north of the Spruce Tree House site. At the site of the test walls, a series-filtration system collects particulate and gaseous pollutants. Passive gas absorption samplers are mounted within 4 m of the walls. All meteorological parameters are digitally recorded as hourly averages. In addition, temperature and humidity are independently measured and continously recorded on a stripchart. A totalizing anemometer gives weekly wind run. Precipitation and actively sampled air chemistry are collected weekly, while the passive gas monitors are exposed for minimum one-month intervals. The continuous stripchart noting the temperature and humidity, mechanical anemometer, and passive gas samplers are collectively referred to as the non-intensive method. Results of the non-intensive approach will be compared with the more detailed data from the active air sampling and digitally recorded meteorological monitoring. If the comparison is favorable, it is ultimately planned to study microclimate variability on a ruin-specific basis with a network of several non-intensive sites.

It is important to note that while the sandstone deterioration measurements are done monthly, the aerometric data are of much finer temporal resolution. This is done to allow the investigation of variations in deterioration rates with respect to not only averaged aerometric data, but also extremes and the range of diurnal variations within those averages.

The meteorological sensors are mounted at the top of a 7 m tower, located in a clearing 40 m south of Spruce Tree House. The sensors are thus about 5 m above the elevation of the test walls, and just above the forest canopy within the canyon. The exposure of the instruments was chosen to reflect accurately the specific environment experienced by the test walls and Spruce Tree House. The non-intensive experiments are located in the same clearing, on a separate mast about 2.5 m tall. In addition, temperature transducers and surface time-of-wetness sensors have been mounted on the test walls. These parameters are also recorded on digital tape.

The active air sampling system is located on the meteorological tower. A constant flow pump draws air through a four-stage series filter pack at 10.0 L/min. The filter holder is suspended face-down beneath a 40 cm diameter polyethylene funnel, 2.5 m above ground. Four 47 mm diameter filters are mounted in a polycarbonate holder. The first filter is 8.0 μm pore diameter Nuclepore, to collect large particles (diameter > 2.5 μm) (Cahill, et al, 1977). The second filter is 1.0 μm Teflon, used to collect the remaining small particles. Third, a 1.0 μm Nylon filter selectively adsorbs nitric acid vapor (Goldan, et al, 1983). The sum of particulate nitrate from the first two filters and nitric acid from the third filter should represent well the total ambient nitrate concentration. The fourth and last filter is a double layer of cellulose fiber paper which has been doped with glycerol and potassium carbonate, to absorb sulfur dioxide (Johnson and Atkins, 1975). Loading of clean filters and the extraction of exposed ones are done at the laboratory at the Illinois State Water Survey. The filterpacks are sealed in several layers of polyethylene for shipments between the field site and laboratory. Analysis of the extracts is done by ion chromatography. About twenty percent of this project's total sample load is reserved for quality assurance purposes, including procedural, reagent, and field blanks.

Discussion

Although there is some variability, the sandstones found in Mesa Verde National Park are relatively uniform in terms of general composition and microstructure when compared to the broad spectrum of sandstone types. Principally, they consist of a narrow size distribution of quartz grains (100-150 microns) that account for about 80 to 90% of the mineral content. Geologists classify this type as a quartz arenite where the grain shapes are subrounded to sub-angular. The angularity is caused by authigenic quartz overgrowths on initially rounded grains. The remainder of the mineral content consists of 5 to 10% clay minerals, 3 to 8% potassium feldspar, a combined 5% of hematite (Fe_3O_4) and calcite ($CaCO_3$), and a combined 1% of such accessory phases as zircon ($ZrSiO_4$), tourmaline, apatite and barite ($BaCO_3$).

A major component of the sandstone is its porosity which ranges between 21 to 36 volume percent. Because of this, its structural integrity owes much to the interlocking packing of the grains. The dominant cementing material is the authigenic quartz. Although normally a very strong and chemically stable bonding material, it does not provide an extensive matrix of bonding material and thereby leaves the stones in a friable state. Other cementing materials, of secondary importance on a volume percentage basis, may be important with respect to the resistance of the stone to weathering process. Clay mineral bonding occurs to a much lesser extent and generally appears in interstitial patches. Normally, calcite and hematite phases appear as discrete grains, however, some stones have been found to contain appreciable amounts as a cement.

The variability of the sandstones found in the Park seems to be a function of the relative proportions of the secondary bonding material. This has an apparent influence on the integrity of ruins stones exposed to natural weathering processes. Frequent observations have been made where individual masonry units in a ruins wall deteriorates rapidly whereas the neighboring blocks are relatively unaffected.

The erosion mechanism of greatest concern to this project is the chemical alteration of the bonding matrix of the sandstone. If cementing material is dissolved, the individual quartz grains on the surface are laid bare to mechanical elements of the erosion process, namely wind and water action. The time dependent element in this process is currently unclear although one would presume that the dissolution of the bonding material is sufficiently slow to dominate. By itself, this is a feasible mechanism of natural erosion. The important question is whether it is accelerated by chemical pollutants. Dry deposition of nitrates, sulphates and chlorides which subsequently are dampened by dew and rain can yield solutions of high acidity on a localized scale. This would act particularly strongly on such compounds as the carbonates and have decreasingly less effect on feldspars, clays, ferrous compounds and the accessory phases. Least affected would be quartz.

At first glance, it would seem that acidic environments should not be particularly important considering the preponderence of loose quartz cementation. However, there is some evidence that these sandstones are susceptible at least to some degree. Previous preliminary investigations have indicated that the cohesive strength of sandstone cores soaked in mild acid solutions is 10 to 90% lower than those soaked in neutral solutions. This would indicate that bonding material is weakened particularly in those cases where significant quantities of carbonates are present. However, it is indeed difficult to apply such observations to the current situation since the manner of delivery of acidic components and degree of uniform exposure is quite uncertain.

The project is presently in the first full year of data collection at the Mesa Verde National Park field site. The constraints of multivariate statistical analysis require that at least several years of sampling be completed before any interpretation can be stated regarding possible direct effects of acid deposition on the sandstone masonry.

Literature Cited

Cahill, T.A., Ashbaugh, L.L., Barone, J.B., Eldred, R.A., Feeney, P.J., Flocchini, R.G., Goodart, C., Shadoan, D.J., and Wolfe, G.W., 1977: Analysis of respirable fractions in atmospheric particulates via sequential filtration. J. Air Poll. Control Assoc., 27, p. 675.

Goldan, P.D., Kuster, W.C., Albritton, D.L., Fehsenfeld, F.C., Connell, P.S., Norton, R.B., and Huebert, B.J., 1983: Calibration and tests of the filter-collection method for measuring clean-air ambient HNO^3. Atmos. Environ., 17, p. 1355.

Johnson, D.A. and Atkins, D.H.F., 1975: An airborne system for the sampling and analysis of sulphur dioxide and atmospheric aerosols. Atmos. Environ., 9, p. 825.

RECEIVED January 2, 1986

19

Effects of Atmospheric Exposure on Roughening, Recession, and Chemical Alteration of Marble and Limestone Sample Surfaces in the Eastern United States

C. A. Youngdahl[1] and B. R. Doe[2]

[1]**Argonne National Laboratory, Argonne, IL 60439**
[2]**U.S. Geological Survey, 12201 Sunrise Valley Drive, Reston, VA 22092**

Marble and limestone surfaces were exposed to atmospheric conditions at four eastern U.S. sites and were monitored for changes in surface chemistry, surface roughness/recession, and weight. The effect of acid deposition, to which calcareous materials are especially sensitive, was of particular interest. Results are described for the first year of testing, and aspects of a preliminary equation to relate damage to environmental factors are discussed. Thus far, findings support that acid deposition substantially damages marble and limestone surfaces.

Improvements in methods to reduce uncertainties in the quantitative contributions of important chemical species are outlined for the ongoing effort.

The deterioration of marble and limestone exposed both to anthropogenic acid deposition from the environment and to natural weathering is being assessed as one of the major activities of the Materials Effects Task Group of the National Acid Precipitation Assessment Program (NAPAP). There is much concern for the calcareous stone materials because of their widespread use as the exterior structure of commercial, institutional, and private buildings as well as in valued monuments and memorials. These calcium carbonate materials are especially sensitive to an acid environment.

Much of the environmental damage to the stone results in surface material loss, usually by reaction and dissolution processes, but also by accumulation of surface and subsurface reaction products that subsequently spall away together with some unreacted stone (1). An example of material loss is shown in Figure 1: loss of engraved detail was nearly complete on the marble tombstone after 65 years of exposure at Arlington National Cemetery. The rate of damage is enhanced at the edges and corners of engraving and sculpture, where the surface area is increased by carving such features.

The environmental damage to these stone materials has been the subject of several previous studies (2-6). The past studies have been more limited in scope, e.g., they have been concerned with a

a

b

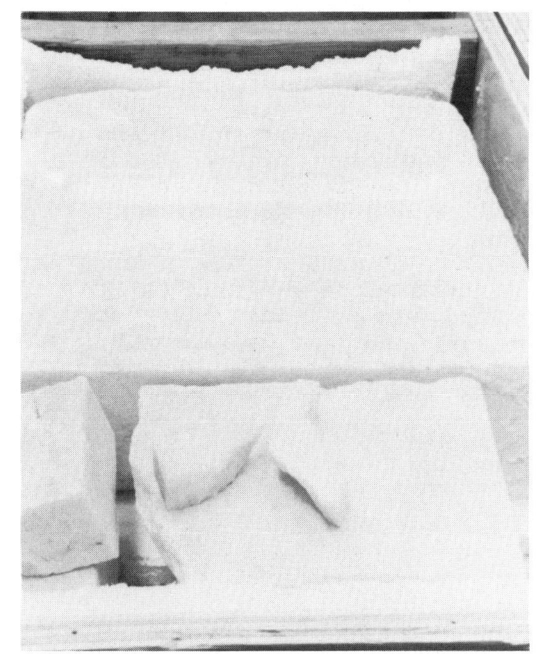

Figure 1. Tombstones of similar varieties of marble. (a) unexposed to atmospheric conditions and (b) exposed for 65 years at Arlington National Cemetery, Arlington, VA. Comparison of engraved inscriptions shows the extent of material dissolution caused by exposure (names of individuals were manually removed).

single environmental pollutant, have measured only one mode of materials damage, or have been of short duration. The previous studies often have been concerned with damage to existing structures for which environmental and maintenance histories are difficult to establish. Therefore NAPAP began an experimental study that is relatively comprehensive with respect to (1) monitoring of environmental factors throughout the duration of stone exposure, (2) damage modes and mechanisms in a variety of geographic locations, and (3) time of stone exposure -- up to ten years in the scheduled plan. Synergism among damaging agents also can be comprehensively investigated in the field tests, which form the focus of the NAPAP study of stone deterioration.

The field program of the Materials Effects Group currently exposes well-characterized specimens of marble and limestone, as well as metals and other materials, to outdoor conditions at sites that are instrumented to monitor meteorological variables, rain chemistry, and air quality factors. Further descriptions of the environmental conditions at the sites are given by Flinn and by Reddy et al. in this volume. Stone samples have been installed at sites near Raleigh, NC; Chester, NJ; Newcomb, NY; and in Washington, DC, and have been monitored for one year.

The damage to stone exposed at the field test sites is assessed by several methods. This paper describes the monitoring of surface roughness, recession, and chemical alteration as well as stone specimen weight changes. Chemical analyses of rainwater and of runoff from relatively large specimens in slab form are being conducted by M. Reddy, and the results are discussed elsewhere in this volume. Tests of changes in color and mineralogical state and of sulfation detected by infrared sensing are also being made and will be reported in future by the pertinent investigators.

Although the results reported below are not yet sufficient in themselves to produce a materials damage function (i.e., an equation relating damage to one or more environmental factors), they can be used to help evaluate and perhaps modify functions that have been proposed on other bases. This paper tentatively concludes that stone material is lost from skyward-facing surfaces at a rate equivalent to about 15 micrometers of surface recession per year at the field test sites and that an approximate agreement with the results of Reddy's runoff chemistry is found where comparison is now possible, i.e., at sites having low ambient concentrations of sulfur dioxide.

Experimental Procedures

Shelburne Marble from Vermont, used for example in the Jefferson Memorial, and Salem Limestone from Indiana, as used in the National Cathedral, were obtained as sample stock materials. Each of the two stone stocks was selected as a monolithic block and was cut into slabs measuring 610 x 305 x 51 mm, under the supervision of NAPAP personnel (7-8). Selected slabs were used to fabricate briquettes, which were employed for several of the damage studies. Briquette dimensions are shown in Figure 2. Surface finishes typical of those provided on stone used for exteriors of buildings were employed on what would become the skyward surfaces of slabs and briquettes: an 80-grit ground surface was produced on marble, and a "smooth planar finish" was used on limestone. All slabs and briquettes were labeled systematically to preserve information on their relative locations and orientations in the source blocks: the geological bedding planes of the stone are parallel to the broad faces of briquettes, and the more recent sedimentary layers are nearer the upper faces. Specimens were water-rinsed, allowed to dry, and stored in polyethylene envelopes until use. Samples were individually protected by plastic bubblewrap during transportation to and from the field sites.

Exposure racks for stone samples at field sites (Figure 3) are designed to support the briquettes at a 30° angle to the horizontal, inclined toward the south. The rack and wedge-shaped sample designs ensure that the samples are securely protected from displacement (e.g., by wind), while avoiding binding forces and providing a practical minimum of contact between specimens and the Lucite rack material (9). Rack locations are in secure areas, and frequent inspections are made and reported by site operators, who brush away interfering material such as bird droppings and who also periodically remove and replace specimens according to a Site Management Plan (9). Samples removed are sealed, padded, and shipped by ground transportation to participating laboratories for specific tests.

Surface Chemistry. The schedule of sample exposure during the first year of the test program was as follows: for each three-month period, briquettes were exposed in racks and then replaced with fresh samples for the following three-month period. An additional set of briquettes was exposed for the entire year, and other samples present in the racks are part of a planned multiyear exposure program. Quarterly withdrawals for chemical analysis are to be made for two years, after which a cumulative annual frequency is to be maintained. One set of samples is continuously exposed for posterity or for unforeseen developments. Three samples of each stone type are employed in each cycle of tests. The plan is subject to change in the light of analysis results as they are produced.

The surface chemistry analysis protocol utilizes methods developed for the subject program (10). Layers 0.25 mm thick are sequentially removed from the stone briquettes at the surface locations illustrated in Figure 2. The process of removal, shown in Figure 4, produces powder samples weighing approximately one gram per layer. Weighed portions of each powdered layer sample are subjected to an extraction procedure to recover the leachable anions for analysis by ion chromatography (IC). A separate portion of each powder can be dissolved and analyzed by inductively coupled plasma spectroscopy (ICP); however, because the ICP analyses have so far indicated little cation change at the stone surfaces, this type of analysis is currently being more selectively applied.

The exposed and analyzed briquettes, as well as residual powders and solutions, are archived for further study, if needed.

Surface Roughening/Recession. Two approaches to measurement of surface recession were explored. One method applied a classical form of optical interferometry to optically smooth samples of stone in an attempt to measure results after brief exposure of the surfaces to rain; this approach was only partially successful and will be only

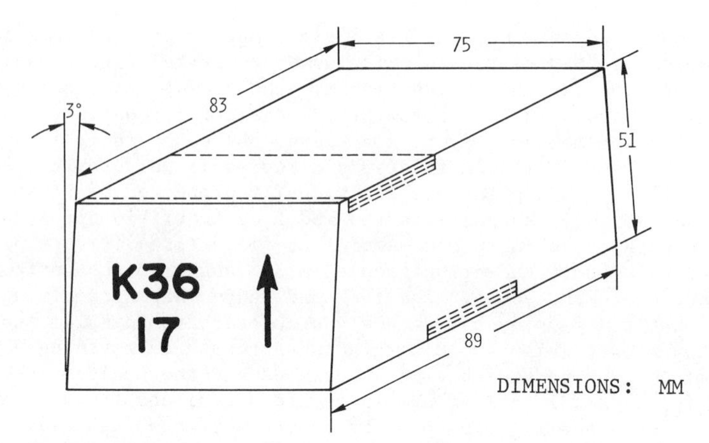

Figure 2. Dimensions of wedge-shaped briquettes employed as specimens of marble and limestone. Also shown are locations of areas sampled in layers to assess depth of chemical alteration caused by atmospheric exposure.

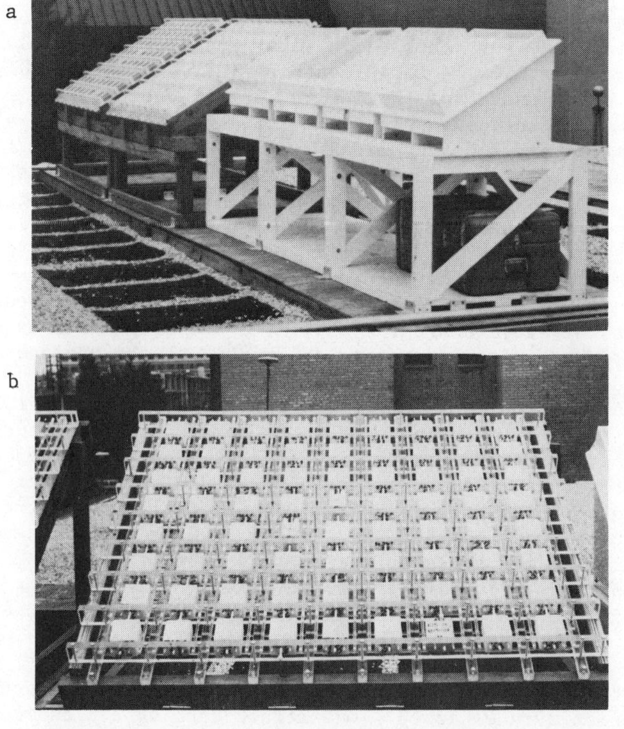

Figure 3. Stone briquettes arrayed for exposure to atmospheric acid deposition and weathering at field test site. (a) side and (b) front views.

briefly summarized. A second method employed laser-holographic moire profiling, developed by Dr. C. Sciammarella of the Illinois Institute of Technology (IIT) and ANL (11-12), and was successfully applied even to the relatively unsmooth, commercial surfaces, for tests in which a substantial exposure to rain is experienced by samples.

In the first approach noted above, the prepared stone specimen surfaces were protected over a portion of their areas by means of a Parafilm M seal which was held in place by a plate of inert material. The assemblies were then exposed briefly in the North Carolina test rack until 38 mm of rain (pH 4) had occurred during exposure. The assemblies were returned to the laboratory, and the samples were examined in a Twyman-Green interferometer in an attempt to measure the height of the step expected between protected and unprotected areas. Some variations in elevation were observed, but results were extremely irregular (an average height of 0.25 micrometers was roughly estimated for marble, and limestone seemed unaffected). For a following test, the sample design was modified to provide two protected zones on each sample's upper surface, separated by a broad and initially flat zone to be exposed to rain. It was planned to employ a Fizeau interferometer to measure changes in surface profiles rather than step heights; however, following a prolonged exposure of assemblies at the test site to accumulate 38 mm of rain exposure, it was found that a substantial upward bowing of the upper surface of both marble and limestone had occurred. This effect was not readily accommodated in data reduction. The dimensional instability of the material has not yet been overcome to enable recession measurements of such high sensitivity (13).

The design of sample assemblies used for the laser-moire approach is illustrated in Figures 5 and 6. Portions of the test surface are protected by Teflon bars that have machined recesses to avoid contact with reference (unexposed) areas of the sample. For some samples, the Teflon is supplemented with Parafilm M gaskets. Nylon fasteners are used where required near specimens, and stainless steel screws fasten together the parts of the Lucite holders. Although some exchange of liquid water between unprotected and protected zones is unavoidable for porous samples having unsmooth surfaces, the liquid is uncontaminated by materials of the holders and may be equilibrated with the specimens; no difficulty has been apparent.

Annual and posterity sets of laser-moire specimens of both stone types have been provided for all four test sites. One year of exposure has been accumulated at the North Carolina and New Jersey sites, and several months' exposure at the New York and District of Columbia sites.

Samples returned to the laboratory after exposure are measured by micrometer prior to laser-moire evaluation; micrometer measurements are reproducible only to within 13 micrometers because surface roughness produces an unusually high sensitivity of measurements to placement of the micrometer. Following laser-moire measurements, the samples have in some instances been checked by an electronic dial gage profiling technique. Because there is risk of damage to samples by the stylus of the gage, its use is minimized; thus, the statistics are fewer than desired and far less comprehensive than those of the completely nondestructive laser-moire method. Samples are returned to test sites for continued exposure after measurements.

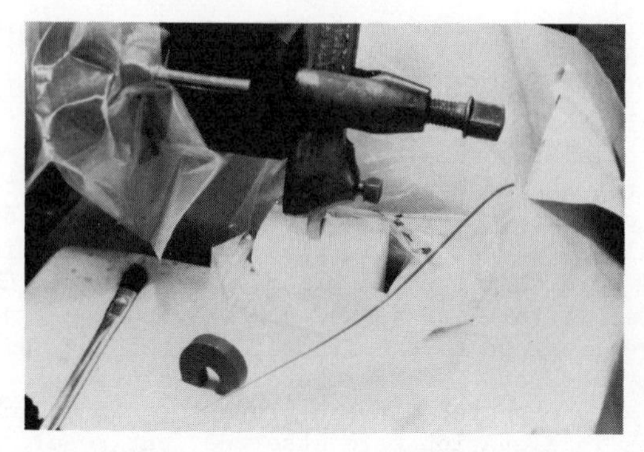

Figure 4. Stone specimen mounted in shaping machine for surface sampling. Tungsten carbide tool bit reciprocating horizontally precisely removes layers of stone, producing powders for chemical analysis.

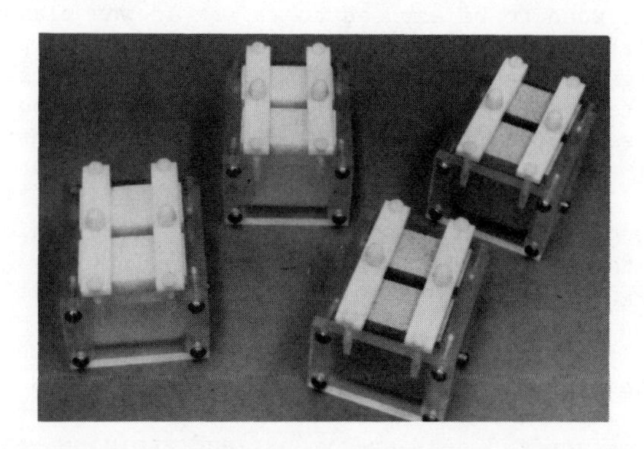

Figure 5. Sections of briquettes in Lucite holders for atmospheric exposure in test racks prior to measurements of stone surface roughness and recession. Teflon bars protect portions of surfaces for reference in measurements by laser holographic moire profiling method.

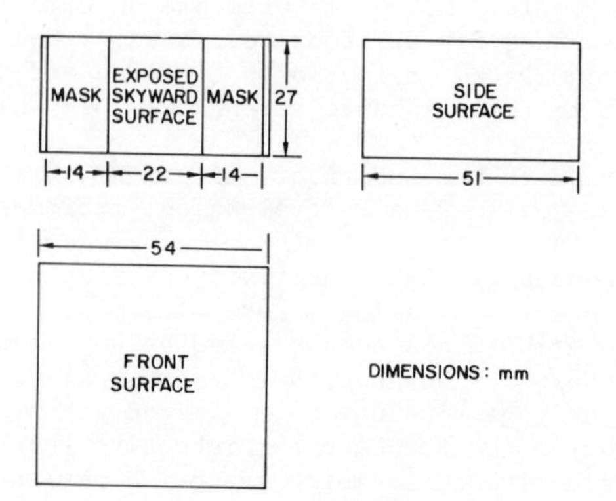

Figure 6. Dimension drawing of stone specimens shown in Figure 5.

The laser-moire method employs a helium-neon source to produce a double exposure hologram from which an interferogram is later made. The angle of incidence of the light on the specimen is changed between the two exposures by an accurately measured amount. Optical filtering is used to eliminate light randomly scattered from the rough sample surface, providing a final profile of the average surface (12). The amount of sample area considered can be varied by changing the magnification used, and this capability is utilized in roughness monitoring. For recession measurements, the entire upper surface of the sample is profiled by evaluating phase changes of the interference fringes as a function of location. In evaluations of both roughness and recession, the interferograms are digitized, and data reduction is performed automatically by computer routines developed especially for this program.

Dimensional instability of the stone materials on a microscale, mentioned above, has been observed by the laser-moire method as well as by the Fizeau and electronic dial gage methods. However, the difficulty has not been apparent in many of the samples, and the relative effect of the instability may diminish as the magnitudes of roughness and recession increase with exposure time.

Weight Change. Briquettes used for monitoring weight changes were prepared with rounded edges and corners in order to minimize errors caused by inadvertent damage during handling. After fabrication, the samples were rinsed under flowing distilled water as surface powder was removed with a soft bristle brush and then placed in a 45°C drying oven for one week. Briquettes were then allowed to equilibrate with air (at 22°C and <40% relative humidity) overnight in a balance room prior to weighing on a Mettler Type B5 C1000 balance that can resolve 0.2 mg. After sample exposure at field sites, the drying and equilibration procedures were repeated prior to reweighing. Although the procedure appears to be adequate, some further study of the drying technique and the use of a drybox in weighing are planned. However, the use of conditions that do not greatly exceed those of natural drying will be maintained.

After weights are determined, the samples are sent from Argonne National Laboratory (ANL) to the National Bureau of Standards (NBS) for measurements of color change; they are then returned to field sites for further exposure.

Sample sets are provided for one-, five-, and ten-year cycles of continuous exposure, with an extra set for posterity. Another set is shipped, handled, dried, and weighed with the test samples but is never exposed outdoors, as a test of possible errors caused by the manipulations. Three briquettes of each stone type comprise each set. Control samples are also retained at the laboratories; they are processed and measured with the sets that have been exposed.

Results

Initial results are presented below. In general, correlations with environmental conditions for results of short-term tests are deferred to a future report. Typical conditions of previous years for the test sites are assumed, pending the availability of environmental data pertinent to the subject exposures.

Surface Chemistry. Stone surface chemistry tests began on these dates in 1984: May 25 in North Carolina, June 5 in New Jersey, June 19 in New York State, and August 11 in the District of Columbia.

Results of IC analyses of limestone samples after the first three-month exposure are plotted in Figure 7, a-f. Concentrations of sulfate, nitrite, nitrate, chloride, and fluoride anions are given in micrograms per gram of stone. Results from skyward surfaces of three briquettes are shown in each of Figures 7 a-e: HU4-06, IU2-01, etc. are briquette names. For each briquette, results from three layers (Figure 2) are given, with the outermost layer designated #1. Consistent trends in concentration with stone depth were not evident for NC and NY, which have the lowest levels of ambient sulfur dioxide and nitrogen dioxide among the four sites (e.g., 3 ppb each of the two pollutants on an annual average for 1983 at NY). An incipient profile of sulfate concentration may be seen in the graphs for NJ (7 ppb ambient sulfur dioxide, average 1983), and clearly higher values of sulfate were found toward the outer surfaces of the briquettes from DC (16 ppb ambient sulfur dioxide, average 1982). Substantial profiles of sulfate were also present on the unexposed controls (Figure 7 e), although sample GU7-07 is regarded as spurious -- the high chloride levels show exceptional contamination. (Note: by a consensus decision of the stone investigators, surface chemistry samples were installed in the "as-received" condition, i.e., were not precleaned of anions.)

Results from controls imply that the skyward surfaces of samples from NC, NJ, and NY were cleaned by the exposure. This cleaning does not suggest that no damage was experienced, but that the sulfate profiles are dynamic features resulting from simultaneous reaction with acid sulfate in rain, rinsing of product from the briquette, flushing of product through the porous limestone, and subsequent precipitation of product from liquid at surfaces during drying. Additional sulfation is thought to occur from ambient sulfur dioxide during drying (14). Figure 7 f shows results from the back surface of one briquette from DC: concentrations of all anions (except nitrite) were exceptionally high. This effect may arise from the lack of rinsing by precipitation and as a result of the gradual flushing-through and drying processes, as well as from reaction with ambient acid species. Several analyses of back surfaces, done subsequent to those reported here, have shown results similar to those of Figure 7 f; however, generalizations cannot yet be made. The sulfate results are qualitatively in accord with those obtained by the U.S. Geological Survey (USGS) at these field sites by means of infrared sensing (15). The appearance of brown stains on limestone back surfaces after some (but not all) quarterly exposures at a given site is a feature worthy of note and continued attention.

Results of IC analyses for skyward surfaces of marble briquettes are shown in Figure 8, a-d. The format of Figure 8 is similar to that of Figure 7. No significantly elevated concentrations of anions were found except in outer surface layers of some of the samples. Sulfation levels above the detection limit were found only on the NJ and DC samples. One briquette from DC was exceptional, showing increased nitrate levels in the outermost two of the three layers. Vehicle emissions undoubtedly contribute to the relatively high nitrogen dioxide level (36 ppb annual average in 1982) in the

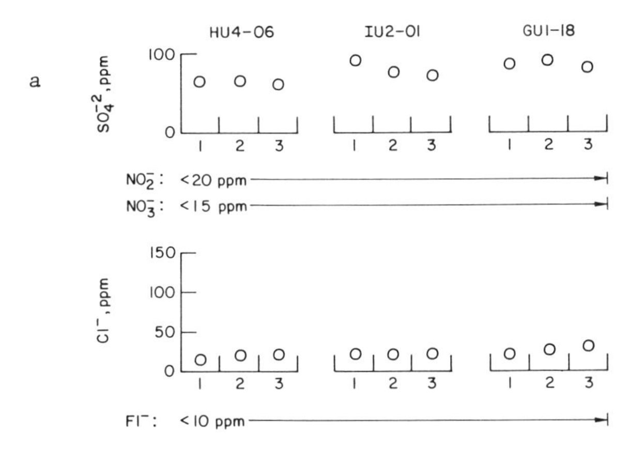

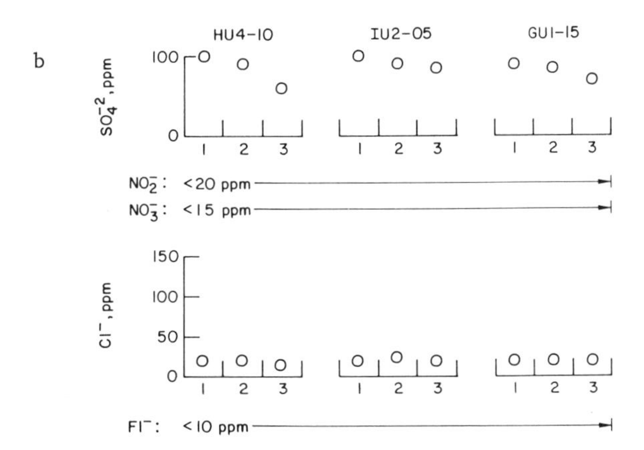

Figure 7 a and b. Results of ion chromatographic analyses of limestone briquette surfaces after three-month exposure at field sites in (a) North Carolina and (b) New Jersey.

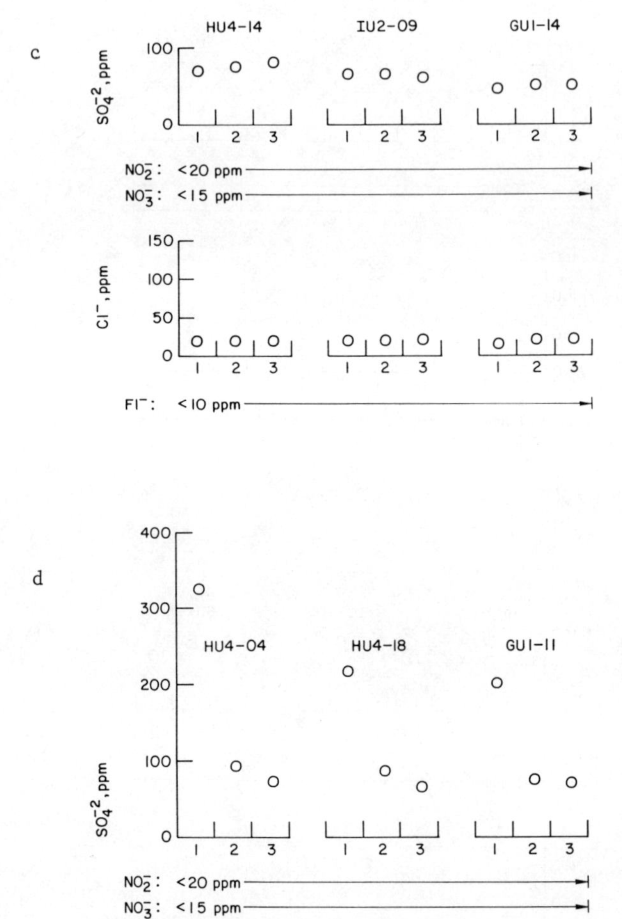

Figure 7 c and d. Results of ion chromatographic analyses of limestone briquette surfaces after three-month exposure at field sites in (c) New York State and (d) District of Columbia.

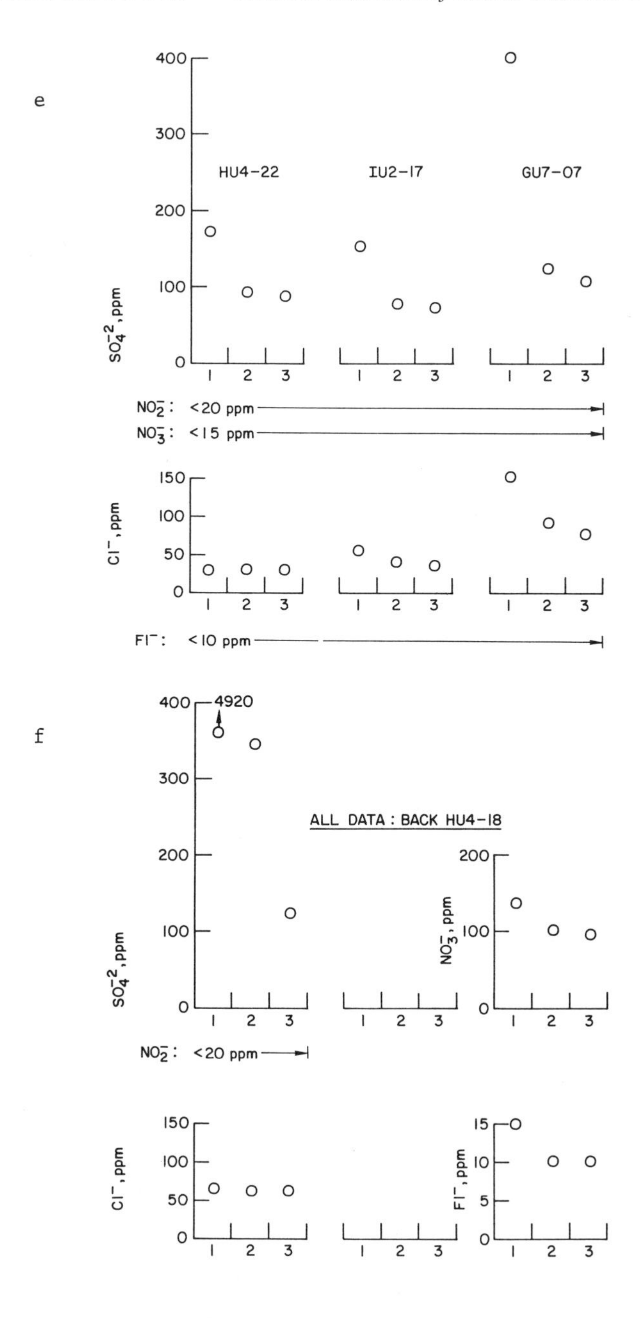

Figure 7 e and f. Results of ion chromatographic analyses of limestone surfaces (e) from control samples not exposed to outdorr conditions and (f) from back surface of briquette from District of Columbia.

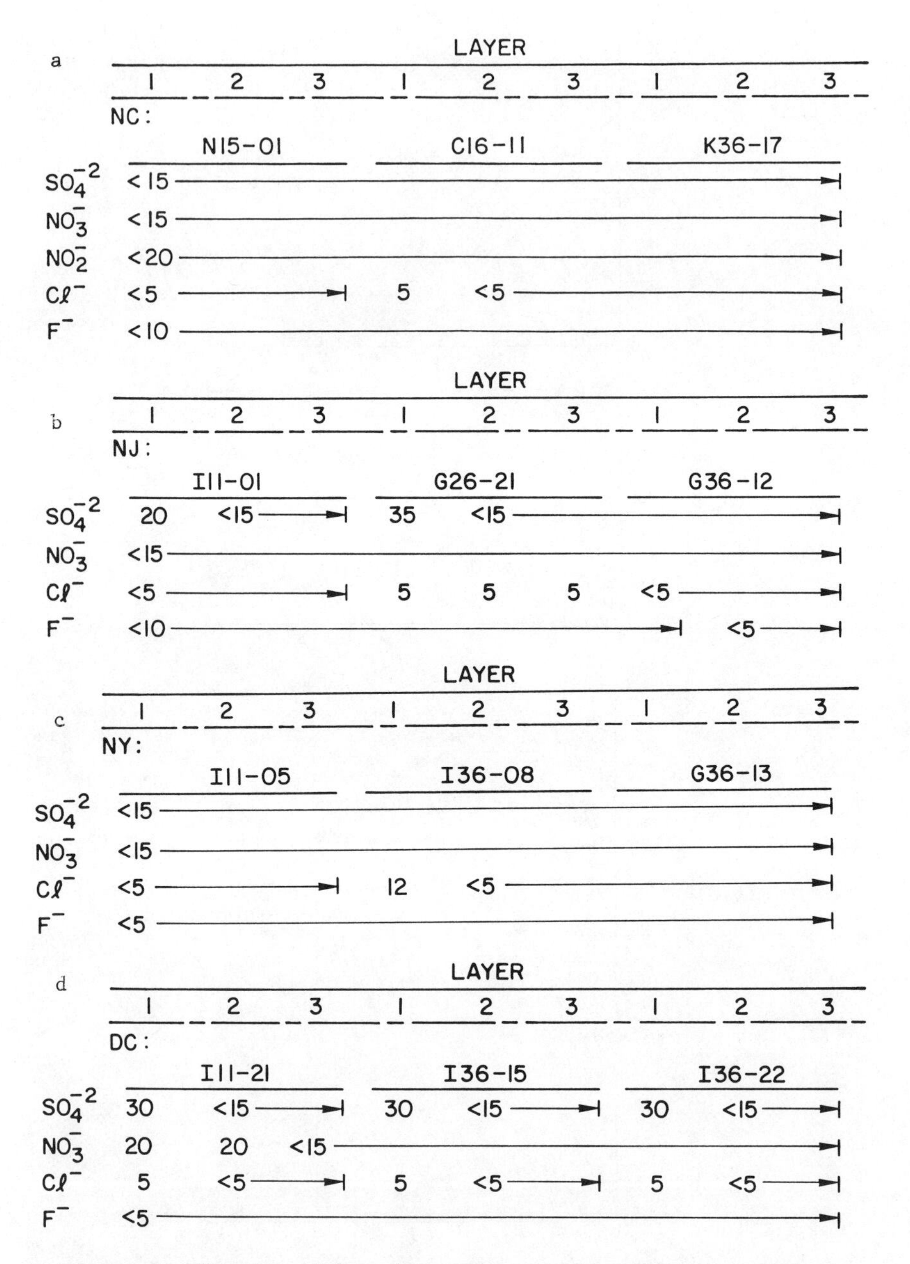

Figure 8. Results of ion chromatographic analyses of marble briquette surfaces after three-month exposure of samples at field sites indicated. (a) North Carolina, (b) New Jersey, (c) New York State, (d) District of Columbia.

atmosphere of DC, which is the only urban site of the four. Marble control samples showed no anion levels above the detection limits.

The frequency of occurrence of sulfate concentrations between 20 and 40 ppm on the skyward surfaces of marble is pointed out for discussion at a later point. Analyses subsequent to those illustrated continue to show values typically in this range. One analysis of a back surface layer after three months at DC, however, has shown 242 ppm sulfate; and a back surface layer after one year in NJ showed 2710 ppm sulfate.

The ICP analyses for Mg, Al, Cd, Mn, Ba, Be, Ni, Pb, Co, Cr, Cu, Fe, Sr, V, and Zn showed the levels of these constituents originally present in the stone; in results thus far available, little enhancement or depletion of these cations was noted after the exposures.

Surface Roughness/Recession. Results of laser-moire measurements of samples from NC and NJ after one year of exposure are given in Tables I and II, respectively. Although the aforementioned problems with dimensional instability appear to have affected at least two of the eight specimens, the remainder show values of recession comparable to expectations, based on weight-change monitoring (see below) and on the runoff chemistry. Roughness values are shown as information only and remain to be compared with initial values to be derived from early holograms.

Tables III and IV show results obtained for marble returned from NY and DC after exposures of less than one year. (The early returns were required owing to difficulties with baseline data previously taken.) Average values of recession are somewhat higher than might be expected on a time basis, when compared with annual results from NC and NJ; but the comparison will be more meaningful after correlation with environmental and rainfall factors is available. Again, dimensional instability was an interfering factor for marble from location M in the source block (compare specimen numbers containing an "M" in Tables II and III).

Electronic dial gage measurements of recession were made on the samples of Tables III and IV in an attempt to verify the results of the relatively new laser-moire method. Although the values found were somewhat lower than those from laser-moire, the statistics of the latter are far more voluminous and the sample area surveyed much more comprehensive. Efforts to improve verification are continuing.

Weight Change. Gravimetric results for samples from the four sites after the first year of test exposure are listed in Table V. Exposure periods were initiated coincidentally with those listed above for surface chemistry samples. For marble, the range of recession values implied by the weight changes is somewhat greater than that found by direct recession measurements and by the runoff chemistry. Weights of marble "simulation" samples were unchanged within 3 mg. For limestone, the weight losses were far in excess of expectations based on direct measurements of recession and on runoff chemistry. It is thought that limestone powders retained in the porous stone were gradually flushed away during test; some originally intact grains of the stone also may have been flushed away after becoming detached by dissolution of cementatious material and by mechanical effects of the environment. A lesser change is anticipated for the

Table I. Stone Surface Recession and Roughness after First Year of Exposure in North Carolina*

Specimen	Recession Avg, µm	Std Dev	Confidence (T-Test)	RMS Roughness Unprotected Avg, µm	RMS Roughness Protected Avg, µm
Marble					
1A K36-21	10.5	27	.995	6.09	6.01
1A C16-17	13.7	27	.995	4.41	5.09
Limestone					
1A HL4-21	8.8	26	.975	13.14	15.26
1A IU5-23	-0.2	5	.0	14.00	13.96

*Exposed June 25, 1984 to May 16, 1985.

Table II. Stone Surface Recession and Roughness after First Year of Exposure in New Jersey*

Specimen	Recession Avg, µm	Std Dev	Confidence (T-Test)	RMS Roughness Unprotected Avg, µm	RMS Roughness Protected Avg, µm
Marble					
2A M15-17	1.5	6	.95	7.37	6.15
2A C16-17	14.0	18	.995	4.04	4.86
Limestone					
2A HL4-02	13.2	22	.995	11.1	12.02
2A IU5-23	16.6	22	.995	13.5	15.58

*Exposed July 4, 1984 to June 7, 1985.

Table III. Marble Surface Recession and Roughness after Three Months of Exposure in New York State*

Specimen	Recession Avg, µm	RMS Roughness Unprotected Avg, µm	RMS Roughness Protected Avg, µm
3A K15-15	4.2	6.19	5.56
3A L26-06	4.46	9.36	7.59
3P H11-20	6.12	6.41	5.96
3P M15-08	(-2.45)	5.27	4.95

*Exposed July 9 to October 17, 1984.

Table IV. Marble Surface Recession and Roughness after Two Months of Exposure in District of Columbia*

Specimen	Recession Avg, µm	RMS Roughness Unprotected Avg, µm	RMS Roughness Protected Avg, µm
4A H36-09	4.11	5.19	4.9
4A L26-14	4.63	6.52	6.42
4P L15-09	2.36	5.82	7.71
4P L26-06	10.5	9.81	7.45

*Exposed August 11 to October 11, 1984.

Table V. Weight Loss* Results after First Year of Exposure

Test Site	Exposed Briquettes		Field Control Briquettes	
	Marble	Limestone	Marble	Limestone
NC	0.2785	1.3480	-0.0025	0.0501
	0.2500	1.5674	-0.0026	0.0323
	0.2915	1.6392	-0.0016	0.0383
NJ	0.4520	1.8482	0.0002	0.0436
	0.4040	1.7099	0.0000	0.0294
	0.4483	1.8569	-0.0002	0.0215
NY	0.2908	1.5769	0.0012	0.0651
	0.2757	1.5006	-0.0002	0.0468
	0.2965	1.7524	0.0004	0.0571
DC	0.3172	1.4169	0.0013	0.0667
	0.3091	1.2900	0.0009	0.0703
	0.3070	1.5560	0.0005	0.0726

*Units = grams/briquette. Upper surface area: 63 sq. cm.

second year's exposure of limestone. Measurements of simulation samples indicated that up to 5% error in the weight losses shown for limestone might be due to differences in briquette dryness at the times of weighing. The effect of retained reaction products on sample weights appeared to be very small, as indicated by the results of surface chemistry (given above).

Discussion

Emphasis in the field tests of stone briquettes is on the long-term damage suffered by exposed marble and limestone. Because the pH variations for the existing test sites are principally variations between one rain and the next rather than from site to site, the separation of the pH effect cannot presently be achieved (except possibly in runoff chemistry on individual rains). The annual average acidity of rain at each site is near pH 4. Measures discussed toward isolating the effect of rain pH on materials damage over the long term have included (1) the addition of a "clean" test site or (2) the provision of additional samples under roof but otherwise exposed to the environment of an existing test site and showered with simulated rain at pH 5.6 during external rainfalls, in the quantity and at the time and rate of each rain. Cooperation with Canadian researchers is being explored to arrange exposures under "clean" conditions; and the use of sheltered samples, already being instituted for metal samples, is planned for stone briquettes.

An objective of the program is to derive for each stone type a damage function in which the materials effects of the various components of acid deposition are quantified separately and apart from other damaging agents (such as natural rain at pH 5.6). The general form of a function might be as follows:

$$\text{DAMAGE} = [G(AH+B+\ldots) + C(SOX\ldots) + D(\ldots)]K$$

where G (gage) = rain quantity, H = concentration of excess hydrogen ion (beyond pH 5.6), B = the dissolution by natural rain, and K = a factor greater than one to account for loss of solids as the more readily soluble portions of the stone preferentially dissolve; the other quantities are open for discussion (below). An ultimate function may be less simple if it is found that synergism is significant; indeed, synergistic effects with mechanisms of mechanical damage from freeze-thaw-cycles and from moisture- and temperature-cycling of retained reaction products are expected. Perhaps an acid fog component will be needed, and a term for nitrogen dioxide is likely to be desired. A term under discussion currently is that for sulfur dioxide, C(SOX...), above.

For the sulfur dioxide term in a marble damage function, attention is given at present to the effect during the period between successive rains. In a formulation that is currently under consideration, the authors attempt to account for the frequent finding of about 30 ppm of sulfate on skyward surfaces and the finding of much greater concentrations on ground-facing surfaces (see Surface Chemistry Results, above). Additionally, the authors are cognizant of Spedding's results with limestone in a moist atmosphere containing sulfur dioxide, where the total sulfur uptake was reportedly rather insensitive to concentration in the gas and to time after brief exposure at room temperature (14). Moisture in the stone might be expected to mitigate the blocking effect somewhat, and rain would be expected to dissolve the product from skyward surfaces of stone samples. A two-part term for marble damage by ambient sulfur dioxide is thus contemplated, to account for (1) initially rapid reaction on freshly rinsed surfaces during dryout of the stone following a rain, and (2) slower reaction at surfaces having acquired a film of calcium sulfate. If the two parts of this term are linearized and fit to the data initially obtained, tentative estimates of the effects of ambient sulfur dioxide can be made: 30 ppm of sulfate might be formed on freshly rinsed surfaces during the initial period of drying following each rain; and subsequently, sulfate may be gained at a rate of the order of 1 ppm/day per ppb of sulfur dioxide in ambient air (estimated from surface chemistry results for ground-facing surfaces of marble, and assuming a linear dependence on gas concentration during the period of slower reaction). Approximately two micrometers of stone surface recession per year (if K = 1) may thus be caused by ambient sulfur dioxide acting on skyward surfaces of marble in DC under typical conditions, and less recession from this cause at the other field sites (and on ground-facing surfaces at any of the sites, unless mechanical damage mechanisms are operative). It appears that the effect on skyward surfaces, even in DC, should be much smaller than the effect of the acid rainfall.

It is emphasized that the foregoing discussion applies to skyward, (i.e., well-rinsed) surfaces, and not to those which are primarily the sites of evaporation of liquid from adjacent areas nor to those more completely sheltered from incoming liquid water. At surface areas where substantial amounts of reaction products are retained in the stone surface, mechanical mechanisms of damage are thought to apply (1). Effects on sheltered, partially sheltered, and vertical surface areas are subjects of planned future work. Studies of the possible synergism of nitrogen dioxide effects with those of sulfur dioxide dry deposition also are being planned.

Conclusions

Measurements of marble and limestone damage in the form of surface recession have shown results in approximate agreement with those of runoff chemistry (circa 15 μm/y). Marble weight changes indicated material loss in a range comparable to that shown by recession and runoff chemistry. Limestone weight losses were anomalously high during the first year of exposure. Surface chemistry indicated that weights of retained reaction products were relatively small but that surface concentrations of reaction products were in accord with expectations based on air pollution concentrations at the stone field test sites. Surface chemistry also provided evidence to help resolve issues related to models of damage, and verified that an increased emphasis on partially sheltered areas of stone is needed.

That acid deposition contributes substantially to the marble and limestone damage has been demonstrated in runoff studies. This finding is given additional support by the briquette surface chemistry, and is in accord with previous work as well as *a priori* expectations. Uncertainties in the quantitative contributions of the important species of deposition remain large at present, and efforts to reduce these uncertainties continue. It is apparent that materials damage rates are worthy of attention but are sufficiently gradual to allow time for development of further understanding as a basis for control strategies.

Acknowledgments

The authors are grateful to C. Moore (NC), P. Fiechter (NJ), and R. Masters (NY) for participation at field test sites and to J. Sieler (NBS) and N. Veloz (National Park Service) for installation of stone-exposure racks at the sites. We wish to thank also F. Williams and E. Huff, who performed the IC and ICP analyses under the leadership of K. Jensen (ANL), and R. Lee (ANL), who participated in development of drying procedures and in sample preparation. Computer program development by Mansour Ahmadshahi (IIT), under C. Sciammarella's leadership, was of much value in the work. The projects described in this paper were monitored at ANL by D. Kupperman, who contributed to the initiation of surface roughness/recession monitoring, and K. Reimann of the Materials Science and Technology Division. The valuable advice and support of the Stone Study Committee organized by NPS and of S. Sherwood and R. Herrmann, Research Coordinator and Chairman, respectively, of the NAPAP Materials Effects Group, also are gratefully acknowledged.

Literature Cited

1. Lewin, S. Z.; Charola, A.E. Scanning Electron Microsc. 1978, 1, 695-703.
2. Proc. 3rd Intl. Congr. Deterioramento E Conservazione Della Pietra, 1979.
3. Preprints Intl. Symp. on the Conservation of Stone, 1981.
4. "Conservation of Historic Stone Buildings and Monuments," National Materials Advisory Board, National Academy Press: Washington, D.C., 1982.
5. Amoroso, G. G.; Fassina, V. "Stone Decay and Conservation"; Materials Science Monographs No. 11, Elsevier: New York, 1983.
6. Livingston, R. A.; Baer, N. S. Proc. 6th World Congr. on Air Quality, 1983.
7. Ross, M; Knab, L. "Selection, Procurement, and Description of Salem Limestone Samples Used to Study Effects of Acid Rain"; Natl. Bur. Stand. NBSIR 84-2905, 1984.
8. Ross, M. "Description, Selection, and Procurement of Shelburne Marble Used to Study Effects of Acid Rain"; USGS Open-File Report 85-XXX, in press.
9. Doe, B., et al. "Acid Rain Site Management Plan"; NPS Preservations Assistance Div., 1984.
10. Jensen, K.; Huff, E.; Williams, F.; Youngdahl, A. "Developmental Procedure of Stone Surface Chemical Analysis"; Report to NPS from ANL, in press.
11. Sciammarella, C. Opt. Eng. 1982, 21(3), 447-57.
12. Sciammarella, C; Youngdahl, A. "Application of Laser Holographic Interferometry to Monitor Stone Surface Roughness and Contour Changes"; Report to NPS from ANL, in press.
13. Primak, W. "Dimensional Stability, Summary in OM85 Basic Properties of Optical Materials"; NBS Special Publ. 697, 1985; p. 36.
14. Spedding, D. Atmos. Environ. 1969, 3, 683.
15. Kingston, M., USGS, personal communication, 1985.

RECEIVED March 3, 1986

Elemental Analysis of Simulated Acid Rain Stripping of Indiana Limestone, Marble, and Bronze

K. M. Neal, S. H. Newnam, L. M. Pokorney, and J. P. Rybarczyk

Department of Chemistry, Ball State University, Muncie, IN 47306

Recent work has shown that the Midwest and specifically Indiana is experiencing significant acidic wet deposition. While the natural abundance of limestone in Indiana soil inhibits the major aquatic and forestry damage observed in the Northeastern United States, other less dramatic yet significant effects are being observed in Indiana, primarily degradation of structural materials. Samples of structurally popular Indiana limestone together with marble, micrite limestone, and bronze have been laboratory treated to the equivalent of ten years of leaching at pH=3.0, 4.0, and 5.6 rainfall. The run-off has been elementally analyzed to determine the leaching rates. Also, the effects on limestone and bronze engraving and embossing caused by similar long-term acidic exposure are also provided. These results are then correlated to actual field observations within Indiana.

The State of Indiana has been closely linked to the acid rain controversy for two primary reasons. The first association arose because of Indiana's ranking as the third largest sulfur dioxide emitting state and the seventh greatest nitrogen oxide emitter, both substances being the apparent precursors of acid rain. In addition, Indiana contains seven of the largest 50 sulfur dioxide emitting stationary power plants in the U.S. (1,2).These facts have led to Indiana being labeled as a primary geographic acid rain source. However, besides being merely a source of acid rain, Indiana also receives significant acid precipitation. The effect of this precipitation on one of Indiana's most commercial products is its second link to acid rain. Indiana building limestone has historically been a popular construction material not only across the nation, but particularly within the State of Indiana. Severe degradation of this popular material has recently been dramatized with the initiation of a $10.5 M restoration project of the most visible symbol of the City of Indianapolis, Monument Circle. This circa-1900 limestone structure, the heart of the metropolitan area, showed severe structural decay, which was attributed primarily to acid rain (3).

0097–6156/86/0318–0285$06.00/0

The majority of studies examining the effects of acid rain have focused on the Northeastern U.S. and have largely ignored the Midwest. Fish kills, dying lakes, and forest damage in the Northeast have been fairly dramatic and widely publicized (4-15). Much work has been concentrated on ecosystem perturbations, such as the leaching of nutrients and trace metals from soils (16-19). Concern for the forest canopy has resulted in a plastic polymer coating to be sprayed on individual urban trees for protection (20). Only recently has work been concentrated on the effects of acid deposition on human health (21,22) and on structural materials (23-25). Ongoing work by the Federal Interagency Task Force on Acid Precipitation centers on acid effects on materials (26,27). The study primarily focuses once again on effects on the East Coast, with the effects on materials in the Midwest being largely ignored. The purpose of this work was to provide data to fill that void.

Recent precipitation parameters common to Indiana have been used to establish the experimental criteria (28). These included such data as acid content, meteorological data and conditions, chemical composition of the precipitation, etc. Only one degradation parameter was focused upon: the cause and effect interaction of pure H^+ in leaching structural materials common in Indiana, including bronze, marble, and Indiana limestone. Quantitative data on the contribution of H^+ can then be selectively isolated from the overall effects of the complex environmental matrix. With this initial step, the effects of further matrix components (whether cations, anions, or compounds), acting both individually and/or synergistically, can then be pursued.

Environmental Factors Governing This Experiment

A three year study monitoring acid precipitation across Central Indiana has just been completed by the authors (28) and the local International Rotary Club who donated the manpower for sample collection. Over 3000 individual samples were collected during approximately 300 individual precipitation events. Using acceptible analytical and statistical protocol, the samples were characterized by meteorological, physical, and trace chemical composition data to determine possible sources of the acidity. The study yielded pertinent environmental data that set the parameters for this work.

The volume-weighted pH average for precipitation of all types (rain, snow, fog) across Central Indiana for the 1983-85 study was 4.02. The average winter pH was 4.27, and the average summer pH was 3.73. Summer rains showed the most severe acidity, with numerous individual events as low as pH=2.8 and routine values in the 3.1 to 3.2 range. Reasonable experimental pH limits utilized for this work were therefore set at pH=3.0 and pH=4.0. A third background control at pH=5.6 was also utilized, this being the generally accepted yet highly controversial pH of pure water caused by the dissolution of carbon dioxide gas at standard temperature and pressure conditions with the resulting carbonic acid/bicarbonate buffer equilibrium system (29).

The total water volume of all moisture measured at the Ball State University Weather Station for 1984 was 86.79, with 90 cm per year considered "typical" for this region of Indiana (30). The predominant wind direction for the above time period was South-

Southwest, with 71% of the total precipitation originating from the Southern (90°-270°) compass hemisphere. Thus, the maximum directional precipitation received by a structural surface would be 64 cm from the Southern hemisphere or 26 cm from the Northern. This work utilized 60 cm as a typical yearly directional exposure value.

The average sulfate ion concentration for these Indiana rains was in the 1.0 to 10 ppm range, while the nitrate ion levels were in 0.1 to 5 ppm range, with about a 2:1 ratio of sulfate to nitrate being typical. Since the relative composition of the acid constituentsvaried so widely and seasonally across Indiana, it would be erroneous to experimentally utilize a particular level of sulfuric or nitric acids. Therefore,to test only the effects of acid leaching on structural materials, and not the effects of the various acid anion salts, nitric acid was the only acid source in the simulated acid rainwater. By deleting the sulfuric acid, gypsum formation ($CaSO_4 \cdot 2H_2O$) on the limestone and marble was avoided. While it is probably the major catalyst of limestone degradation, the formation and subsequent peeling of the soft sponge-like gypsum was beyond the scope of this work. The removal of each gypsum layer exposes a virgin limestone surface to the acid leaching process, which would merely multiply the individual acid leaching processes of relatively stable surfaces that were studied in this work.

Experimental

Reagents & Equipment. All reagent solutions were prepared with doubly distilled and de-ionized water that was also filtered through a Gelman Sciences (Ann Arbor, MI) Water I purification system to an ion conductivity of 5 μmho/cm. The simulated acid rain solutions of pH=3.0, 4.0, and 5.6 were prepared in this water from concentrated nitric acid, Fisher ACS Certified Reagent Grade. The pH was adjusted using a Corning Model 5 pH Meter (Medfield, MA) that was continuously calibrated with a pH=4.01 Mallinkrodt Buffar buffer solution. Five standards each for calcium, magnesium, copper, and zinc were prepared from 1000 ppm Fisher Scientific Atomic Absorption standards. All standards, samples, and blank solutions at each pH were analyzed on a Perkin Elmer 306 Atomic Absorption Spectrophotometer (Norwalk, CN) according to standard acceptable methods.

Samples. Three standard stone samples were obtained from the collection of the Ball State University Geology Department. The primary sample was a specimen of Salem Limestone (calcite, $CaCO_3$), commonly known as Indiana Building Limestone, which was obtained in the Bloomington-Bedford area. Because of its popularity as a construction material, the Salem Limestone was the focus of this study. It was subdivided into three equal sections for the testing of leaching rates at pH=3.0, 4.0, and 5.6. Henceforth, these three samples will be labeled Salem 3.0, Salem 4.0, and Salem 5.6. The somewhat porous crystalline face of each surface was cut smooth and square, but not polished.

The second stone sample was common white Georgian marble (metamorphic calcite, $CaCO_3$). Since marble is a rearranged and stronger crystal modification of limestone, it should provide a contrast in acid leaching rates. This sample was square cut, and its face of relatively large crystalline grains was highly polished to a glass-like finish. It was only exposed to pH=3.0 water, and henceforth will be labeled as Marble 3.0.

The final stone sample was a fine-grained and densely-packed dolamite-like grey limestone Micrite from the Onondaga formation in New York. The large facial area of this sample was smooth-cut into which the words "ACID RAIN" were mechanically engraved. The average depth of these letters was 0.218 cm, which is comparable to the average depth of 0.268 cm for the letters engraved on a 1985 Veterans' Administration marble gravestone in Muncie. Micrometers and machinists' depth gauges were used to measure the letter depths periodically in the study. This sample was labeled Micrite 3.0.

The bronze samples were obtained in 30.5 cm x 30.5 cm sheets from the C.R. Hills Co. (Berkley, MI). The bronze alloy was cast as Alloy #220 with 90% copper and 10% zinc by weight, with trace lead and iron impurities. It was in sheet rolled form at 18 gauge (1.00 mm thickness) and with a density of 8.80 g/cm^3. Historically, bronze was classified as copper/tin alloy with a minimum 10% by weight tin, while brass was a copper/zinc alloy with at least 10% zinc. The contemporary classification of "bronze" has been expanded to encompass a much wider range of copper alloys with bronze-like structures, properties, and color, but which contain no tin. Thus, although it is commercially classified as bronze, the sample utilized in this work was historically brass.

This bronze sheet was machine cut into four strips, each with dimensions of 15.25 x 3.75 cm. The letters script "A" and block "T" were engraved into each strip, with the "T" having an average depth of 0.033 cm. Two of the bronze strips were leached with pH=3.0 water as duplicates, a third sample received pH=4.0 exposure, while the final strip as a control received pH=5.6 water. These samples will be labeled Bronze 1-3.0, Bronze 2-3.0, Bronze 3-4.0, and Bronze 4-5.6, respectively.

Procedure. The Micrite 3.0 was suspended vertically on a ringstand, with a 15 cm polyethylene funnel and 100 ml bottle assembly directly beneath to collect the water run-off. To simulate rain, the pH=3.0 solution was sprayed on the engraved stone using a conventional 400 ml handpump aerosol sprayer, which was set to deliver a medium droplet size mist at a distance of 20 cm from the stone, to ensure complete coverage of the face by the circular spray pattern. Only direct run-off from the stone was collected, with the remainder of the spray pattern disposed of as waste. The stone was sprayed over a ten minute "exposure" period or until 65 ml were collected in pre-calibrated bottles (the amount being equivalent to a 1.05 cm rain on the surface area of the Micrite). The calcium and magnesium levels were measured on the Atomic Absorption instrument. The stone was allowed to air dry between collections, with two hours being typical due to the porosity of the stone. This procedure was used for 300 simulated rains. The engraved "ACID RAIN" letters and all other dimensions were measured after every 30 events. Sample weight changes on this and other stone samples were not measured due to inconsistent water retention.

The Marble 3.0 and all three Salem Limestone samples were leached by a modified procedure. The spraying was replaced by setting each stone into a 14 cm petri dish together with 100 ml of the respective pH water for the standard 10 minute exposure time. The stone sides, in addition to the face, were thus also partially exposed to the water, and these side areas were included in the final exposure area calculations. Since each stone had different exposed surface areas but identical 100 ml rain volumes, each stone received differing simulated rain depths as follows: Marble 3.0 = 1.59 cm, Salem 3.0 = 1.73 cm, and Salem 4.0 & 5.6 = 1.96 cm. All leaching results were normalized to 1.00 cm to compensate for the variations.

The four bronze samples were each sprayed with simulated rain according to the procedure for the Micrite 3.0 above. Each bronze was sprayed until 30 ml was collected in the collection bottles, being equivalent to 1.0 cm of simulated rain. These water samples were then analyzed for copper and zinc on the AA instrument. An air dry time of 30 minutes was typical. Each Bronze 3.0 was subjected to 300 simulated rains, while the Bronze 4.0 & 5.6 received only 150. Engraved letter depths and overall dimensions of each were measured after every 30 events.

Results and Discussion

Physical Observations. Micrite: The Micrite 3.0 exhibited a noticable color change after 16 cm of rain, from a dark grey to a grey-white which whitened gradually throughout all 300 rains. No gross dimensional changes were detectable, with the exception of smoother and more rounded corners. After 50 rains, the surface of the engraved "ACID RAIN" letters was broadening, with the edges of each letter becoming less sharp and defined. This letter broadening progressively increased throughout the experiment. Table I summarizes the engraved letter depth changes as a function of simulated rain depth (60 cm being equivalent to one year of directional rain):

Table I. Depths of Engraved "ACID RAIN" Letters in Micrite 3.0

Total Rain Depth,cm	Average Letter Depth,cm	Std. Dev.
12	0.0194	0.0143
26	0.0575	0.0332
34	0.0746	0.0379
42	0.0879	0.0334
60	0.1016	0.0333
84	0.1026	0.0330
114	0.1051	0.0328
138	0.1064	0.0323
173	0.1099	0.0336
213	0.1137	0.0343
243	0.1162	0.0346
273	0.1187	0.0349

The letter depth measurements were made repeatedly at the same representative location on each letter. Relatively large but consistent standard deviations probably resulted from crystal structure inhomogeniety, variations in letter engraving, etc. Figure 1 illustrates two distinct rates of leaching. The greatest rate occured in the first 1½ year, with a 50% overall increase in depth. This is probably the result of the pulverization and weakening of the interior walls and floor of the letters caused by the engraving process. This weakened material could then be leached at a significantly more rapid rate than the surface Micrite. Following this stage, the depth rate increase slowed appreciably for the next 3½ year period to a consistent but very slight increase. At this point , the Micrite letter floor was probably virtually identical to the surface Micrite with very similar stripping rates. The engraved letters were still stress points, as shown by the continued letter broadening, resulting in a slight rise in the stripping plateau. In addition, the letter surfaces were receiving two-dimensional exposure on side and top versus only one-dimension for the normal surface. This could also contribute to the slight depth increase long-term.

From the above, it is projected that the letter depth change would eventually reach zero and become negative as the letters broadened and the surface leached slightly faster than the letter interiors, resulting in the letters "fading". This process is readily observed on U.S. Government Military Veterans' gravestones and other naturally "weathered" monuments. Depths of V.A. marble stones in Muncie's Beech Grove Cemetery were measured for the period 1930-1985. The results are summarized in Table II:

Table II. Depths of Engraved Marble Letters in V.A. Stones

Date	# Letters	Average Depth,cm	Std. Dev.
2-1985	13	0.2688	0.0489
6-1983	14	0.2778	0.0408
1978	14	0.3097	0.0231
1-1975	11	0.3380	0.0384
12-1969	11	0.2803	0.0131
1-1967	12	0.2646	0.0576
4-1961	14	0.2329	0.0357
5-1951	14	0.2054	0.0414
8-1947	12	0.1947	0.0460
11-1940	12	0.1770	0.0515
1-1935	9	0.0923	0.0172
11-1931	11	0.1259	0.0327

Random stones all with a westerly exposure were chosen at approximately five year date intervals. Standard deviations were probably large due to the wide variation of response to environmental factors and to the marble's own structural variations. At the same location, a granite family gravestone also offered a unique stripping record. Vertically arranged on the western face of this stone were embossed name and date lines: 1913, 1939, 1957, 1941, respectively top to bottom. Table III summarizes the average embossed letter heights above the stone face for the last nine letters of each line:

Table III. Heights of Embossed Granite Letters, Muncie, IN.

Date	# Letters	Average Height,cm	Std. Dev.
1913	9	0.0751	0.0021
1939	9	0.1270	0.0142
1941	9	0.1778	0.0210
1957	9	0.2362	0.0365

Figure 2 illustrates the rates of letter changes for both the V.A. marble engraved letters and the granite embossed letters. The correlation between the two rates was surprising in view of the greater bond strength of the granite crystal. The similarity could result partially from the greater stress on the protruding embossed letters. Of note are the accelerated changes for both the marble and granite in the 1935-40 period, possibly due to a local environmental perturbation. A close comparison of the experimental Figure 1 with the 1985-75 V.A. marble period of Figure 2 shows that the marble letters also increased in depth over the initial 10 year exposure, which is virtually the same as the experimental Micrite 3.0. However, as discussed above, the depth increase was then reversed after 10 years, and the letters broadened and faded due to surface decay.

Salem Limestone: After 9 cm of rain, both the Salem 3.0 & 4.0 limestone changed color slightly from light tan-brown towards a white-brown, becoming more pronounced with continued exposure. Pitting of the surface was obvious after 43 cm of rain and became progressively deeper and more numerous, with the Salem 3.0 showing the greatest effect. Surface area dimensions of the Salem 3.0 remained relatively constant, again with the rounding of corners. Sample depth decreased slightly by an average of 0.1 cm. The Salem 5.6 showed virtually no observable changes.

Marble: Marble 3.0 showed very light pitting surrounding the individual marble crystals, which was minimal compared to both the Salem 3.0 & 4.0. The Marble 3.0 showed no color change, and the surface remained relatively mirror-like. No changes in the physical dimensions of area or depth were detectable.

Bronze: The duplicate Bronze 3.0 samples both exhibited numerous pitting and discoloration areas. The random pitting may be the result of the zinc dissolving to provide cathodic protection for the copper, due to zinc's large 0.76 V oxidation potential. The pitting occured across the entire surface and inside the letters. Since the strips were leached vertically, the solution accumulated at the bottom while drying, causing a grey-green area of severe discoloration. This material was probably a combination of green $Cu_2(OH)_2CO_3$ and blue-grey $Zn_2(OH)_2CO_3$, both of which are natural oxidation products. The overall color of both samples changed from a dark brown-gold to a light white yellow-gold. No detectable changes in dimensions of the strips or engraved letters were detected, except at the bottom of the block "T" where a slight depth increase of 0.002 cm was observed in both samples. This probably resulted from the vertical positioning which allowed the base of the "T" to retain the water.

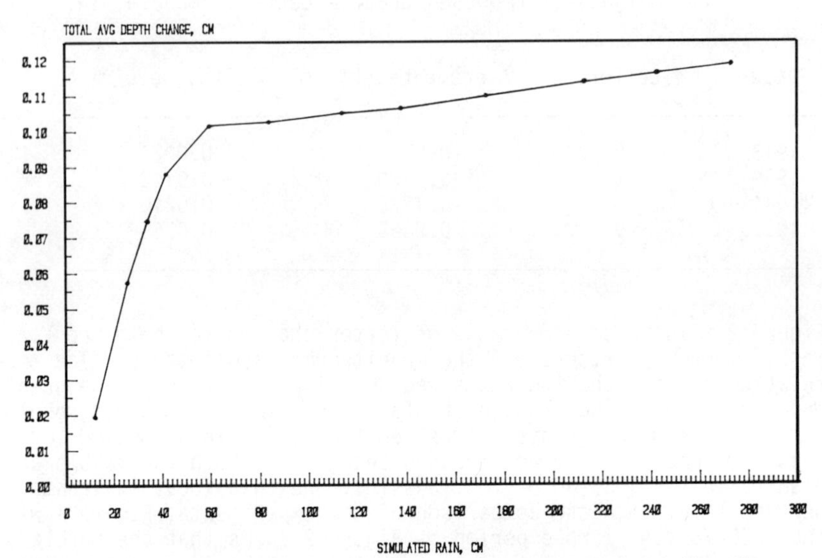

Figure 1. Letter depth changes for engraved letters for the Micrite sample at pH=3.0.

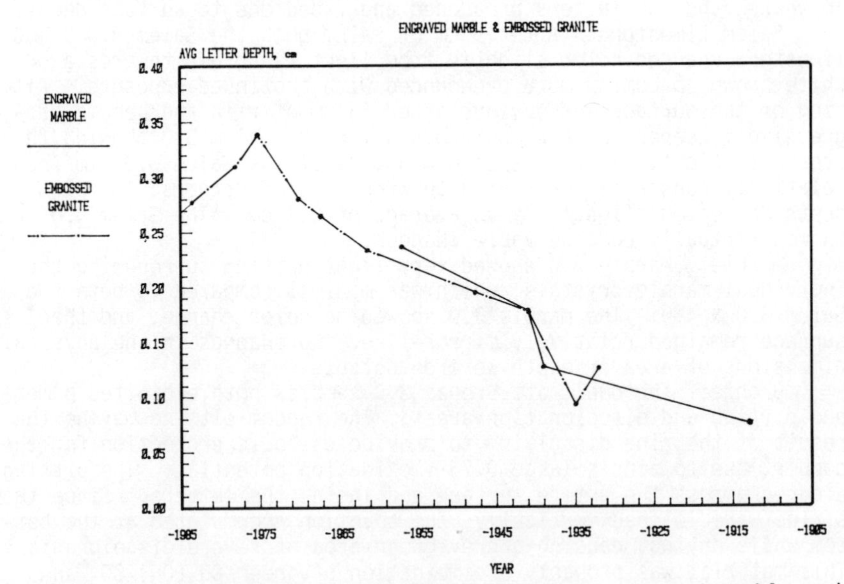

Figure 2. Changes in letter depths for V.A. engraved marble and changes in the heights of embossed granite as a result of natural weathering over time.

The Bronze 4.0 showed no dimensional or letter depth changes. It did change color towards a medium gold hue, with very slight pitting. The Bronze 5.6 showed no color, dimensional, or letter changes. It was a virtual match to an untreated bronze sample.

Leaching Results. Figure 3 is a comparison between the calcium leaching rates for the Micrite 3.0, the Marble 3.0, and the Salem 3.0 samples. As expected, the Salem exhibited the highest rate, with Marble being similar but less easily stripped. Micrite with its higher magnesium content yielded very little calcium. All three materials showed large initial stripping rates, followed by a rough exponential decrease to a steady minimal rate. No explanation was immediately obvious for the somewhat sinusoidal variation in all three samples at approximately 60 cm (1 year) cycles. This pattern was also observed in the succeeding Figures of this work. Of interest was the extreme calcium stripping for Salem 3.0 at 240 cm rain depth. This may have been an artifact of inhomogeniety in the particular crystalline sample.

Magnesium leaching rates for the same three samples at pH=3.0 showed the same rapid initial leaching, the exponential decrease, and the sinusoidal fluctuations in each case. Similarly, the Salem and Marble samples showed nearly identical magnesium rates, while the Micrite leached at slightly higher levels, probably due to the higher percentage of magnesium in the Micrite rather than a more efficient leaching process.

Figure 4 illustrates the calcium leaching rates for the three Salem limestone samples at pH=3.0, 4.0, and 5.6. As above, the magnesium stripping rates were virtual mirror images of the calcium rates. The general shapes of all the curves resembled the pattern discussed above. Salem 5.6 leached only a very small amount of both calcium and magnesium, with most occuring in the first 180 cm (3 years), since this is considered "neutral" water or a "blank". The Salem 3.0 leached the highest levels of both metals, being 1x to 3x higher than Salem 4.0 at most rain depths until about 360 cm (6 years), where the rates became nearly identical. The calcium rate for Salem 4.0 even slightly surpassed the Salem 3.0 at about 420 cm rain before merging with it. This tailing of the curves at low leaching levels and the apparent pH independence of the rates at long exposure times suggests that the leaching process was partially diffusion controlled, eventually reaching a diffusion-limited steady state. The severe pitting of these samples supports this theory in that the "soft" surface limestone dissolved more rapidly, causing the observed pits and leaving the "harder" crystals or fossilized material to dissolve at a much slower rate. As the acid penetrates into the interior through these pits, the leaching then becomes a diffusion-controlled process that depends on the presence of a moderate acid, but not necessarily its exact concentration. This process should dominate as long as the original pitted surface remains relatively intact. If it should be transformed into gypsum with subsequent crumbling or else degraded by other chemical agents, an entirely new soft surface could be exposed, beginning the large initial stripping rate process completely again. This synergistic interaction would lead to the more rapid degradation of the material than would be observed from simple acid leaching.

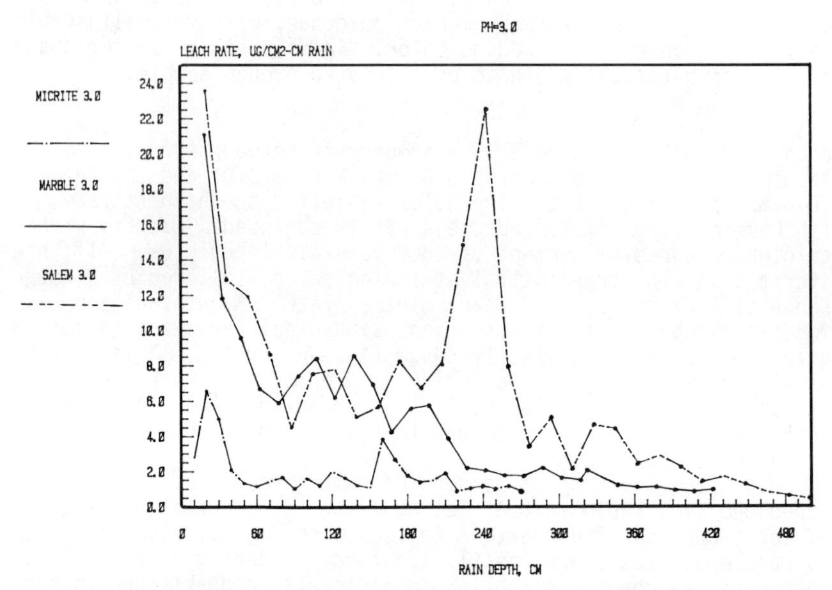

Figure 3. Calcium leaching rates of Micrite, Marble, and Salem Limestone at pH=3.0.

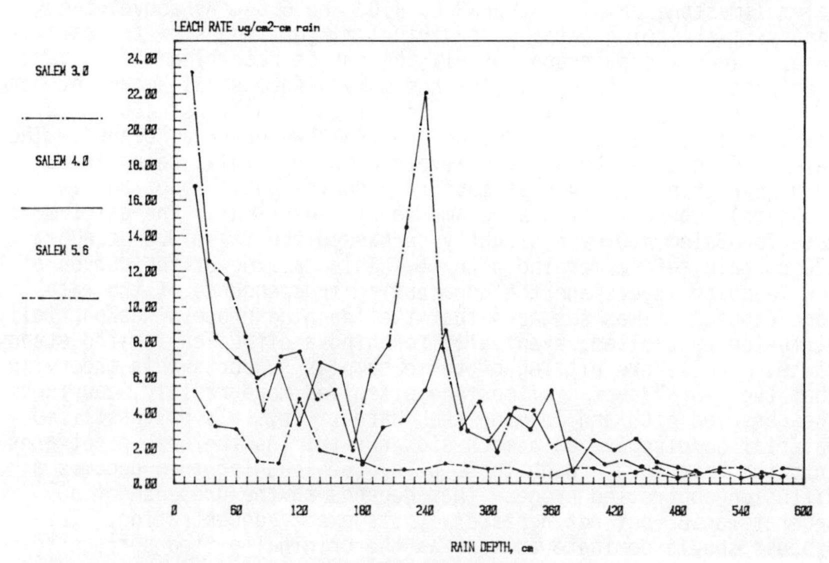

Figure 4. Calcium Leaching rates of Salem Limestone at pH= 3.0, 4.0, and 5.6.

Table IV is a summary based upon the total elemental recovery concentrations of calcium and magnesium leached from the five stone samples, their equivalent carbonate weights, and the total average sample leaching rate (normalized for surface area and rain depth):

Table IV. Total Acid Leachate for Stone Samples

	Micrite 3.0	Marble 3.0	Salem 3.0	Salem 4.0	Salem 5.6
Depth of Each Rain, cm	1.05	1.59	1.73	1.96	1.96
Total # of Rains	279	274	287	290	300
Total Rain Depth,cm	293	435	496	568	588
Year Equivalent	4.88	7.26	8.27	9.47	9.80
Total Ca Leachate,g	0.0306	0.1348	0.1800	0.1130	0.0398
Equivalent $CaCO_3$,g	0.0764	0.3366	0.4494	0.2823	0.0993
Total Mg Leachate,g	0.0048	0.0049	0.0056	0.0051	0.0014
Equivalent $MgCO_3$,g	0.0168	0.0169	0.0195	0.0178	0.0049
Total $CaCO_3$+$MgCO_3$,g	0.0933	0.3535	0.4689	0.3000	0.1042
Total Average Leach Rate ug/cm rain-cm^2 area	5.13	12.94	16.38	10.37	3.60
Weight Ratio, $CaCO_3/MgCO_3$	4.55	19.92	23.05	15.86	20.26
Calc % $CaCO_3$	81.98	95.22	95.84	94.07	95.30
Calc % $MgCO_3$	18.02	4.78	4.16	5.93	4.70

As expected, the Salem 3.0 had the greatest total average stripping rate, followed closely by the stronger crystalline Marble 3.0. The fine-grained Micrite with its higher magnesium content was leached at only 31% the Salem 3.0 rate. Based on these rates, the percent composition was calculated for each sample. These values compared closely (within 2%) with the actual compositions determined after sample digestion. The Marble and Salem had virtually the same composition, while the Micrite contained only 82% $CaCO_3$.

The three Salem samples show increased leaching rates as a function of increased acidity, but not the rate values predicted by simple chemical stoichiometry. A pH decrease from 5.6 to 4.0 is a 39.8x increase in acidity, while a change from 4.0 to 3.0 is a 10x acidity increase. The observed changes were factors of 2.88x and 1.58x, respectively. These discrepancies can be attributed to the complex equilibrium interactions involved in the solubilities of the metal carbonates. These two solubility equilibria are further complicated by the two acid equilibria for the carbonic acid/bicarbonate/carbonate system in addition to the equilibrium solubility of CO_2 gas in water. The solution of these simultaneous equilibria processes to determine the relationship between carbonate solubility and acid concentration is a non-trivial one (sixth degree in H^+ concentration). This solubility problem has been approached from several different viewpoints (31-35), the most convenient being a graphical solution of the solubility as a function of initial solution and final solution pH. From this method, it can be theoretically shown that the leaching of $CaCO_3$ by pH=5.6 solution is approximately 3.9×10^{-5}M, while in pH=4.0 solution, the solubility is 1.3×10^{-4}M, or a 3.17x increase. This compares well with the observed change in these experiments of 2.88x for the same pH=1.6 change (or a factor of 2.03x for a pH=1.0 change). The theoretical solution also predicts a solubility of 7.8×10^{-4}M in pH=3.0 solution, or a factor of 6.22x more soluble than pH=4.0. The experimental results show only a 1.58x difference between these two acid solutions. This discrepancy could result from the stripping rates being based on the entire 10 year average, which includes the last five years where the leaching rates for all pH solutions were virtually identical. In this system, it appears that a general conclusion can result that the leaching rate increases by a factor of 2x for each pH=1.0 decrease.

Bronze Samples: Acid leaching of the bronze samples occured at a much lower rate than limestone. Both Bronze 3.0 samples, run in duplicate, exhibited virtually identical stripping rates and weight losses (less than 2% difference), so only Bronze 1-3.0 is discussed here. Figure 5a illustrates the average copper stripping rates for the 10 years of simulated rain. A moderate amount of stripping was initially evident on Bronze 3.0, followed by a rather sharp decrease for 40 cm of rain (0.67 years), possibly due to the slow stripping of surface oxide layers. At 60 cm of rain, a rapid increase was followed by a relatively steady rate of 1.25 g/cm^3 for the duration of the work. The exponential rate decrease over time with the Salem limestone was absent here, probably due to more uniform monolayer arrangement of the bronze alloy that could strip evenly and consistently. The sinusoidal pattern was also again observed. Little copper leaching was evident for both Bronze 4.0 and 5.6, in contrast to Bronze 3.0. No direct pH dependence seems to exist since the Bronze 4.0 & 5.6 had very similar overall rates, with Bronze 4.0 stripping about 1.4x higher than Bronze 5.6, but 11.7x less than Bronze 3.0. However, a certain threshold pH between 3.0 and 4.0 seems to be required to initiate significant copper leaching in a stoichiometric relationship with pH.

Figure 5b illustrates similar behavior for the zinc leaching. As above, the relatively stable rate and sinusoidal pattern for Bronze 3.0 exists, as well as its higher rate values in comparison to Bronze 4.0 & 5.6. However, a significant difference was noted within the first 60 cm rain (1 year) where the stripped zinc from the Bronze 4.0 & 5.6 was quite large before decreasing almost exponentially to a low steady rate. This resembled the limestone behavior, and probably resulted from the high reactivity of zinc in acid solution, in comparison to copper which is relatively inert in acid. The stoichiometric reaction of zinc with acid is strongly thermodynamically favorable, with each pH decrease of 1.0 resulting in a 3.75x increase in zinc stripping. From Figure 5b, the average difference between Bronze 3.0 & 4.0 was 3.19x, which reasonable corresponds to theoretical. The experimental rate response factor between Bronze 4.0 & 5.6 was only 1.26x, as opposed to the predicted value of 14.9x. This implies that the Bronze 5.6 leached more than predicted. At these low levels, this may result from simple water erosion of the sample surface.

Table V is a summation of the stripping losses for the four Bronze samples:

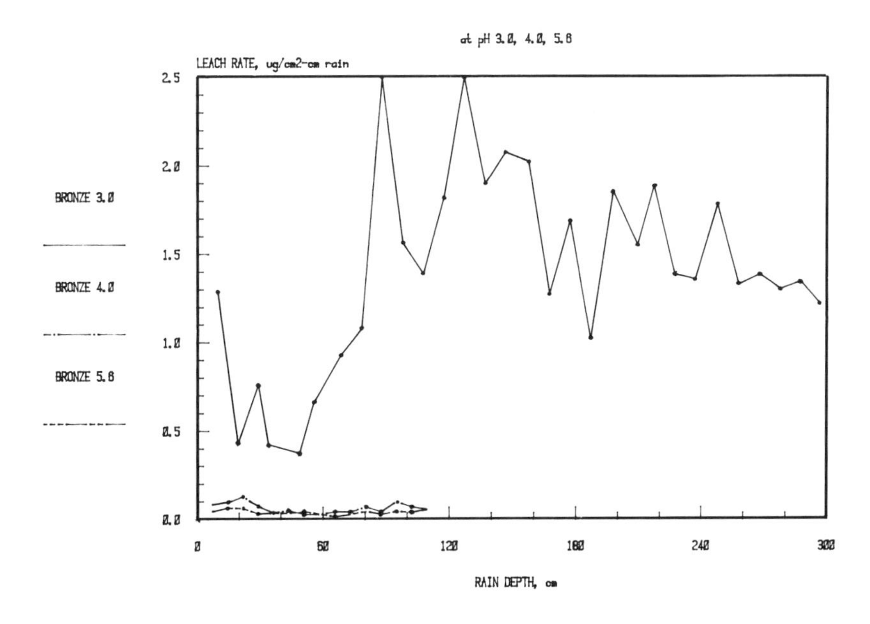

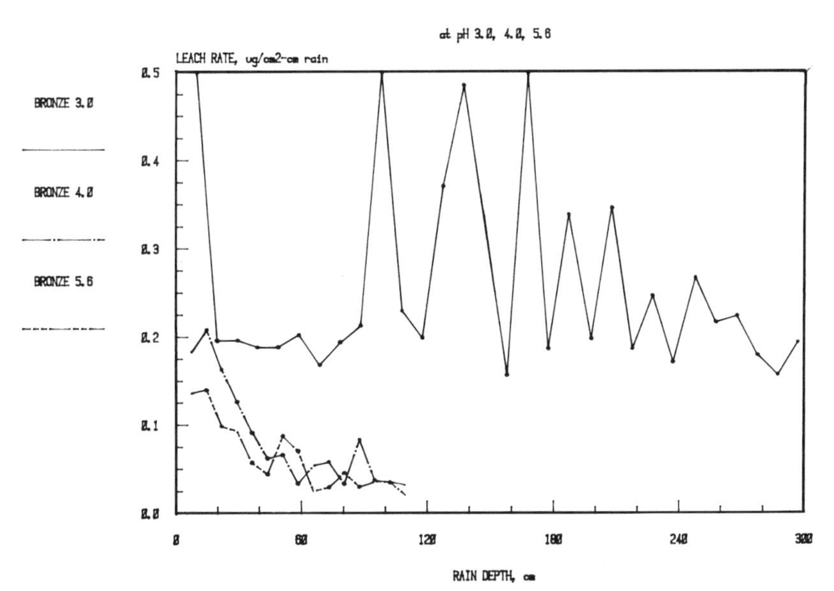

Figures 5a & 5b. Copper and Zinc leaching rates of Bronze at pH= 3.0, 4.0, and 5.6.

Table V. Total Acid Leachate for Bronze Samples

	Bronze 1-3.0	Bronze 2-3.0	Bronze 4.0	Bronze 5.6
Depth of Each Rain,cm	0.9817	1.0081	0.7273	0.7234
Total # of Rains	300	300	150	150
Total Rain Depth,cm	294.5	302.4	109.1	108.5
Total Cu Leachate,g	0.01294	0.01308	0.00028	0.00018
Total Zn Leachate,g	0.00242	0.00244	0.00038	0.00029
Total Weight Loss,g	0.01536	0.01552	0.00066	0.00047
Total Balance Weight Loss,g	0.0192	0.0204	0.0037	0.0016
% Recovery	80.00%	76.06%	17.81%	29.13%
Total Avg Leach Rate $\mu g/cm$ rain-cm^2 area	1.717	1.723	0.146	0.104
Cu/Zn Leachate Ratio	5.35	5.36	0.75	0.62

Since these samples could be fully oven dried and accurately weighed, a comparison between the true balance-measured weight losses and the cumulative experimental stripping rates could be made. For the Bronze 3.0 a 80% chemical recovery was attained, but for the Bronze 4.0 & 5.6, poor recoveries of less than 30% were achieved. No logical explanation can be offered for this large discrepancy.

The Cu/Zn stripping rate ratios for the Bronze 3.0 had a ratio of 5.35 which equates to 84% Cu and 16% Zn by weight (actual percentages from a digested sample were 90% Cu and 10% Zn). However, for both the Bronze 4.0 & 5.6, the Cu/Zn ratio was apparently inverted, with the zinc leached at higher relative rates in comparison to copper. This inversion of relative rates refers to the earlier discussion of the greater zinc reactivity in acid solution and the apparent threshold pH for significant copper leaching.

Summary

Laboratory acid leaching studies of structural materials have clarified the role of one major component in the complex acid rain system. Future work will concentrate on determining the interactions between the individual atmospheric components and how they catalyze each other's degradation contribution. It was shown that after initial leaching rates exponentially decay, the acid leaching process apparently becomes diffusion limited, assuming that a new fresh surface was not exposed by a synergistic component. A pH decrease of 1.0 resulted in a doubling of the acid leaching rates from limestone. Zinc was stoichiometrically stripped from bronze at all pH levels, but copper did not appreciably strip until a threshold between pH= 3.0 & 4.0 was reached. Overall, acid leaching of the various structural materials was fairly consistent and predictable from theory. These results should greatly aid in the study of more complicated interactions of acid rain, such as the formation and loss of gypsum.

Acknowledgments

Ball State University Department of Chemistry
Ball State University Office of Research
District 656 of the International Rotary Club
Alan Bourgault and the Ball State Department of Geology
Carol Bourgault and the Ball State Weather Station

Literature Cited

1. —— Hearings before the Subcommittee on Health and the Environment of the U.S. House of Representatives, 98th Congress, 1983-84, Serial No 98-115.
2. —— Acid Deposition: Atmospheric Processes in Eastern North America, National Academy Press, Washington D.C., 1983.
3. Schmidt, W., Report to the Indianapolis Historical Preservation Commission on Renovation of Monument Circle, 1985.
4. Likens, G.E., Chem. and Engr. News, Nov 22, 1976, 54(48), 29-44.
5. Glass, N.R., Powers, C.F., Rennie, P.J., Environment International, 1980, 4, 443-452.
6. Cowling, E.B., Environ. Sci. & Tech., 1982, 16(2), 110A-123A.
7. Martin, H.C., Materials Performance, 1982, 21(1), 36-39.
8. Baker, J.P., Schofield, C.L., Water, Air, and Soil, 1982, 18, 289.
9. Ember, L.R., Chem. and Engr. News, Sept 14 1981, 59(37), 20-31.
10. Kaplan, R.T., Hode, H.C., Protas, A., Environ. Sci & Tech, 1981, 15, 539.
11. Leivestad, H., Muniz, I.P., Nature, 1976, 258, 391.
12. Evans, L.S., J. Air Pollut. Control Assn., 1979, 29, 1145.
13. Likens, G.E., Borman, F.H., Science, 1974, 184, 1176.
14. Likens, G.E., Borman, F.H., Science, 1975, 188, 958.
15. McFee, W.W., Kelly, J.M., Beck, R.H., Water, Air, and Soil Pollution, 1977, 7, 401.
16. Fairfax, F.A., Lepp, N.W., Nature, 1975, 225, 324.
17. Miegroet, H.V., Cole, D.W., J. Environ. Qual., 1984, 13, 586.
18. Huete, A.R., McCoth, J.G., Environ. Qual., 1983, 13, 366.
19. Johnson, D.W., Richter, D.D., Miegroet, H.V., J.Air Pollut. Control Assn., 1983, 33, 1036.
20. Muir, P., Loucks, O.L., Proceedings on Methods of Evaluating Symptoms of Air Pollution, Butler Univ., in press.
21. —— Indiana State Board of Health, Testimony to U.S. Congressional Hearings on Bill HR3400, Indianapolis, Feb 1984.
22. Kawecki Assoc., Sulfur Oxides and Public Health, American Lung Assn., 1983.
23. Fassina, V., Atmospheric Environment, 1978, 12, 2205-2211.
24. Gauri, K.L., Holden, G.C., Environ. Sci. & Tech., 1981, 15(4), 386-390.
25. Longinelli, A., Bartelloni, M., Water, Air, and Soil Pollution, 1978, 10, 335-341.
26. —— Interagency Task Force on Acid Precipitation, Annual Report, 1982, Washington D.C.
27. —— Interagency Task Force on Acid Precipitation, Annual Report, 1984, Washington D.C.
28. Rybarczyk, J.P., Pokorney, L.M., A COmprehensive Three Year Study of Acid Rain Across Central Indiana, in preparation.

29. —— Electric Power Research Institute Journal, Nov 1983.
30. —— Monthly Reports, Ball State University Weather Station, 1982-85.
31. Butler, J.N., "Ionic Equilibrium: A Mathematical Approach", Addison -Wesley Publishing Co., Reading MA, 1964.
32. Blackburn, T.R., "Equilibrium: A Chemistry of Solutions", Rinehart and Winston, Inc., New York, 1969.
33. Freiser, H., Fernando, Q.,"Ionic Equilibria in Analytical Chemistry", John Wiley & Sons, Inc., New York, 1963.
34. Morel, F.M., "Principles of Aquatic Chemistry", John Wiley & Sons, Inc, New York, 1983.
35. Manahan, S.E., "Environmental Chemistry", Willard Grant Press, Boston, MA, 1980.

RECEIVED January 13, 1986

Effects of Acid Rain on Deterioration of Coquina at Castillo de San Marcos National Monument

D. G. Rands, J. A. Rosenow[1], and J. S. Laughlin[2]

Department of Chemistry, Southern Illinois University, Edwardsville, IL 62026

Synthetic acid rain samples have been allowed to percolate through pieces of coquina, the material of construction at Castillo de San Marcos National Monument, St. Augustine, Florida. Chemical analyses of the solutions are used to determine the extent of dissolution of the coqiuna, a limestone material, by acid rain. Because of the location of the Castillo on the Atlantic coast, the effect of salt spray on weathering of coquina appears to be as significant as the influence of acid rainfall.

Castillo de San Marcos is a National Park Service Monument located in St. Augustine, Florida. The monument was constructed of coquina blocks during 1672 to 1695. St. Augustine was considered by the Spanish to be the keystone in the defense of Florida. The Castillo was designed to preserve Spain's sovereignty over Florida and protect Spanish shipping routes [1]. Coquina is composed of sea shells (Donax shells) and calcareous sandstone cemented together by their own lime (calcium carbonate) [2,3]. The coquina has a spongy texture and is grainy and extremely porous, providing ducts for the passage of water.

Park officials in St. Augustine are concerned about the deterioration of the internal and external walls of the monument caused by moisture. The exterior of Castillo shows degradation due to rainfall and visitor traffic. Increasing acidity levels in the rainfall of Florida have led officals to believe that this may be a source of additional damage to the monument. In this study we consider the possible effects acid rain may have on coquina.

Materials and Methods

Rectangular blocks of coquina of 4 x 4 cm cross-section ranging in length from 5 to 20 cm were enclosed with sheet plastic and glass

[1]Current address: U.S. Gypsum Corporation, 700 North Highway 5, Libertyville, IL 60048
[2]Current address: Bruker Instrument, Inc., Manning Park, Billercia, MA 01821

0097-6156/86/0318-0301$06.00/0

tubing affixed at either end to permit flow of solutions. To determine the effects of various solutions on the deterioration of coquina, one liter samples of deionized water, HNO_3 at pH = 4, H_2SO_4 at pH = 4, and representative synthetic acid rain samples (Table 1) were allowed to trickle through the coquina column. The resulting solutions were analyzed for pH, alkalinity by titration, phosphate by the colorimetric method, sulfate by turbidimetry, calcium and magnesium by atomic absorption spectrometry and sodium and potassium by flame photometry.

Using a computer model developed from solution equilibrium techniques discussed by Lindsay [4], analytical data were used to calculate the activities and molarities of 29 species in solutions recovered from the coquina columns. A typical computer print-out shown in Figure 1 lists the solution species as well as their activities and molarities. The resulting data were used to interpret the effects of initial pH, ionic strength, acid rain composition and column length on degradation of coquina.

Results and Discussion

A visual examination of Castillo de San Marcos showed the deterioration of coquina occurs primarily through the dissolution of calcium carbonate which serves as the bonding agent for shells and particulate matter. Because coquina is extremely porous material, a great amount of rainfall seeps through the walls and drips into interior rooms of the fort [5].

Because of its location on Matanzas Bay at St. Augustine, Florida, Castillo is subject to the effects of sea spray carried down in precipitation falling along the Florida seashore. In our study we have chosen representative precipitation compositions (Table 1) taken from NADP reports for Everglades National Station and Cape Canaveral, Florida [6].

In our study we have not included a consideration of dry deposition or salt spray deposited on the surfaces of the walls of the fort.

pH and Calcium Concentration. In Figure 2 are plotted the results of total calcium analyses of all solutions recovered from percolation through coquina columns. We have plotted the logarithm of total calcium concentration in moles per liter versus pH of the recovered solution. An examination of our experimental data showed no relationship between the initial pH of solutions or of column length and calcium concentration. This phenomenon is attributed to the heterogeneity of natural coquina.

The recovered solutions displayed pH values between 7.0 and 9.1 indicating that these solutions had not reached equilibrium with calcium carbonate and atmospheric CO_2. The equilibrium system attains a pH of 8.3 [7].

Those eight solutions have pH values greater than 8.5 resulted from partial exclusion of atmospheric CO_2 and approach the theoretical pH of 10.0 for the equilibrium system $CaCO_3$-H_2O in the absence of atmospheric CO_2.

Linear regression analysis of the data points in the pH range of 7.0 to 8.5 shows an increase in calcium concentration as the pH increases (Figure 2). This calcium increase indicates a progressive

```
ORIGINAL CONDITIONS:

PH :8.025
IONIC STRENGTH :4.56278964E-03
TOTAL CARBONATE :1.227E-03 MOLAR
TOTAL PHOSPHATE :4.85E-06 MOLAR
TOTAL SULFATE :5.25E-06 MOLAR
TOTAL CALCIUM :3.72E-04 MOLAR
TOTAL MAGNESIUM :8.61E-05 MOLAR
TOTAL SODIUM :3.174E-04 MOLAR
TOTAL POTASSIUM :5.83E-05 MOLAR

RESULTS:    MOLARITY            ACTIVITY

[CO3 2-]   :7.31713015E-06     5.47395435E-06
[HCO3-]    :1.18732862E-03     1.1042338E-03
[H2CO3]    :2.39154538E-05     2.39154538E-05
[CAHCO3+]  :4.28268467E-06     3.98296231E-06
[CACO3]    :2.06461025E-06     2.06461025E-06
[MGHCO3+]  :8.6110624E-07      8.00841986E-07
[MGCO3]    :5.88615338E-07     5.88615338E-07
[NAHCO3]   :5.66400099E-07     5.66400099E-07
[NACO3-]   :3.15616547E-08     2.93528221E-08
[PO4 3-]   :2.52240621E-10     1.31288264E-10
[H2PO4-]   :4.46171623E-07     4.1494644E-07
[HPO4 2-]  :3.7072792E-06      2.77342027E-06
[CAHPO4]   :4.16844065E-07     4.16844065E-07
[CAPO4-]   :1.11235315E-07     1.03450546E-07
[MGHPO4]   :1.42621557E-07     1.32640231E-07
[MGPO4-]   :2.57434264E-08     2.39417807E-08
[SO4 2-]   :4.98413654E-06     3.72863886E-06
[CASO4]    :2.07892255E-07     2.07892255E-07
[MGSO4]    :5.04714263E-08     5.04714263E-08
[NASO4-]   :5.91800398E-09     5.5038343E-09
[HSO4-]    :3.60476238E-12     3.3524842E-12
[KSO4-]    :1.53903427E-09     1.43132543E-09
[CAOH+]    :6.22570003E-09     5.78999634E-09
[MGOH+]    :2.55746801E-08     2.3784844E-08
[NAOH]     :1.96925041E-10     1.96925041E-10
[CA 2+]    :3.65327352E-04     2.73301855E-04
[MG 2+]    :8.45484889E-05     6.32508318E-05
[NA+]      :3.16795923E-04     2.94625059E-04
[K+]       :5.8298461E-05      5.42184612E-05
```

Figure 1. Typical computer print-out.

TABLE 1: Composition of Synthetic Acid Rain Samples

	SOLN 1	SOLN 2
$MgCl_2$, m/l	2.1×10^{-5}	7.4×10^{-6}
Na_2SO_4, m/l	1.7×10^{-5}	1.1×10^{-5}
NH_4Cl, m/l	2.8×10^{-5}	4.4×10^{-6}
KCl, m/l	7.7×10^{-6}	2.6×10^{-6}
$CaSO_4$, m/l	2.5×10^{-5}	6.7×10^{-6}
Na_3PO_4, m/l	5.3×10^{-5}	-----
NaCl, m/l	7.7×10^{-5}	3.7×10^{-5}
HNO_3, m/l	3.2×10^{-5}	1.3×10^{-5}
HCl, m/l	1.1×10^{-6}	-----
pH	4.57	4.85
Ionic Strength	5.4×10^{-4}	1.4×10^{-4}

neutralization of the acid solutions by calcium carbonate in the interstices of the coquina and an approach to atmospheric equilibrium. Data points at pH values between 8.5 show a reversal in the solubility trend indicating that, in those experiments, partial pressures of CO_2 were lower than normal atmospheric values resulting in lower solubility of $CaCO_3$.

Our results show that in the event that solutions seeping through coquina blocks at Castillo are sufficiently isolated from the atmosphere, calcium carbonate dissolved at the surface can be re-deposited in the interior of the building material when the pH exceeds 8.3.

Ionic Strength and Calcium Concentration. In order to assess the contribution of sea spray to coquina degradation we have plotted total calcium concentration vs ionic strength (Figure 3). Ionic strength of each solution was calculated from the print-out of all solution species. It can be seen from Figure 3 that there is a general increase in dissolution of $CaCO_2$ as the ionic strength of our experimental samples increased. This correspondence is a common phenomenon caused by a general decrease in the value of ionic activity coefficients as solution ionic strength increases as predicted by the Debye-Huckel Theory. The straight line drawn in Figure 3 is the theoretical solubility of $CaCO_3$ in equilibrium with atmospheric CO_2 and shows the expected increase in solubility with ionic strength.

To illustrate the effect of ionic strength on degradation of calcium carbonate we have calculated the solubility of calcium carbonate in deionized water, acid at pH = 4.0 and acid rain at pH = 4.0 with an ionic strength of 7.2×10^{-3} in the absence of CO_2. The results of these calculations are shown in Table 2 and are plotted in Figure 3. These data show that the ionic strength contribution of sea spray and other atmospheric sources are as significant as the neutralization reaction with acid at pH = 4.0 in the degradation of coquina by acid rainfall.

Table 2: Calculated Calcite Solubility Under Varying Conditions
Atmospheric CO_2 Excluded

	Total Calcium
Deionized Water	1.29×10^{-4} m/l
pH 4 Acid	1.62×10^{-4} m/l
pH 4 Acid Rain	2.13×10^{-4} m/l

Using our computer generated data we have observed that only about 1% of the total calcium in solutions occurs as ion-pair species such as $CaSO_4$, $CaCO_3$, and $CaHCO_3$. The remaining calcium occurs in solution as free calcium ions. Consequently, ion-pairing with the substituents of acid rain is not a significant factor in the degradation of coquina.

Formation of gypsum in marble monuments has been cited as a major contributor to degradation [8]. Our data show that the ion

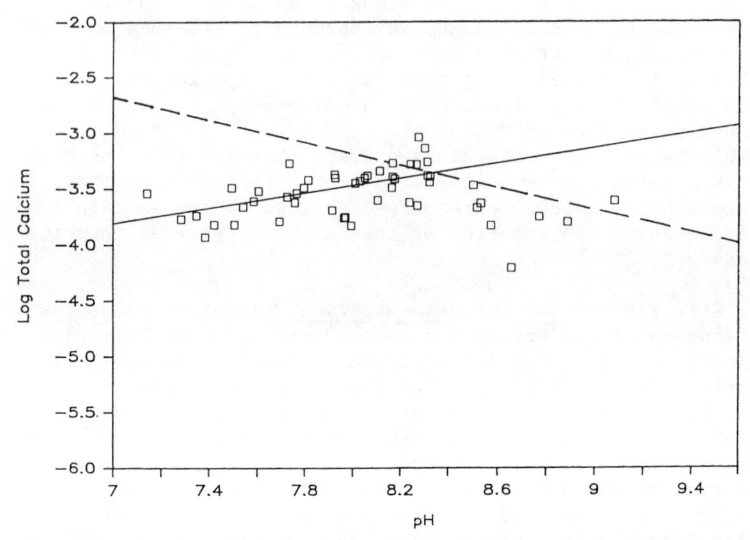

Figure 2. Plot of log total calcium vs pH. Dashed line, calculated solubility as a function of pH. Solid line, linear regression plot for experimental points below pH = 8.5.

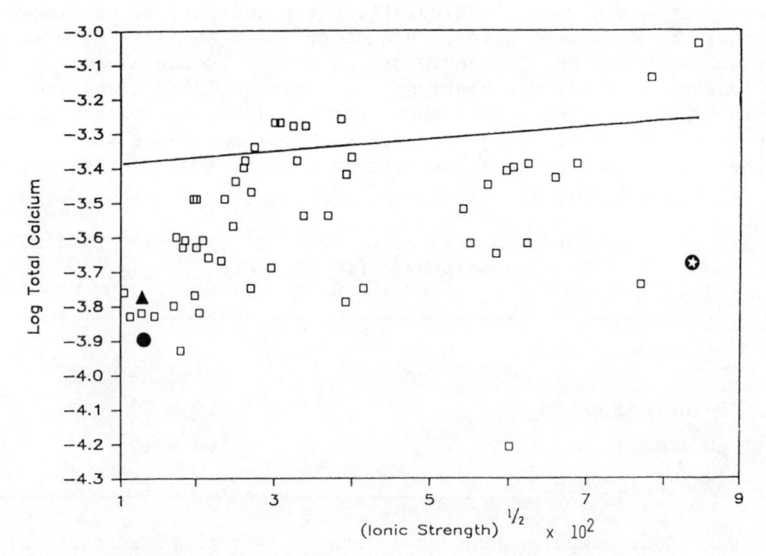

Figure 3. Plot of log total calcium vs ionic strength. Solid line, calculated solubility of calcium carbonate in equilibrium with atmospheric CO_2. Calculated solubility of calcium carbonate in the absence of atmospheric CO_2 for: Circle, deionized water; triangle, acid solution at pH = 4.0; star, acid rain at pH = 4.0.

activity product for Ca^{2+} and SO_4^{2-} in our system is less than the thermodynamic solubility product of gypsum, $CaCO_4 \cdot H_2O$, by two to three orders of magnitude. Our study, however, deals only with dissolution problems and not the secondary effect of gypsum formation by alternate wetting and drying. Consequently, no conclusions can be drawn regarding deleterious effects of gypsum in coquina.

Conclusions

Although rain percolating through the porous coquina of Castillo de San Marcos does not attain saturation with respect to calcium carbonate, this study shows that the acidity and ionic strength of precipitation falling along the Florida coast are significant contributors to degradation of the material of construction. In fact, the ionic strength of the precipitation, enhanced by sea-spray is as significant as increased acidity in causing degradation.

Algal growth on interior walls of the facility cause some degradation [5], however, visual inspection confirms that this effect is insignificant compared to precipitation and visitor traffic.

Acknowledgment

The authors are grateful for the cooperation of National Park Service officials at Castillo de San Marcos National Monument. The financial support of the Graduate School at Southern Illinois University at Edwardsville is gratefully acknowledged.

Literature Cited

1. Arana, L.R. and Manucy, A.: "The Building of Castillo de San Marcos." Philadelphia: Eastern National Park and Monument Association 9, (1977).
2. Cooke, C.W.: "Geology of Florida." Geol. Bull. 29, Florida Geological Survey, Tallahasee, Florida, (1945).
3. Miles, R.D., Junquera, O.M. and Harms, R.H.: "Coquina shells as a supplemental source in layer hen diets." Florida Sci. 45, 142-144, (1982).
4. Lindsay, W.L.: "Chemical Equilibria in Soils." Wiley-Interscience, New York (1979).
5. Rands, D.G., Davis, J.S., and Arana, L.R.: "Chemical and Biological Processes in Coquina of Castillo de San Marcos National Monument." Schweiz A. Hydrol. 46 (1), 109-116, (1984).
6. National Atmospheric Deposition Program: NAPD Data Report, Precipitation Chemistry. National Resource Ecology Laboratory. Colorado State University, Fort Collins, (1983-84).
7. Nakayama, F.S.: "Hydrolysis of $CaCO_3$, Na_2CO_3 and $NaHCO_3$ and their combinations in the presence and absence of external CO_2 source." Soil Sci. 109, 391-398, (1970).
8. Fassina, V.: "A Survey on Air Pollution and Deterioration of Stonework in Venice." Atmospheric Environment 12, 2205-2211, (1978).

RECEIVED January 21, 1986

DEGRADATION OF ORGANICS

22

Effects of Acid Rain on Painted Wood Surfaces: Importance of the Substrate

R. Sam Williams

U.S. Department of Agriculture, Forest Products Laboratory, One Gifford Pinchot Drive, Madison, WI 53705-2398

The effects of acid rain on painted materials can be seen in least two phenomena, degradation of the coating and degradation of the substrate. Most research on acid degradation and painted materials has focused on degradation of coatings caused by gaseous pollutants such as sulfur dioxide and nitrogen dioxide--known precursors of acid rain. This work showed that the type of pigment and extenders used in the paint formulation had a direct bearing on paint performance in an acid environment. The degradation of the substrate also has a direct bearing on coating performance. This substrate degradation may involve different failure mechanisms; therefore, future acid rain research should include the reaction to acid rain of the substrate-coating interface. Preliminary work at the Forest Products Laboratory has shown minor increases in weathering rate of uncoated wood specimens that were dipped in dilute acid periodically during xenon-arc accelerated weathering. The effect of this wood degradation on subsequent coating performance is unknown but is the topic of continuing research.

The mere mention of the term "acid rain" or acid deposition (or acid precipitation) can usually generate lively discussion, controversy and even confrontation among friends, scientists, and nations. One aspect of acid rain now being discussed is its role in the degradation of paints and painted surfaces, even though ultraviolet light is certainly a much more severe factor in degrading many polymers, including the polymer (binder) in paints.

The term "acid rain" itself is not easily defined; however for the purpose of this discussion, let us include under acid deposition those solids, liquids, gases, or aerosols of man-made origin. The degree of contamination, pH levels, type of anions, or ecological significance are not defined, although continued emphasis in these areas of research is important. This report will focus on the effect

Published 1986, American Chemical Society

of acid deposition on wood and painted (finished) wood. Finishing problems and degradation of finishes depend on the substrate and, for this reason, considerable discussion of the material properties of wood are included.

Normal Weathering of Wood

Before discussing acid deposition effects on wood, a brief review of normal wood weathering may be helpful. (A glossary of wood-related terms is included at the end of this paper.) As mentioned previously, UV degradation can be a severe problem with many polymers. From a polymer chemistry viewpoint, wood is similar to other polymer composites. And as is true of many synthetic polymer composites, both the matrix and fiber are organic polymers. In wood, the fibers are composed of cellulose crystallites surrounded by amorphous cellulose and hemicelluloses. The matrix polymer is a highly cross-linked aromatic polymer called lignin. Lignin, with its abundance of phenolic, methoxy phenolic, and ketone groups, has sufficient chromophores to be an exceptionally good UV light absorber (1). The absorption of this energy by the lignin is the main cause of UV degradation of wood. This UV degradation is manifest in an initial color change followed by the gradual erosion of the wood surface. This erosion or weathering is not to be confused with decay. Decay is caused by fungi and can lead to rapid deterioration throughout the volume of the wood. Weathering, on the other hand, is a surface deterioration; and, although the initial color changes can be seen within days or even hours, the surface erosion proceeds very slowly. The erosion rate for solid wood in temperate zones is in the order of 1/8 to 1/2 inch per century and depends mainly on amount of UV exposure and the wood species (1, 2). Other degrading factors include moisture, mechanical abrasion, temperature, and pollution (3).

Wood Properties That Affect Coating Performance

Wood and wood-based materials such as plywood, fiberboard, paper-overlayed panels, and flakeboard, have specific properties which must be taken into account when formulating finishes for them.

Wood changes dimension as the moisture content varies between approximately 30% (fiber saturation) and 0% (ovendry). Moisture contents above fiber saturation occur as water in the cell lumen and do not produce further dimensional change. Within the range from fiber saturation to ovendry the amount of dimensional change depends on species (Table I), individual trees (Figure 1), and the type of cut (Table I). Because the dimensional changes differ in the radial and tangential directions of wood, various cuts distort differently (Figure 2). Note the uniform shrinkage in the radial (vertical-grained) cut piece (Figure 2). The painting characteristics of various species are listed (Table II). Vertical-grained cut western redcedar and redwood lead as the most paintable substrates. Comparison of the radial shrinkage values for these species show them to be among the most dimensionally stable. In addition to these gross features, the microstructure at the wood-paint interface is extremely important. Because moisture effects dimensional changes in wood, moisture is the biggest enemy of painted wood. How acid

Table I. Shrinkage Values of Domestic Softwoods (30)

Species	Shrinkage from green to ovendry moisture content[1]		
	Radial	Tangential	Volumetric
	- - - - - -	Percent	- - - - - -
SOFTWOODS			
Baldcypress	3.8	6.2	10.5
Cedar:			
Alaska-	2.8	6.0	9.2
Atlantic white-	2.9	5.4	8.8
Eastern redcedar	3.1	4.7	7.8
Incense-	3.3	5.2	7.7
Northern white-	2.2	4.9	7.2
Port-Orford-	4.6	6.9	10.1
Western redcedar	2.4	5.0	6.8
Douglas-fir:[2]			
Coast	4.8	7.6	12.4
Interior north	3.8	6.9	10.7
Interior west	4.8	7.5	11.8
Hemlock (con.):			
Western	4.2	7.8	12.4
Larch, western	4.5	9.1	14.0
Pine:			
Eastern white	2.1	6.1	8.2
Jack	3.7	6.6	10.3
Loblolly	4.8	7.4	12.3
Lodgepole	4.3	6.7	11.1
Longleaf	5.1	7.5	12.2
Pitch	4.0	7.1	10.9
Pond	5.1	7.1	11.2
Ponderosa	3.9	6.2	9.7
Red	3.8	7.2	11.3
Shortleaf	4.6	7.7	12.3
Slash	5.4	7.6	12.1
Sugar	2.9	5.6	7.9

Table I. Shrinkage Values of Domestic Softwoods (30)--con.

Species	Shrinkage from green to ovendry moisture content[1]			Species	Shrinkage from green to ovendry moisture content[1]		
	Radial	Tangential	Volumetric		Radial	Tangential	Volumetric
	- - - - - - Percent - - - - - -				- - - - - - Percent - - - - - -		
			SOFTWOODS--con.				
Fir:				Virginia	4.2	7.2	11.9
Balsam	2.9	6.9	11.2	Western white	4.1	7.4	11.8
California red	4.5	7.9	11.4	Redwood:			
Grand	3.4	7.5	11.0	Old-growth	2.6	4.4	6.8
Noble	4.3	8.3	12.4	Young-growth	2.2	4.9	7.0
Pacific silver	4.4	9.2	13.0	Spruce:			
Subalpine	2.6	7.4	9.4	Black	4.1	6.8	11.3
White	3.3	7.0	9.8	Engelmann	3.8	7.1	11.0
Hemlock:				Red	3.8	7.8	11.8
Eastern	3.0	6.8	9.7	Sitka	4.3	7.5	11.5
Mountain	4.4	7.1	11.1	Tamarack	3.7	7.4	13.6

[1]Expressed as a percentage of the green dimension.

[2]Coast Douglas-fir is defined as Douglas-fir growing in the States of Oregon and Washington west of the summit of the Cascade Mountains. Interior West includes the State of California and all counties in Oregon and Washington east of but adjacent to the Cascade summit. Interior North includes the remainder of Oregon and Washington and the States of Idaho, Montana, and Wyoming.

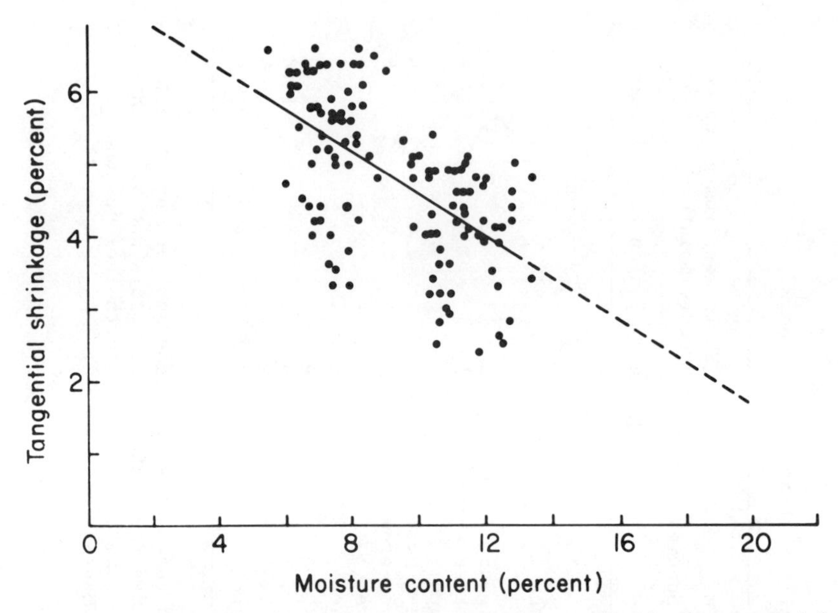

Figure 1.--An illustration of variation in individual tangential shrinkage values of several boards of Douglas-fir from one locality, dried from green condition (30). (ML85 5194)

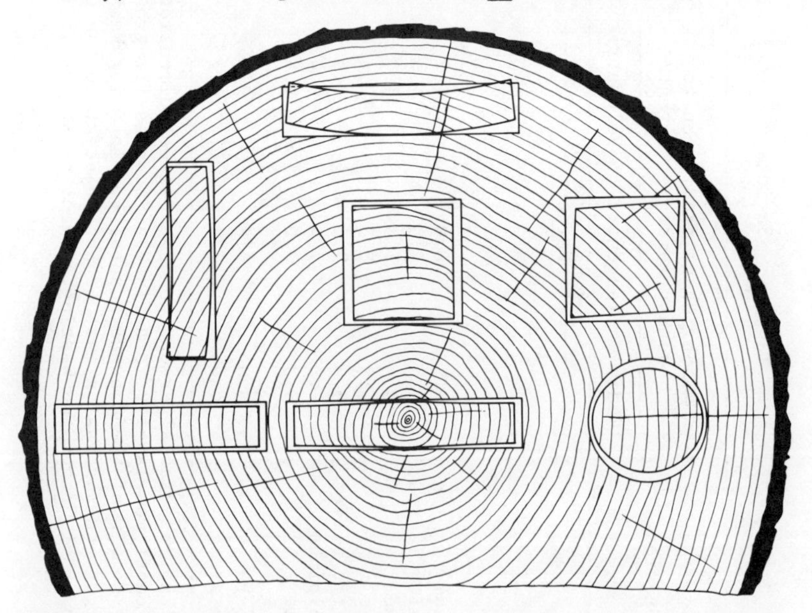

Figure 2.--Characteristic shrinkage and distortion of flats, squares, and rounds as affected by the direction of the annual rings. Tangential shrinkage is about twice as great as radial (30). (ZM 12494F)

Table II. Characteristics of Woods For Painting and Finishing (omissions in the table indicate inadequate date for classification) (30)

Wood	Ease of keeping well painted; I--easiest, V--most exacting[1]	Weathering		Appearance	
		Resistance to cupping; 1--best 4--worst	Conspicuousness of checking; 1--least, 2--most	Color of heartwood (sapwood is always light)	Degree of figure on flat-grained surface
SOFTWOODS					
Cedar:					
Alaska-	I	1	1	Yellow	Faint
California incense-	I	--	--	Brown	Faint
Port-Orford-	I	--	1	Cream	Faint
Western redcedar	I	1	1	Brown	Distinct
White-	I		--	Light brown	Distinct
Cypress	I	1	1	Light brown	Strong
Redwood	I	1	1	Dark brown	Distinct
Products[2] overlaid with resin-treated paper	I	--	1	--	--
Pine:					
Eastern white	II	2	2	Cream	Faint
Sugar	II	2	2	Cream	Faint
Western white	II	2	2	Cream	Faint
Ponderosa	III	2	2	Cream	Distinct
Fir, commercial white	III	2	2	White	Faint
Hemlock	III	2	2	Pale brown	Faint
Spruce	III	2	2	White	Faint
Douglas-fir (lumber and plywood)	IV	2	2	Pale red	Strong
Larch	IV	2	2	Brown	Strong

Continued on next page

Table II. Characteristics of Woods For Painting and Finishing (omissions in the table indicate inadequate date for classification) (30)--con.

Wood	Ease of keeping well painted; I--easiest, V--most exacting[1]	Weathering		Appearance	
		Resistance to cupping; 1--best 4--worst	Conspicuousness of checking; 1--least, 2--most	Color of heartwood (sapwood is always light)	Degree of figure on flat-grained surface
		SOFTWOODS--con.			
Lauan (plywood)	IV	2	2	Brown	Faint
Pine:					
Norway	IV	2	2	Light brown	Distinct
Southern (lumber and plywood)	IV	2	2	Light brown	Strong
Tamarack	IV	2	2	Brown	Strong
		HARDWOODS			
Alder	III	--	--	Pale brown	Faint
Aspen	III	2	1	Pale brown	Faint
Basswood	III	2	2	Cream	Faint
Cottonwood	III	4	2	White	Faint
Magnolia	III	2	--	Pale brown	Faint
Yellow-poplar	III	2	1	Pale brown	Faint
Beech	IV	4	2	Pale brown	Faint
Birch	IV	4	2	Light brown	Faint
Cherry	IV	--	--	Brown	Faint
Gum	IV	4	2	Brown	Faint

Table II. Characteristics of Woods For Painting and Finishing (omissions in the table indicate inadequate date for classification) (30)--con.

Wood	Ease of keeping well painted; I--easiest, V--most exacting[1]	Weathering		Appearance	
		Resistance to cupping; 1--best 4--worst	Conspicuousness of checking; 1--least, 2--most	Color of heartwood (sapwood is always light)	Degree of figure on flat-grained surface
		HARDWOODS--con.			
Maple	IV	4	2	Light brown	Faint
Sycamore	IV	--	--	Pale brown	Faint
Ash	V or III	4	2	Light brown	Distinct
Butternut	V or III	--	--	Light brown	Faint
Chestnut	V or III	3	2	Light brown	Distinct
Walnut	V or III	3	2	Dark brown	Distinct
Elm	V or IV	4	2	Brown	Distinct
Hickory	V or IV	4	2	Light brown	Distinct
Oak, white	V or IV	4	2	Brown	Distinct
Oak, red	V or IV	4	2	Brown	Distinct

[1]Woods ranked in group V for ease of keeping well painted are hardwoods with large pores that need filling with wood filler for durable painting. When so filled before painting, the second classification recorded in the table applies.

[2]Plywood, lumber, and fiberboard with overlay or low-density surface.

deposition interacts with moisture and affects painted wood is unknown at this time.

Dimensional changes are a function of density; therefore the higher density latewood changes more than the lower density earlywood. Thus the large areas of latewood in flat-grained lumber hold paint poorly (Figure 3). This is particularly true of the large latewood bands often found in plywood. Miniutti (4, 5) reported different swelling of earlywood and latewood in vertical-grain lumber. He also showed that the differential swelling of poorly machined flat grain siding developed severe strains in the coating over these areas (Figure 4). This raised grain developed cracks over the latewood portions of the swelled wood (Figure 5).

Acid Deposition Effects on Coatings

A comprehensive review of the effects of pollution on coatings applied to wood and many other materials was published in 1979 (6) and it is not my intent to duplicate this effort. I will however, review some of the work that impacts directly on wood and wood finishing.

Of finishes used outside, the polymers that can offer the best protection against acid deposition are those not containing acid-sensitive groups such as esters. Inclusion of acid resistant paint binders such as vinyls, urethane, and epoxies would produce acid resistance only if the other components are also acid resistant. Saponification of esters catalyzed by hydroxyls formed during corrosion of steel substrates may also degrade polyesters (7).

Virtually all of the research on acid deposition effects on finishes has dealt with coating degradation and did not include the effect on the substrate. In early laboratory experiments of SO_2 effects on paint, Holbrow (8) showed that some oil-based paints dried more slowly in a SO_2 contaminated atmosphere. The effect was dependent upon SO_2 concentration, type of oil, and pigment. Paints based on linseed oil, bodied dehydrated castor oils, and tung oil, with titanium dioxide pigments, were more susceptible to drying retardation than unbodied dehydrated castor oil and basic pigments such as white lead or zinc oxide. The greatest effect occurred within the first day or two of cure and the effect was more pronounced under moisture condensing conditions. The SO_2 exposure during the early stages of oil cure rendered the films moisture sensitive and they wrinkled under further exposure to moist conditions. The drying of latex paints has not been evaluated.

The soiling of paints by various particulates has been documented. These particulates include sulfates and chlorides of iron, calcium, and zinc (8) as well as dust from alkaline mortar (9). Spence and Haynie (10) discussed two surveys by Michelson and Tourin, and Booz *et al*. in which an attempt was made to correlate concentration of atmospheric particulate matter with the frequency of repainting. Although these surveys showed a correlation, there are many unanswered questions. The effects of other pollutants, the type of paint used, and the social and economic factors affecting painting frequency were not taken into account.

Figure 3.--Paint failure over latewood of flat-grained siding (30). (M147 211-12)

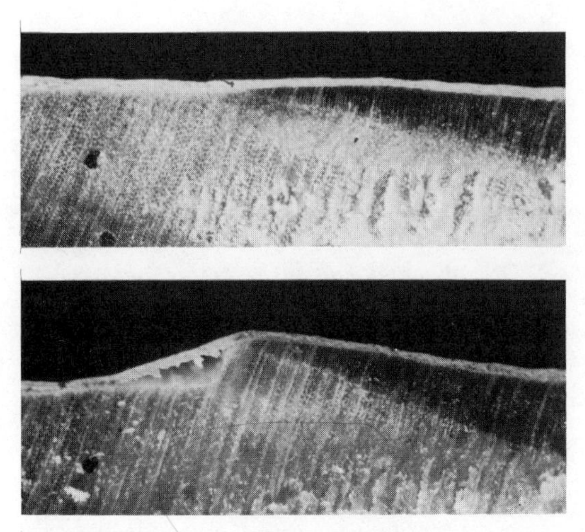

Figure 4.--Enlarged view of the end grain of painted wood before (top) and after (bottom) swelling with water. The paint is one thick coat of solvent-borne primer. (Upper): Dry. (Lower): Wet (4). (M121 550)

Figure 5.--Enlarged views of the painted face (top) and end grain (bottom) of a piece of wood similar to that in Figure 4. The crack in the thick coat of solvent-borne primer is over the earlywood-latewood junction and developed after water was placed on the end grain to swell the crushed springwood cells beneath the band of latewood (4). (M121 551)

Gutfreund (11) used exposure of paint films to O_3 as a means of predicting paint performance. Higher grades of paint embrittled less during exposure to this pollutant.

In the early 1970's a series of experiments were begun by Campbell, Spence, Schurr, and other collaborators to assess the effects of acid deposition on paint films. The work was a logical continuation of previous work by Holbrow, Tice, and Gutfreund. In the first of these experiments, using stainless steel substrates, the accelerated erosion of the paint surface by SO_2 and O_3 was evaluated by attenuated total reflectance infrared spectroscopy (ATR IR), scanning electron microscopy (SEM), and by measuring paint erosion or loss gravimetrically. In subsequent work, the panels were evaluated only gravimetrically.

A 1974 study by Campbell *et al.* (12) involved both laboratory and field exposure of five coating systems on stainless steel panels: oil-based house paint; latex house paint; industrial maintenance paint; coil-coating paint; and nitrocellulose/acrylic automotive paint. The four field exposure sites were north central North Dakota; Los Angeles, California; Chicago, Illinois; and Valparaiso, Indiana.

The research focused on accelerated effects of pollution on erosion, the normal degradative mechanism for a good paint system, and did not include catastrophic failures caused by loss of paint adhesion. The weight loss of specimens at low SO_2 concentrations seemed to correlate well with the paints having $CaCO_3$ extender pigments (Table III). At higher levels of SO_2 (1.0 ppm), the erosion rates were substantially higher. The greatest increase in erosion rate occurred with oil-based house paint having $CaCO_3$ extender pigments. ATR IR spectroscopy evaluation of the surface showed more rapid breakdown of the binder in the oil/alkyd paints than in the acrylic-based latex paints. Shaded specimens showed similar trends but less degradation. Surface evaluation using SEM substantiated the results from the weight loss measurement. ATR IR spectroscopy showed that oil-based house paints have the greatest surface degradation. Latex house paints also showed severe surface degradation. The results with O_3 were less clear although the oil-based paint was more severely affected than the latex paint. Results of laboratory exposure to SO_2 correlated well with the outdoor exposure; the highest erosion rate occurred for coatings having acid-sensitive extender pigments (mainly $CaCO_3$) in areas of high pollution.

The laboratory exposure values obtained for unshaded specimens were generally higher than for the low concentration SO_2 (0.1 ppm) and almost as high as the high concentration SO_2 (1.0 ppm). Oil-based house paint containing $CaCO_3$ was an exception, having twice as high an erosion rate for 1.0 ppm SO_2 exposure (shaded) as for the 0.1 ppm SO_2 (unshaded). In most cases the effect of light and SO_2 appeared additive except for the oil-based house paint. Here the erosion was more than twice as fast as the sum of SO_2 and light

Table III. Slope of Erosion Data (mils loss x 10^{-5}/hr with 95% confidence limits) Accompanied by a T-test Probability (%) that a Statistical Difference Exists Between the Respective Slope for a Given Pollutant Type and Level vs. the Zero Pollutant Level, Based on Accelerated Weathering Data Collected at 400, 700, and 1,000 Hours Only (used, with permission, from Campbell et al. (12))

Coating	Controls	SO_2 concentration		O_3 concentration	
	0 ppm	0.1 ppm	1.0 ppm	0.1 ppm	1.0 ppm
			SHADED		
Automotive refinish	1 ± 1.3	1 ± 2.2	3 ± 1.4 95%	1 ± 2.3	3 ± 1.4 96%
Latex	1 ± 1.8	2 ± 2.2	8 ± 2 99%	2 ± 4.4	2 ± 1.6 95%
Industrial maintenance	11 ± 4.6	8 ± 9.4	13 ± 2.9 50%	6 ± 8.8	21 ± 15 87%
Coil	3 ± 1.9	3 ± 5.4	19 ± 10.5 99%	1 ± 6.1	4 ± 4.2 55%
Oil	0 ± 6.4	9 ± 4.3	47 ± 10.6 99%	8 ± 6.8	11 ± 3 30%

Table III. Slope of Erosion Data (mils loss x 10^{-5}/hr with 95% confidence limits) Accompanied by a T-test Probability (%) that a Statistical Difference Exists Between the Respective Slope for a Given Pollutant Type and Level vs. the Zero Pollutant Level, Based on Accelerated Weathering Data Collected at 400, 700, and 1,000 Hours Only (used, with permission, from Campbell et al. (12))

Coating	Controls	SO_2 concentration		O_3 concentration	
	0 ppm	0.1 ppm	1.0 ppm	0.1 ppm	1.0 ppm
			UNSHADED		
Automotive refinish	2 ± 0.8	4 ± 5	3 ± 2.6 75%	2 ± 1.7	5 ± 1.3 99%
Latex	4 ± 1.5	3 ± 13.4	11 ± 1 99%	2 ± 0.3	9 ± 5.9 93%
Industrial maintenance	19 ± 5.1	12 ± 3.3	22 ± 7 66%	10 ± 14.1	28 ± 14 85%
Coil	12 ± 2.3	9 ± 1.7	34 ± 4.7 99%	6 ± 3.3	15 ± 2.5 94%
Oil	20 = 7.2	22 ± 2	141 ± 19 99%	22 ± 17.2	45 ± 10.5 99%

exposure. Although this result might suggest some synergism between these two effects with the oil-based house paint, the data do not support extending this to other paints. It is likely an isolated effect caused by combining a highly UV-sensitive binder with $CaCO_3$ extender pigments. The results from these experiments should be viewed with some caution because of possible complications in measuring weight loss. Addition or condensation type reactions of pollutants with paint components, diffusion of pollutants into the film, outgassing of the paint, and accumulation of dirt (particularly in the field exposures) can all bias the weight-loss measurements.

In a later study (13), a controlled environment chamber was used to identify direct and possible synergistic effects of SO_2, NO_2, and O_3 on an oil-based house paint, an acrylic latex house paint, a vinyl coil coating, and an acrylic coil coating, all on aluminum substrates. Variables in chamber conditions included temperature, relative humidity (RH), pollutants (SO_2, NO_2, and O_3), dew, and light. The chamber was constructed such that the paint surfaces were subjected to moisture-condensing conditions. The result of the study, based only on weight loss, indicated that oil-based house paints having silicate extenders were affected by SO_2 and RH. Degradation to the acrylic and vinyl coil coatings was very slight. It is significant that the latex house paint failed because of corrosion of the aluminum substrate. In the case of porous films such as latex paints, the diffusion of SO_2 through the film was sufficient to bring about catastrophic paint failure.

In a more recent 30-month outdoor exposure study involving nine sites in the St. Louis, Missouri, area, the performance of good-quality oil- and latex-based house paints on stainless steel were evaluated by weight-loss measurements (14). The paints did not contain $CaCO_3$ extender pigments. Spencer and Haynie reported no effects caused by SO_2, but erosion rates were a function of time of wetness, temperature, and sunlight. No explanation was given for change in SO_2 effects in comparison with the earlier investigations. Because the paints that performed poorly in the earlier studies contained acid sensitive extenders, the higher resistance of the paints used in the later study may be attributed to the lack of these extenders. Spence and Haynie reported significantly lower erosion rates for oil- versus latex-based paints. However, erosion rates alone may not be a good criteria for evaluating latex- versus oil-based paints. The major cause of paint failure on wood is not erosion but catastrophic failure of the wood-paint interface and subsequent peeling of the paint. Reductions in erosion rates through formulating highly crosslinked paints may be counter productive.

Acid Deposition Effects on Wood and Cellulosic Materials

The effects on wood and other cellulosic materials of acid rain and the oxides of nitrogen and sulfur have been reported.

Raczkowski (3) found that exposing strips of microtomed spruce (*Picea abies* Karst.) to sunlight, wind, and rain resulted in decreased tensile strength compared with unexposed controls. Twelve sets of specimens were exposed for 1-month periods over 1 year. The loss in tensile strength along the grain was generally directly related to the amount of sunlight during the summer and fall (Figure 6); however, the loss in strength for winter and spring seemed higher than could be accounted for on the basis of sunlight alone. The acidity of the rain and SO_2 levels were higher during the winter and spring, and this higher loss was attributed to these higher acidity levels. Initial inspection of the data indicated that the main effect was caused by sunlight and that there may be an acid effect particularly during the winter months. However the summer acid levels seem rather low. The pHs ranged from 6 to 7.5 (Figure 7). The pH of rain caused only by atmospheric CO_2 should be 5.6. If the study had included another control that was exposed to sunlight but not to acid conditions, it might have been possible to separate these two effects.

In a series of three papers, Arndt and Gross (15-17) reported on color and weight changes caused by outdoor exposure and accelerated weathering of wood. In one of the papers, painted steel was also evaluated. In both the wood and painted steel, only color changes could be observed following outdoor and accelerated weathering. Conclusions based on color change must be viewed with some caution because, in the case of paint, minor pigment instability could show color change without seriously affecting the binder integrity. With most wood, color changes are related to the instability of extractives, which have little to do with the integrity of the wood. In addition, the gravimetric measure of weathering was complicated by several covariables; the changes in SO_2 levels in different locations also had different moisture exposure, solar exposure, and dry deposition rates.

The degradation of wood surfaces by ozone (O_3) was also studied (18). Exposure of wood at two moisture contents (6% and saturated) to O_3 concentrations of 0.5, 1.0, and 1.5% resulted in a weight loss of wood. However, the specimens showed no loss of crystallinity or strength parallel to the grain. It appeared that the loss in weight was caused by degradation of the easily accessible hemicelluloses. The crystalline cellulose was not degraded and therefore the strength was unaltered.

Evans and Banks (19) exposed microtomed sections of Lime (*Tilia vulgaris*) and Corsican pine (*Pinus nigra*) to dilute sulfuric, sulfurous, nitric, acetic, and formic acids, pH 2.0 to 6.0, for up to 12 months at 40°C. Controls were soaked in water at the same temperature. At a pH of 2.0, all acid treatments caused greater loss of strength and toughness compared with the controls. The amount of strength loss was pH dependent. The most interesting result, however, was that the strength loss caused by sulfurous acid was greater than all other acids at similar pHs. They noted that the strength losses may be caused by a combination of both hydrolytic degradation of the hemicelluloses and a sulfonation reaction of the lignin. Scanning electron microscopy (SEM) data showing failure at the lignin-rich middle lamella further supported the speculation that lignin degradation is partially responsible for the decrease in strength and toughness.

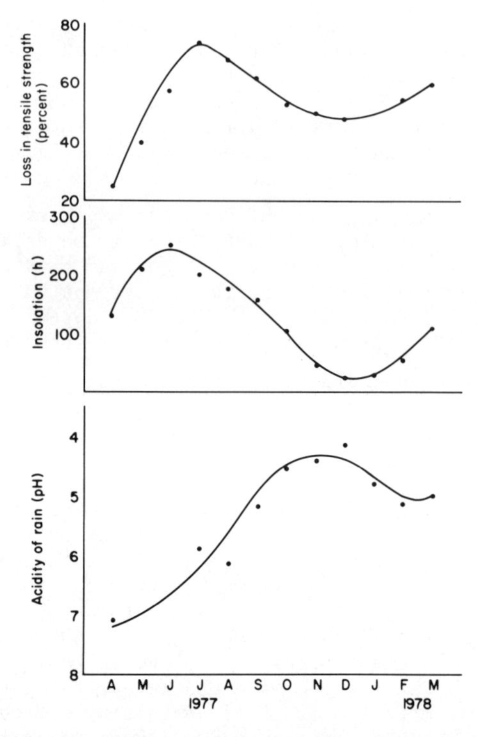

Figure 6.--The loss of tensile strength of spruce microsections, the mean monthly insolation, and the acidity of the rain water (3). (ML85 5193)

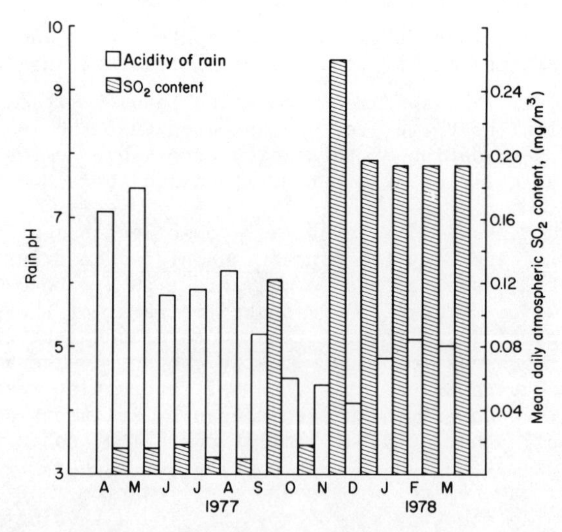

Figure 7.--Mean monthly pH of rainwater and sulphur dioxide concentration in the air (3). (ML85 5192)

Feist (20) and Williams (21) have used xenon arc accelerated weathering to determine the efficacy of surface treatments of wood. The degradation of the surface is manifest as erosion and can be measured microscopically (2). In recent work at the Forest Products Laboratory (22), these techniques were used to determine the effect of acid treatment on the erosion rate of western redcedar (*Thuja plicata* Donn ex D. Don).

Five replicate specimens of western redcedar and their longitudinally end-matched controls were artificially weathered (xenon arc Weather-Ometer, approximately 24-h light and 4-h deionized water spray daily). The Weather-Ometer was shut down for about 1 h Monday through Friday in order to soak the specimens in dilute acid for 15 min. Six types of acid soaks were used: nitric and sulfuric acids at pHs of 3.0, 3.5, and 4.0. I found similar results with both acids. Compared with the unsoaked controls, the 3.0 pH acid caused a 10% increase in erosion rate. At a pH of 3.5 the effect was a 4% increase due to the acid, and no effect was found at a pH of 4.0.

Although only a limited amount of work has been done on air pollution effects on wood, the effects on other cellulosic material, such as cotton yarn and fabric, have been studied by several groups (23-28). With minor differences as to the importance of nitrogen oxides, the general results of these studies showed that there was an accelerated degradation effect caused by pollution. These studies compared the strength of fabric following various exposures of cotton cloth to outdoor environments having varying amounts of pollution. The results indicated a strong correlation between strength loss and SO_2 concentration, with soiling also having an impact. Also as pointed out by these researchers, the solar effects cannot be neglected, particularly during the summer months. As for the mechanism of degradation or the interactions of the various factors, (SO_2, soiling, and light), these aspects of the problem have not been addressed. In addition, the effect of wet deposition versus dry deposition has not been adequately addressed. The work showing the effect of soiling covers only one aspect of the dry deposition problem. The results of the work with cellulose clearly demonstrate an effect of acid deposition. Further work is needed to delineate the various factors and the relationship of the various factors with one another.

Final Comments

There are several avenues for future research in the resistance of wood and wood coatings to degradation by acid rain.

Previous research on acid deposition effects on wood coatings did not take into account the effect of the substrate but linked poor performance in acid environments to the extenders in the paint. It should be relatively easy to formulate paints to overcome this erosion problem. However, in wood it is catastrophic failure that causes most poor-paint performance. Thus, the effect of acids on this interface failure should be addressed in future work.

It is important to test coatings on the substrate for which they are designed. Research on painted steel will not likely be applicable to painted wood because the properties of the paint must be different in order to accommodate the vastly different properties of steel and wood. A highly cross-linked moisture barrier applicable to steel will actually trap moisture at the wood-paint interface and could cause peeling. These coatings are brittle and are prone to cracking, which leads to further moisture penetration. Paints for wood products should be more flexible and porous.

In the research done so far, specimens have not usually been exposed to the specific locations where acid deposition is most likely to occur on a painted structure--protected areas such as under eaves and soffits. In fact, intercoat peeling of zinc-containing paints under eaves has been linked to the formation of moisture-sensitive salts through the action of pollutants (12). The link is rather circumstantial and further research is needed.

The discoloration or "frosting" of latex paints is reported to be caused by the concentration of pollutants such as SO_2, NO_2, or H_2S in protected areas such as under eaves and soffits (12, 29). When these pollutants are absorbed by paint films under moist condensing conditions, the acids that form may penetrate the film and react with some pigments. The salts formed by this reaction leave a grayish-white deposit on the paint after the water evaporates. The problem is more noticeable with darker colors.

It is conceivable that acid or SO_2 would diffuse through the film over an extended period of time and thus have an impact on the wood-paint interface. It should be kept in mind, however, that the pH of most wood species is already in the range of wet-acid deposition. Under dry deposition conditions, higher buildup of acid (SO_2) may occur at this interface. It would be worthwhile to design an experiment to measure the acid conditions under the film.

Future research on acid degradation should determine which components of wood are most affected by acid and what effect this degradation has on paint performance. Assuming there is an acid environment at the wood-paint interface, some hydrolysis of the hemicelluloses is probable. It is also likely that amorphous cellulose would be hydrolyzed. Recent work by Banks (19) linked lignin degradation to sulfurous acid. Crystalline cellulose should remain unaltered; but if hydrolysis of the hemicelluloses does occur, it is not necessarily a sufficiently serious condition to weaken the paint adhesion. We know from recent work at the Forest Products Laboratory that brief exposure of wood to ultraviolet light (natural sunlight) prior to painting drastically decreases paint adhesion. This exposure affects mainly the lignin. Thus, it seems that of the three main constituents of wood (cellulose, hemicellulose, and lignin), lignin is the most critical for paint adhesion. Evaluating lignin susceptibility to degradation by acid deposition may be the key to understanding the effects of acid deposition on wood and finished wood.

Disclaimer Notice

The use of trade, firm, or corporation names in this publication is for the information and convenience of the reader. Such use does not constitute an official endorsement or approval by the U.S. Department of Agriculture of any product or service to the exclusion of others that may be suitable.

Literature Cited

1. Feist, W. C.; Hon, D. N.-S. In "Chemistry of Solid Wood;" Rowell, R. M., Ed.; ADVANCES IN CHEMISTRY SERIES No. 207, American Chemical Society: Washington, DC, 1984; pp 401-51.
2. Feist, W. C.; Mraz, E. A. Forest Prod. J. 1978, 28(3), 38-42.
3. Raczkowski, J. Holz als Roh-und Werkstoff 1980, 38(6), 231-34.
4. Miniutti, V. P. Offical Digest 1963, 35(460), 451-71.
5. Miniutti, V. P. Forest Prod. J. 1964, 571-76.
6. Gross, H. Effects of air pollution on surface finishes and similar coatings," Erich Schmidt Verlag, Berlin, 1979.
7. Wicks, Z. W., Jr. In "Corrosion Control by Coatings," Leidheiser, H., Jr., Ed.; Science Press: Princeton, 1979; p. 29-34.
8. Holbrow, G. L. J. Oil Colour Chemists' Assoc. 1962, 45(10), 701-18.
9. Tice, E. A. J. Air Pollution Control Assoc. 1962, 12, 533.
10. Spence, J. W.; Haynie, F. H. J. Paint Technol. 1972, 44(574), 70-74.
11. Gutfreund, K. J. Paint Technol. 1966, 38(503), 732-39.
12. Campbell, G. G.; Schurr, G. G.; Slawikowski, D. E.; Spence, J. W. J. Paint Tech. 1974, 46(593), 59-71.
13. Spence, J. W.; Haynie, F.; Upham, J. B. J. Paint Technol. 1975, 47(609), 57-63.
14. Haynie, F. H.; Spence, J. W. J. Air Pollution Control Assoc. 1984, 34, 941-44.
15. Arndt, U.; Gross, U. Staub Reinhaltung der Luft. 1974, 34(6), 225-27.
16. Arndt, U.; Gross, U. Staub Reinhaltung der Luft. 1976, 36(10), 405-10.
17. Arndt, U.; Gross, U. Staub Reinhaltung der Luft. 1977, 37(2), 53-55.
18. Dick, J. L.; Murphey, W. K. Research Briefs. 1972, 6(2), 4-6.
19. Evans, P. D.; Banks, W. B. The International Research Group on Wood Preservation, Working Group III, Preservatives and Methods of Treatment, Sixteenth Annu. Meet., Brazil, 12-17 May 1985. Document No: IRG/WP/3326, 21 Feb. 1985.
20. Feist, W. C. "Protection of wood surfaces with chromium trioxide;" Res. Pap. FPL 339, Madison, WI, U.S. Department of Agriculture, Forest Service, Forest Products Laboratory; 1979.
21. Williams, R. S. J. Appl. Polym. Sci. 1983, 28, 2093-103.
22. Williams, R. S. Unpublished.
23. Race, E. J. Soc. Dryers Colourists 1949, 65, 55-63.
24. Bogaty, H.; Campbell, K. S.; Appel, W. D. Textile Res. J. 1952, 22, 81-83.
25. Morris, M. A.; Young, M. A.; Molig, T. A. Textile Res. J. 1964, 34, 563-64.
26. Brysson, R. J.; Trask, B. J.; Upham, J. B.; Booras, S. G. J. Air Pollution Control Assoc. 1967, 17(5), 294-98.
27. Brysson, R. J.; Trask, B. J.; Cooper, A. S., Jr. Am. Dyestuff Rep. 1968, 57, 512-17.
28. Brysson, R. J., Walker, A. M.; Cooper, A. S., Jr. Textile Res. J. 1975, 45, 154-59.

29. Olympic Stain, A division of Clorox Company. Olympic Technical Information Bulletin No. 14, Bellevue, WA; 1977.
30. Forest Products Laboratory. Wood Handbook: Wood as an engineering material; Agric. Handb. 72, rev., Washington, DC; U.S. Department of Agriculture; 1974.
31. Feist, W. C.; Oviatt, A. E. "Wood siding--installing, finishing, maintaining," Home and Garden Bull. 203, Washington, DC; U.S. Department of Agriculture; 1983. 23 p.

Glossary (Extracted from the Forest Products Laboratory, Wood Handbook, 1974.)

ANNUAL GROWTH RING. The layer of wood growth put on a tree during a single growing season. In the temperature zone the annual growth rings of many species (e.g., oaks and pines) are readily distinguished because of differences in the cells formed during the early and late parts of the season. In some temperate zone species (black gum and sweetgum) and many tropical species, annual growth rings are not easily recognized.

CELLULOSE. The carbohydrate that is the principal constituent of wood and forms the framework of the wood cells.

DECAY. The decomposition of wood substance by fungi.

Advanced (or typical) decay. The older stage of decay in which the destruction is readily recognized because the wood has become punky, soft and spongy, stringy, ringshaked, pitted, or crumbly. Decided discoloration or bleaching of the rotted wood is often apparent.

Incipient decay. The early stage of decay that has not proceeded far enough to soften or otherwise perceptibly impair the hardness of the wood. It is usually accompanied by a slight discoloration or bleaching of the wood.

DIMENSIONAL STABILIZATION. Special treatment of wood to reduce the swelling and shrinking that is caused by changes in its moisture content with changes in relative humidity.

EARLYWOOD. The portion of the annual growth ring that is formed during the the early part of the growing season. It is usually less dense and weaker mechanically than latewood.

EDGE-GRAIN LUMBER. Another term for vertical-grain lumber.

EXTRACTIVE. Substances in wood, not an integral part of the cellular structure, that can be removed by solution in hot or cold water, ether, benzene, or other solvents that do not react chemically with wood components.

FIBER SATURATION POINT. The stage in the drying or wetting of wood at which the cell walls are saturated and the cell cavities free from water. It applies to an individual cell or group of cells, not to whole boards. It is usually taken as approximately 30 percent moisture content, based on ovendry weight.

FINISH (FINISHING). Wood products such as doors, stairs, and other fine work required to complete a building, especially the interior. Also, coatings of paint, varnish, lacquer, wax, etc., applied to wood surfaces to protect and enhance their durability or appearance.

FLAT-GRAINED WOOD. Lumber that has been sawed parallel to the pith and approximately tangent to the growth rings. Lumber is considered flat grained when the annual growth rings make an angle of less than 45° with the surface of the piece.

HEARTWOOD. The wood extending from the pith to the sapwood, the cells of which no longer participate in the life processes of the tree. Heartwood may contain phenolic compounds, gums, resins, and other materials that usually make it darker and more decay resistant than sapwood.

HEMICELLULOSE. A celluloselike material (in wood) that is easily decomposable as by dilute acid, yielding several different simple sugars.

LATEWOOD. The portion of the annual growth ring that is formed after the earlywood formation has ceased. It is usually denser and stronger mechanically than earlywood.

LIGNIN. The second most abundant constituent of wood, located principally in the secondary wall and the middle lamella, which is the thin cementing layer between wood cells. Chemically it is an irregular polymer of substituted propylphenol groups, and thus no simple chemical formula can be written for it.

LUMEN. In wood anatomy, the cell cavity.

MOISTURE CONTENT. The amount of water contained in the wood, usually expressed as a percentage of the weight of the ovendry wood.

RADIAL. Coincident with a radius from the axis of the tree or log to the circumference. A radial section is a lengthwise section in a plane that passes through the centerline of the tree trunk.

QUARTERSAWED LUMBER. Another term for vertical-grained lumber.

RELATIVE HUMIDITY. Ratio of the amount of water vapor present in the air to that which the air would hold at saturation at the same temperature. It is usually considered on the basis of the weight of the vapor but, for accuracy, should be considered on the basis of vapor pressures.

SAPWOOD. The wood of pale color near the outside of the log. Under most conditions the sapwood is more susceptible to decay than heartwood.

SPRINGWOOD. (See EARLYWOOD.)

SUMMERWOOD. (See LATEWOOD.)

TANGENTIAL. Strictly, coincident with a tangent at the circumference of a tree or log, or parallel to such a tangent. In practice, however, it often means roughly coincident with a growth ring. A tangential section is a longitudinal section through a tree or limb perpendicular to a radius. Flat-grained lumber is sawed tangentially.

VERTICAL-GRAINED LUMBER. Lumber that has been sawed so that the wide surfaces extend approximately at right angles to the annual growth rings. Lumber is considered vertical-grained when the rings form an angle of 45° to 90°with the wide surface of the piece.

WEATHERING. The mechanical or chemical disintegration and discoloration of the surface of wood caused by exposure to light, the action of dust and sand carried by winds, and the alternate shrinking and swelling.

RECEIVED January 2, 1986

23

Effect of Acid Rain on Woody Plants and Their Products

Ellen T. Paparozzi

Department of Horticulture, University of Nebraska, Lincoln, NE 68583-0724

Acid rain has been shown to effect woody plants at all stages of their development. Seedlings, immature through older, fully expanded leaves and needles, as well as annual rings of woody plants have been shown to be injured by simulated acid rain. There has been research that suggests that the growth of trees could be affected if acid rain has a long term acidifying effect on soils. The ultimate implication of the effects of this pollutant as it impacts tree growth will be in terms of tree productivity and wood quality. This could influence products such as paper and wood for furniture, blight our landscape and affect roofs, decks, telephone poles, and fences. Little work has been done on the effect of acid rain on wood products. However, work on wood chemistry indicates that acid rain has the potential and the characteristics needed to be a destructive force to these materials.

Of all the organic materials mentioned during this conference, trees are one of the most multidimensional. They impact our lives visually, as well as physically. They are a source of recreation, the key to a vital industry employing over a million people, and provide a haven and/or food source for wildlife. Until recently trees have been taken for granted in our everyday lives. The renewed public interest in woody plants, particularly trees, is largely due to atmospheric pollutants, of which acid rain is one, which have been causing stress and damage to forests here and in Europe. Acid precipitation is composed of both dry and wet deposition. As dry deposition is difficult to measure, most collection studies have centered on the wet deposition. Quantity, elemental position and pH are routinely recorded by weather stations participating in the National Atmospheric Deposition Program. Rain is made acidic by the addition of CO_2,SO_2 and NO_x, with the SO_2 and NO_x components being of greatest concern. The proportions of SO_2 to NO_x vary across the United States and are projected to favor the increasing of the NO_x component of acid rain in the future.

0097-6156/86/0318-0332$06.00/0

Trees can be impacted by acid rain in any number of ways and these effects can be classified as either direct or indirect(1). The following discussion will present an overview of the many ways acid rain could affect woody plants specifically as well as unfinished wood products.

POLLEN GERMINATION AND POLLEN TUBE GROWTH

Pollen of nine forest trees was collected and cultured on a medium that had been adjusted, using dilute sulphuric acid to pH levels as low as 2.6. A reduction in pollen tube length was observed in sugar maple (*Acer saccharum*) and quaking aspen (*Populus tremuloides* clone III) at a medium pH level of 4.0. At pH 3.6 and below sugar maple, all three clones of quaking aspen and paper birch (*Betula papyrifera*) showed a decrease in pollen germination as well as a reduction in pollen tube length, while white pine (*Pinus strobus*) showed only a decrease in pollen germination. At pH 3.0, pollen tube length of white pine and Canadian hemlock (*Tsuga canadensis*) was decreased, with all species (including *Prunus pennsylvanica*, *Pinus resinosa*, *Picea mariana*, and *Pinus banksiana*) showing reduced pollen germination and pollen tube growth (2). Sidhu (3) found that simulated acid rain of pH 3.6 could inhibit pollen germination of white spruce by up to 30% as well as pollen tube growth.

In a field study it was found that, while simulated acid rain at pH 2.5 damaged 'Empire' apple blssoms and reduced pollen germination, it did not influence fruit set (4).

WOODY PLANT SEED GERMINATION

Germination of woody plant seed has been found to be inhibited or stimulated by simulated acid rain. Inhibition of red maple (*Acer rubrum*) seed germination was found at pH 4.0 and 3.0, while staghorn sumac (*Rhus typhina*) seed germination was inhibited at pH 3.0 (5). However, seed from flowering dogwood (*Cornus florida*), sugar maple (*Acer saccharum*), red alder (*Alnus rubra*), American beech (*Fagus grandifolia*), tulip-poplar (*Liriodendron tulipifera*), and shagbark hickory (*Carya ovata*) did not germinate any differently when exposed to pH levels of 3.0, 3.5, 4.0 or 5.7 (5). Raynal et al. (6) also found that while seed germination overall was unaffected, when seeds were exposed to pH 3.0 simulated acid rain, radical elongation was reduced, mostly due to bacterial infection. Germination of white pine (*Pinus strobus*) (5, 7), Eastern red cedar (*Juniperus virginiana*) and Douglas fir (*Pseudotsuga menziensii*) was stimulated by pH levels of 3.5 and 3.0 simulated acid rain (5).

SYMPTOMS OF INJURY

Several woody plant species will react to exposure of simulated acid rain by forming foliar lesions. While the pH level of simulated acid rain that produced these lesions varies from species to species, the lesions themselves are generally yellow to brown necrotic spots or regions (see Table 1). Additionally, on two clones of poplar, galls were formed in response to simulated acid rain. Surface characteristics such as stomatal presence and density, trichome density, type and amount of epicuticular wax, leaf surface wettability and buffering capacity have all been shown to influence foliar injury by simulated acid rain. However, it may be the interaction of any or all of these factors, in addition to leaf/needle age that determines the individual response by species to pH levels of simulated acid rain (8, 9, 10, 11).

Leaf injury during one growing season may not be as critical to deciduous species as it is to coniferous species. Paparozzi (12) found that, despite extensive leaf injury to young birch trees from simulated acid rain of pH 2.8 and 3.2, the rate of abscission of injured leaves was not accelerated. The following spring, the same previously injured plants leafed out and grew comparably to the control plants and plants which had received simulated acid rain of pH 4.0 and 5.6. Coniferous or evergreen species, however, which would hold their injured needles for up to three years could show a gradual decline in growth due to a reduction in photosynthesis.

The previous studies and almost all of the simulated acid rain experiments discussed herein were designed to quickly produce symptoms and obtain threshholds for plant injury. To date there has been no documented field identification of ambient acid rain injury to forest trees or woody plants (8). Among many potential reasons, this may be due either to the fact that ambient acid rain is often accompanied by other atmospheric pollutants, thus injury symptoms are different, or that changes are occurring subtly over time.

TABLE 1. A partial list of woody plants which developed foliar lesions after exposure to simulated acid rain.

PLANT	pH	CITATION
Acer saccharum Marsh	3.0	(13)
Acer rubrum L.	2.0	(14)
Betula alleghaniensis Britt.	3.0	(15, 11)
Carya illinoensis (Wang.) K. Kalt	2.0	(14)
Cornus florida L.	2.0	(14)
Liriodendron tulipifera	2.0	(14)
Malus hypehensis (Pamp.)	3.25	(16)
Pinus strobus L.	1.0	(16)
Populus spp. hybrids	3.1	(17)
Quercus phellos	3.2	(18)
Quercus prinus L.	2.0	(14)
Quercus rubra L.	2.8	(19)
Rhododendron spp.	2.8	(19)
Robinia pseudoacacia L.	2.0	(14)

LEACHING OF ELEMENTS FROM FOLIAGE

Rain and mist, acid or otherwise, have been shown to leach nutrients from the foliage of woody plants (19). Wood & Bormann (13) found that when sugar maple seedlngs were exposed to pH 3.0 (leaf injury), 3.3, and 4.0 of simulated acid rain, there were significant increases in leaching of K^+, Mg^+ and Ca^+ ions. Cronan (21) found that in coniferous throughfall which was exposed to ambient rain of pH 4.0, there was an increase in H^+, Ca^+, Mg^+, K+, Mn^+, SO_4^-, NO_3^- and Cl^- as compared to bulk precipitation. In contrast, throughfall from hardwoods showed increases in Ca^+, Mg^+, K^+, SO_4, NO_3^-; no change in Mn+ and $Cl^-$and a decrease in H^+ compared to bulk precipitation. While the exact ion concentrations differed, generally other researchers (22, 23, 24), also made similar observations in their field studies. Scherbatskoy and Klein (25) using simulated acid mist, found that, in addition to leaching of K^+, Ca^+ and NO_3^-, amino acids were also leached at pH 2.8. Leaching of chlorophyll, $H_2PO_4^-$, and carbohydrates was not affected by varying the mist pH level. Additionally, it was found pH 4.3 simulated acid mist, after it passed through yellow birch and white spruce trees, became more acidic. They suggested that this may be due to removal of dry acidic deposits from the leaves. Alcock and Morton (26) also found a decrease in the pH level of rainfall as it passed through field sites of European birch (Betula pendula) and Scotch pine (Pinus sylvestris). However, they suggest that, in addition to the previous explanation, the increase in acidity could be related to the leaching of organic acids from the leaves. Direct nutrient availability from leaves to the tree terminates when leaves abscise. However, leaf litter around the base of the tree can contribute nutrients back to the soil as the litter decomposes.

Hovland et al. (27) used acidified water on Norway spruce needles to test its effect on nutrient leaching. They found that K^+ leaching was increased as water quantity increased the leaching, while Mg+, Mn+ and Ca+ leaching was increased by using water solutions of increased acidity. Acidified water decreased phosphate leaching during the first fourteen weeks of the experiment, but then increased leaching after that time. Hagvar and Kjondal (28) using white birch (Betula verrucosa) leaves, also found that Ca+, Mg^+ and Mn^+ were effectively removed from leaf litter as the acidity of the water solution increased. Initially, pH 3.0 and 2.0 increased the decomposition rate. However, later in the experiment it was found that the decomposition rate was reduced at pH 2.0. Both groups of researchers commented that leaching may be dependent on the substance that is being decomposed at that time and that this will change over time as the litter decomposes.

Additional information on how leaf litter potentially interacts with acid rain was contributed by Lee & Weber (29). In their experiments, acid rain was simulated in a field situation on sugar maple and red alder. Rain as throughfall was allowed to interact with leaf litter and the leachate was collected. Litter leachate was found to be higher in SO_4^-, Ca+ and Mg+, and the pH was found to have increased. Thus, they hypothesized that the litter was neutralizing the simulated acid rain, with red alder litter being more effective than sugar maple.

SOIL EFFECTS

The effect of acid rain on leaf litter should be viewed in conjunction with the effect of acid rain on the underlying soils.

If the chemistry of soils is affected by acid rain directly or through interaction with the leaf litter, this in turn will affect elements available to woody plants. Of most concern are the macronutrients N, P, K, S, Mg and Ca, as well as elements that may be toxic to plants, such as Al. According to Krug and Frink (30,31) there are many misconceptions about soil acidity. One of these is that acid rain will increase the acidification of soil and water.

This comment concurs with the observations of Richter et al. (32) in two forest ecosystems. They found at both sites that strong acids in bulk precipitation were neutralized by forest canopies and the surface soil layers. Tabatabai (33), agreeing that additions of acid rain to soils are probably insignificant, notes that soil formation is an acidifying process. Soils alone produce sources of acidity by mineralization of organic N & S, carbonic acid formation and nitrogen fixation. Soils also receive sources of acidity in the form of organic acids from litter decomposition.

Johnson et al.(34) in their review on cation leaching by natural processes and acid deposition, state that it did not appear that atmospheric inputs caused significant losses of base cations. This is probably due to the fact that many soils have a substantial buffering capacity. If that buffering capacity is exceeded by acid inputs and if leaching losses of cations are greater than those produced by the weathering of minerals, then there will be an increase in soil acidification and weathering. However, though this occurrence may be rare, an example where it could occur would be poorly buffered soils receiving large inputs of acid deposition (35).

However, as Krug and Frink (31) and Johnson et al. (34) respectively point out, the influence of changing land use and successive vegetation and forest management practices will need to be accounted for. Johnson et al. (35) go on to suggest that generalizations about acid rain effects will have little meaning. Site conditions, amounts and types of inputs will vary and the result may be a positive, negative or no effect on the forest ecosystem. An extensive review of this area and interesting reading is found in Evans et al. (36).

ACID RAIN AND WOODY PLANT DISEASES AND INSECTS

There is little published work on the relationship between acid rain and insects. However, Smith et al. (37) point out that a number of damaging forest insects detect and respond to trees under stress. Environmental changes such as acid rain could cause tree stress. The insects that would bear observation would be those that spend all or part of their life cycles on leaves or needles, as that is where injury occurs and substances are leached, or microarthropods inhabit leaf litter and soil.

Research addressing the effect of simulated acid rain on woody host:disease pathogen interactions is also limited. Shriner (38) found that oak trees exposed to pH 3.2 simulated acid rain had 84% less telia, produced by the oak-pine rust pathogens, compared to the control. Bruck & Shafer (39) exposed loblolly pine (*Pinus taeda*) to simulated acid rain of pH 5.6 to 2.4 and inoculated plants with fusiform rust. Needle necrosis occurred, with necrosis formed on 60% of needles exposed to pH 2.4 as compared to no needle necroses at pH 5.6. Six months later they found significantly fewer rust galls on needles exposed to the low pH level. They noted that the basidiospore of fusiform rust penetrates through healthy needle tissue. With fewer areas of the needle left alive due to the exposure to simulated acid rain of pH 2.4, infection was less and thus, fewer rust galls.

These few experiments are not enough, however, to define the effects of simulated acid rain on the woody host:disease pathogen interaction. As Evans (8) points out, as yet there are no experimental field data that show that the host:pathogen interaction significantly changes plant survival or productivity. Certainly, this is one of the more complex study areas with the interaction by the leaf, the pathogen and the rainfall all being multidimensional and ever changing, depending on the type of plant and pathogen, stage of pathogen infection, the leaf surface, the lesion produced and the elemental leaching from the leaf by acid rain.

Woody plants often also have a symbiotic relationship with a mycorrhizal fungus. Mycorrhizae, of which there are two general types; endomycorrhizae and ectomycorrhizae, are thought to enhance nutrient uptake, particularly phosphorus.

Preliminary work using loblolly pine seedlings growing in sand, explored the effect of simulated acid rain on the infection of roots by ectomycorrhizae. Shafer et al. (40) found that simulated acid rain of pH 4.0 and 3.2 was inhibiting ectomycorrhizal infection when compared to roots exposed to pH 5.6. At pH 2.4 there appeared to be stimulatory effects on infection. The authors suggested that increased soil acidity was the cause of the enhanced ectomycorrhizal infection, as other experiments have shown that, if substrate acidity is increased, so will infection. The authors did caution, however, that this short term greenhouse study was just preliminary.

ACID RAIN AS A STIMULATOR AND INHIBITOR OF GROWTH

Some researchers have reported that, despite needle or leaf injury at low levels of simulated acid rain, growth of the plant occurred. Wood and Bormann (7) exposed white pine seedlings to simulated acid rain of pH 2.3, 3.0, 4.0 and 5.6. Plants exposed to pH 3.0 and 4.0 had greater total plant and needle weights than plants exposed to pH 5.6. Plants exposed to pH 2.3 showed needle necroses but had significantly greater total plant and needle weights than any other treatment. Raynal et al. (41) found that sugar maple seedlings under nutrient-limited conditions showed foliar damage and growth stimulation at pH 3.0. Both groups suggest that the NO_3^- component may be responsible for growth stimulation. However, this is questionable, as Wood & Bormann (13) pointed out, growth was also accompanied by leaching of K^+, Mg^+ and Ca^+. One would expect that, for growth to occur, these essential macronutrients would need to be readily available.

Tveite and Abraham (42) also observed a stimulation in height and diameter of Scots pine saplings when pH levels 2, 2.5 and 3.0 of simulated acid rain were supplied. There were no effects on Norway spruce or lodgepole pine. The authors suggested that the growth of Scotch pine may be due to increased uptake of nitrogen from the soil.

The ultimate aim of all the acid rain research was, and is, to help identify and characterize acid rain injury so it can be used to assess injury to woody plants in forests. This process has begun. Unfortunately, we need the key now.

The Pinelands of New Jersey, and the forests in the northeastern and southeastern U.S. and in Central Europe have shown a problem referred to as a decline. In the Pine Barrens, Johnson and co-workers (43) found that two-thirds of the shortleaf, loblolly and pitch pine trees sampled showed either a noticeably abnormal or dramatic decrease in ring increment size. This decreased growth occurred in young and old trees. Pests, ozone and wildfire were absent. Unfortunately, when data were analyzed, pH was significantly correlated with growth when combined with many of the independent variables. Thus, it was difficult to single out any one factor. However, the authors felt that summer drought, a variable that was frequently correlated with growth and pH, could be exacerbated by acid rain. They further pointed out that there was a clear relationship between stream pH and growth rate. Thus, acid rain should still be considered when evaluating this situation.

In the northern Appalachians, large numbers of red spruce have been dying. This has been occurring over the last twenty years without any obvious provocation. Here again, Johnson and Siccama (44) found that there was a dramatic decrease in tree ring increment size in the mid 1960's in 40% of the trees sampled. They suggested that this rather abrupt shift to narrow increments is an indication of red spruce decline. The authors felt that acid rain may be but one of several stresses, probably predisposing the plant to drought stress. Others stresses that could be involved include ozone, SO_2, long-term change in climate, heavy metals and drought. They also noted that drought is one common factor in the tree growth declines in the Pine Barrens, the northern Appalachians and the German forests.

In central Europe the forest decline is called the 'Waldsterben' syndrome. The Waldsterben syndrome is of great concern to the forestry industry and the public, especially in West Germany (45). The forested area that it has affected has increased from about 8% in 1982 to 50% in 1984. As Schutt and Cowling (46) noted, this phenomenon possesses several features that were different than the decline in the U.S., such as the fact that it affects simultaneously both deciduous and coniferous trees and that a rapid decrease in health and vigor occurs over a wide range of environmental conditions. Symptoms include that of water stress, growth decrease and abnormal growth. The stress factors inducing the syndrome are unknown, but acid rain may be one.

ACID RAIN EFFECTS ON UNFINISHED WOOD PRODUCTS

Millions of trees are harvested annually and processed into varying types of wood products from paper to lumber. Unfinished or raw wood products no longer possess the protective outer layer, the bark, that the tree does. Thus, the wood which is normally protected from acid rain in the forest, now, as a wood product, will be directly exposed to the environment. Therefore, it is important to consider the direct effect of acid rain on wood products such as telephone poles, fences, wood siding and shingles.

There was little research available in the literature on the effects of acid rain on wood products. Letters to various utility companies also indicated that there was little work done in this area. However, the available literature will be cited and integrated with current knowledge in the area of wood science.

Extensive research has been conducted utilizing acid hydrolysis on wood in order to investigate the chemistry of wood and convert wood to cellulose and even further to glucose for technological utilization (47). However, concentrations of acid used, even in dilute acid procedures, are generally much higher than found in acid rain.

Hovland and Abrahamsen (in 28) exposed cellulose sheets and small pieces of aspen wood (Populus tremula) to artificial acid "rain" to determine its effect on decomposition. These materials were placed on the leaf litter in three coniferous forest sites. The only result was that application of pH 2.5 water reduced the decomposition rate of cellulose in one of the sites. Hon (48) is currently working on the degredative effect of acid rain on the surface quality of wood and has found that acid rain does deteriorate wood. The effects of acid rain were further pronounced by the presence of UV light and moisture, thus, a synergism. Hon has observed that these factors together appear to affect the exposed surface by changing the color of the wood, its ultrastructure, tensile strength and chemical composition.

Banks et al. (49), using thin and thick sections of pine (Pinus sylvestris) and linden (Tilia vulgaris), found significant strength losses when 100 um longitudinal sections were exposed to dilute acid solutions, particularly sulfurous acid solutions. Using larger, transverse sections significant loss in mechanical properties occurred only to a depth of about 0.5 mm. Thus, the wood surface may be modified and this could influence the susceptibility of surfaces.

When a raw wood surface is exposed to the environment, it will weather. Weathering, which consists of the wood turning color and the gradual breakdown of the cells and surface of the wood, results from the interaction of factors such as light (particularly UV), water and wind. This process is not considered detrimental and often wood siding on buildings remains untreated in order to achieve a gray, natural look (50). However, the potential of normal weathering in combination with acid rain accelerating the exposure of wood surfaces to insects and diseases should not be overlooked.

Under environmental conditions, soft-rot fungi will slowly, gradually and progressively decay wood surfaces. Additionally, some fungi grow best around pH 4.0 to 6.0 with some fungi apparently being able to change the pH of the wood slightly as they grow (50). Thus, it is possible that acid rain may influence the invasion of wood surfaces by some fungi, but this will also depend on the type of wood.

Not all woods may show a predisposition to invasion by disease pathogens and insects. As Scheffer and Cowling (51) pointed out, woods do vary in the extent to which they will inherently resist heartwood decay. Certain types of oak and redwood are resistant to decay while some pines, birches and hickories are slightly or not resistant to heartwood decay. Two of the members of this slightly or not resistant decay category did show surface deterioration in work done by Banks et al. (49). It seems feasible that wood from these trees could be affected by acid rain and possibly other pollutants in combination with light and water. The result of this multiple factor interaction may then be impacted by insects or diseases.

Whether wood products made from trees which have been injured by atmospheric deposition (including decline) will show reduced wood strength and durability to the above factors is unknown. However, an informal working party within the International Association of Wood Anatomists has been formed to help address this topic. Baas

(52) notes that research in this area should increase in the future. As tree growth in relation to wood structure is a very complex area, it is important that experienced researchers get involved so that rash or misleading conclusions are not the result. The first endeavor of this working party was to compile an annotated list of scientists working on acid rain and wood structure. This effort will certainly help bring together interested scientists and further focus research in this area.

ASSESSMENT

The research available to date presents a partial view of the impacts of acid rain on woody plants. Many of the impacts are still only 'potential' impacts, as simulation studies versus field studies present a conflicting view. However, one thing appears quite clear - more research is needed. As many researchers have found, the effect of acid rain is not going to be one of simple cause and effect, but rather one of a multiple factor interaction. Thus, future work should be statistically designed to test the interaction(s) rather than main effects. Work needs to be done over both the short and long term to assess injury. Basic physiological work across disciplines with the standardization of techniques used (e.g. one set type of simulator for all researchers to produce simulated acid rain) must be employed in order for different experimental results to be comparable. If we can discover how plants will react to given combinations of stresses, only then will we be able to propose an appropriate course of action.

Published as Paper Number 7838, Journal Series, Nebraska Agricultural Experiment Station.

LITERATURE CITED

1. Tamm, C.O.; Cowling, E.B. Water, Air, Soil 1977, 503-512.
2. Cox, R.M. New Phytol. 1983, 95, 269-276.
3. Sidhu, S.S. Can. J. Botany 1983, 61, 3095-3099.
4. Forsline, P.L.; Musselman, R.C.; Kender, W.J.; Dee, R.J. J. Amer. Hort. Sci. 1983, 108, 70-74.
5. Lee, J.J.; Weber, D.E. Forest Sci. 1979, 25, 393-398.
6. Raynal D.J.; Roman, J.R.; Eichenlaub, W. Environ. Exp. Bot. 1982, 22, 385-392.
7. Wood, T.; Bormann, F.H. Water, Air, Soil 1977, 7, 479-488.
8. Evans, L.S. Ann. Rev. Phytopath. 1984, 22, 397-420.
9. Evans, L.S. Botan. Rev. 1984, 50, 449-490.
10. Craker, L.E.; Bernstein, D. Environ. Poll. Ser. A. 1984, 36, 375-381.
11. Paparozzi, E.T.; Tukey, H.B. Jr. J. Amer. Soc. Hort. Sci. 1983, 108, 890-898.
12. Paparozzi, E.T. Ph.D. Thesis. Cornell University, Ithaca, NY. 1981.
13. Wood, T.; Bormann, F.H. Ambio 1975, 4, 169-171.
14. Haines, B.; Stefani, M; Hendrix, F. Water, Air, Soil 1980, 14, 403-407.
15. Wood, T.; Bormann, F.H. Environ. Pollut. 1974, 7, 259-267.
16. Forsline, P.L.; Dee, R.J.; Melios, R.E. J. Amer.Hort. Sci. 1983, 108, 70-74.

17. Evans, L.S.; Gmur, N.F.; Da Costa, F. Phytopath. 1978, 68, 847-855.
18. Shriner, D.S. Water, Air, Soil 1977, 8, 9-14.
19. Keever, G.J. Ph.D. Thesis. Cornell University, Ithaca, N.Y., 1982.
20. Tukey, H.B. Jr. Ann. Rev. Plant Physiol. 1970, 21, 305-324.
21. Cronan, C.S. In "Indirect Effects of Acidic Deposition on Vegetation"; Teasley, J. Ed.; Acid Precip. Ser. 1984, 5, 65-79.
22. Eaton, J.S.; Likens, G.E.; Bormann, F.H. J. Ecol. 1973, 61, 495-508.
23. Hoffman, W.A. Jr.; Lindberg, S.E.; Turner, R.R. J. Environ. Qual. 1980, 9, 35-100.
24. Mollitor, A.V.; Raynal, D.J. Soil Sci. Soc. Amer.J. 1982, 46, 137-141.
25. Scherbatskoy, T.; Klein, R.M. J. Environ. Qual. 1983, 12, 189-195.
26. Alcock, M.R.; Morton, A.J. J. Appl. Ecol. 1981, 18, 835-839.
27. Hovland, J.; Abrahamsen, G.; Ogner, G. Plant and Soil 1980, 56, 365-378.
28. Hagvar, S.; Kjondal, B. Pediobio. 1981, 22, 232-245.
29. Lee, J.J.; Weber, D.E. J. Environ. Qual. 1982, 11, 57-64.
30. Krug, E.C.; Frink, C.R. Bulletin 811. The Conn. Ag. Exp. Station. 1983.
31. Krug, E.C.; Frink, C.R. Science 1983, 221, 520-525.
32. Richter, D.D.; Johnson, D.W.; Todd, D.E. J.Environ. Qual. 1983, 12, 263-270.
33. Tabatabai, M.A. Environ. Control 1985, 15, 65-110.
34. Johnson, D.W.; Richter, D.D.; Van Miegroet, H.; Cole, D. J. Air Poll. Con. Assc.1983, 3, 1036-1041.
35. Johnson, D.W.; Turner, J.; Kelly, J.M. Water Res. Res. 1982, 18, 449-461.
36. Evans, L.S.; Hendrey, G.R.; Stensland, G.J.; Johnson, D.W.; Francis, A.J. Water, Air, Soil 1981, 16, 469-509.
37. Smith, W.H.; Gebballe, G.; Fuhrer, J. In "Indirect and Indirect Effects of Acidic Deposition on Vegetation"; Teasley, J., Ed.; Acid Precip. Ser.1984, 5, 33-34.
38. Shriner, D.S., Phytopath. 1978, 68, 213-218.
39. Bruck, R.I.; Shafer, S.R. In "Indirect Effects of Acidic Deposition on Vegetation"; Teasley, J., Ed.; Acid Precip. Ser. 1984, 5, 19-32.
40. Shafer, S.R., Grand, L.F.; Bruck, R.I.; Heagle, A.S. Can. J. For. Res. 1984, 15, 66-71.
41. Raynal, D.J.; Roman, J.R.; Eichenlaub, W. Environ. Exp. Bot. 1982, 22, 385-392.
42. Tveite, B.; Abrahamsen, G. SNSF-contribution FA 29/78. 1978.
43. Johnson, A.H.; Siccama, T.C.; Wang, D.; Turner, R.S.; Barringer, T.H. Environ. Qual. 1981, 10, 427-430.
44. Johnson, A.H.; Siccama, T.G. Tappi J. 1984, 67, 68-72.
45. Steinbeck, K. Forestry 1984, 81, 719-720.
46. Schutt, P.; Cowling, E.B. Plant Disease 1985, 69, 1-9.
47. Wenzl, H.F.J. "The Chemical Technology of Wood"; Academic Press, NY. 1970.
48. Hon, D.N.-S. Amer. Chem. Soc. Abstr. 1985, Cell 0007, 0009.
49. Banks, W.B.; Evans, P.D. Amer. Chem. Soc. Abstr. 1985, Cell 0010.

50. Haygreen, J.G.; Bowyer, J.F. "Forest Products and Wood Science"; Iowa State Univ. Press, Ames, Iowa, 1982.
51. Scheffer, T.C.; Cowling, E.B. Ann. Rev. Phytopath. 1966, 4, 147-170.
52. Baas, P. IAWA Bulletin 1984, 5, 316.

RECEIVED January 2, 1986

Acid Rain Degradation of Nylon

Karen E. Kyllo[1] **and Christine M. Ladisch**

Textile Science, Consumer Sciences and Retailing Department, Purdue University, West Lafayette, IN 47907

Nylon 6,6 fabric exposed to simulated acid rain in light and darkness conditions showed polymer damage particularly when acids of pH 2.0 and 3.0 were used. This study reports the effects of sunlight, aqueous acid, heat and humidity (acid rain conditions) on spun delustered nylon 6,6 fabric. Untreated nylon and nylon treated with sulfuric acid pH 2.0, 3.0 and 4.4 were exposed to light in an Atlas Xenon-arc fadeometer at 63°C and 65% RH for up to 640 AATCC Fading Units. The untreated and acid treated fabrics were also exposed to similar temperature and humidity conditions without light. Nylon degradation was determined by changes in breaking strength, elongation, molecular weight and amine end group analysis. Physical damage was assessed using SEM.

Acidic precipitation is a growing environmental problem. Acid rain, snow or fog is formed when oxides of sulfur and nitrogen from fossil fuel combustion are oxidized in the atmosphere by ultra-violet light and ozone to give sulfuric and nitric acids. These acids then mix with atmospheric water to form acidic precipitation. Since the normal pH of atmospheric water is 5.6-5.7, precipitation having a pH below 5.6 is termed "acid rain" (1, 2).

The pH values of acid rain have steadily dropped over the last 25 years. Areas east of the Mississippi River Valley which once had precipitation of pH 5.0 are now subject to precipita-

[1]Current address: Department of Merchandising, Consumer Studies and Design, University of Vermont, Burlington, VT 05405

0097-6156/86/0318-0343$06.00/0

tion with an average pH of 4.4-4.2. Single incidents of precipitation with pH's as low as 2.1 have also been recorded (2-4).

The deleterious effect of acid rain on lakes, aquatic ecosystems, and vegetation has been widely publicized and is undergoing continual study (2, 4, 5). Damage to buildings and synthetic materials is also thought to be significant but is not as well documented. Deterioration of outdoor textiles by acid rain is reflected by reduced tensile and tear strengths and discoloration or spotting (6,7). Evidence of acid rain damage to textiles is complicated by the synergistic effects of other environmental factors such as sunlight and heat.

Nylon fibers are used extensively in outdoor textiles and as a result are subject to sunlight, varying temperatures and acid precipitation. The degradation of nylon by light, heat, humidity and air polluted with sulfur dioxide has been widely studied (8-13). However, little data is available on the effect of aqueous acid on nylon in the presence of heat, light and moisture (i.e. acid rain conditions). Therefore, the purpose of this work was to determine the effect of acid rain conditions on nylon. The synergistic effects of aqueous acid, light and heat on nylon were also examined.

Weathering of Nylon Textiles

Nylon is readily degraded by ultraviolet radiation from sunlight (10, 14-15). Absorption of light in the 270-280 and 300-340 nm region of the electromagnetic spectrum is responsible for the majority of the degradation of nylon 6,6 (15). Exposure to wavelengths of radiation below 300 nm results in photolysis and and exposure to wavelengths of radiation above 300 nm results in photooxidation.

The free radical reactions of photolysis can result in chain scission at the amide linkage of the nylon polymer or crosslinking between polymeric chains (15, 17). Degradation of nylon by photolysis is independent of oxygen, heat, moisture or additives.

Photooxidation reactions only take place in the presence of oxygen (15, 17). These reactions are the primary source of most of the sunlight damage to textiles. Photooxidation of nylon and model compounds have shown that oxidative attack usually produces free radicals, peroxides, and ultimately polymer chain scission. This results in lower tensile strength and ultimately a shorter useful lifetime of the textile product.

Heat, humidity, atmospheric pollutants and delustrants such as titanium dioxide also highly influence photooxidation (9, 15, 18). Studies on UV light-exposed nylon showed a decrease in tensile strengh as the relative humidity was increased (19) and as the temperature was increased (8). Delustered nylon showed a greater degree of strength loss than non-delustered nylon (10). Exposure of nylon 6,6 fabric to 0.2 ppm SO_2 gas (the concentration representative of SO_2 in a polluted atmosphere) also resulted in tensile strength loss (20).

All of these accelerating effects were found to occur only in the presence of light. Since acid alone is also known to degrade nylon (12, 13) it is reasonable to suspect a synergistic action by light, humidity, heat, and aqueous acid (i.e. acid rain conditions) on the chemical and physical properties of the nylon fabric.

One researcher examined the resistance of nylon to acid hydrolysis (15% HCl for 16 hours at $50^{o}C$) after exposure to light, heat and humidity (13). Acid hydrolysis resulted in polymeric damage comparable to that found in nylon exposed to light and 0.2 ppm SO_2 under similar conditions. The amino end group concentration ($[NH_2]$) (an indicator of cleavage of the amide linkage in the nylon polymer) of the acid hydrolyzed nylon was, however, higher than that of the nylon exposed to light and SO_2. It was concluded that exposure to SO_2 gas in the presence of light in a humid atmosphere did not cause acid hydrolysis and that the SO_2 may not have been converted to H_2SO_4 during the exposure period.

Another study indicated that nylon 6,6 exposed to light and SO_2-contaminated air showed a 13% strength loss after 168 hours of exposure at 40% humidity. The strength loss was 39% when an 18 minute water spray every 2 hours was added to the conditions. Exposure with 0.2 ppm SO_2 resulted in 41% strength loss without water spray and 68% with water spray, indicating the significant role of SO_2 and water in the degradation process. SO_2 is thought to form sulfuric acid in the presence of water and thus may catalyze chain scission of the nylon polymer (8, 10).

In an effort to determine the effect of aqueous acid on nylon in the presence of light, Zeronion et al (12) submerged nylon fabric in 20% sulfuric acid at $50^{o}C$ in a flint glass jar and exposed it to irradiation from a 275 watt sunlamp at a distance of 6 inches from the fabric. The nylon showed more polymer chain scission and greater $[NH_2]$ than nylon degraded by light and SO_2 gas both with and without a water spray. From these experiments, it was concluded that if sulfuric acid was present in the atmosphere, its attack on nylon was accelerated by the presence of light.

Scanning electron miscroscopy of the fibers exposed to SO_2, sulfuric acid and light revealed that photodegradation with SO_2 produced pitting of the fiber surface. Combined acid and light conditions also resulted in fiber surface pitting and roughness, whereas acid hydrolysis alone did not produce pitting. It was concluded that the mode of degradation in the presence of sulfuric acid and light was different from the attack by acid or light alone (12, 21).

The purpose of this study was to examine the effects of light, heat, humidity and aqueous acid on delustered nylon 6,6. The degradation resulting from separate exposure to light, heat, humidity and aqueous acid was compared with degradation resulting from combined exposure to light, heat, humidity and aqueous acid. The exposed nylon was evaluated for degradation by measuring breaking strength, viscosity and $[NH_2]$. Scanning electron microscopy was also used to characterize the physical degradation in the fiber.

Materials and Methods

Fabric. A plain weave fabric (Testfabrics #361) made from spun, delustered nylon 6,6 was used in this study. The fabric was chosen because it was representative of the type of nylon fiber used outdoors and the type of fabric used in previous photodegradation studies (21-23).

The nylon fabric was treated with sulfuric acid because of its nonvolatile nature and because it is considered to be a major contributor to rain acidity (1). Fabric samples measuring 27 cm (warp) by 36 cm (fill) were treated with 7.4×10^{-2}, 0.9×10^{-2} and 0.2×10^{-2} percent H_2SO_4 solutions. These acid concentrations represent pH values of 2.0, 3.0 and 4.4, respectively, and are representative of the range of acid rain pH values found in the U.S. east of the Mississippi River (2). The acid solutions were padded onto the fabric with a 50:1 liquor-to-goods ratio to produce a 78% wet pick-up.

Exposure to Heat, Light and Humidity. The acid-treated and untreated control fabrics were attached to a Teflon screen (280u filaments, 33 filaments/inch, Tekco, Inc., N.Y.), which was then clipped to the specimen rack in an Atlas Xenon-arc Fadeometer model 25-FT. White cardboard was wrapped around the outside circumference of the specimen rack to serve as a solid white, reflective backing for the samples. The screen and the cardboard did not touch one another.

The Fadeometer was equipped with a 2500 watt xenon-arc lamp and borosilicate inner and outer filters. Conditions of exposure were $63 \pm 3^{\circ}C$ and $65 \pm 5\%$ R.H. The fabrics were exposed to light for 40, 80, 160, 320 and 640 AATCC Fading Units (AFU), as measured by AATCC Blue Wool Lightfastness Standards L-5, L-6, L-7, L-8 and L-9, respectively. The fabrics were padded with acid immediately prior to each 40 AFU of light exposure. For example, fabrics exposed for 80 AFU were padded with acid a total of 2 times; fabrics exposed to 160 AFU were padded with acid 4 times.

Nylon fabric was also exposed in darkness to the same temperature, humidity and acid conditions described above. The fabrics were placed in an Atlas Gas Exposure Cabinet model GE-1RC. Filtered air was allowed to flow through the cabinet. For each treatment, fabrics were exposed to light and darkness simultaneously to insure equal amounts of exposure. Following exposure, all fabrics were rinsed in distilled water, neutralized in 1.0% sodium carbonate, rinsed again in distilled water and air dried. The fabrics were then stored at $21 \pm 1^{\circ}C$ and $65 \pm 1\%$ R.H. for at least 24 hours.

Evaluation of Fiber and Fabric Properties. Breaking strengths of the control and treated fabrics were determined according to ASTM D 1682, ravelled strip method (24) using an Instron model 1130 equipped with a 1,000 pound load cell and a gear ratio of 1:1. Five conditioned warp strips from each of two replicates were tested for each experimental condition.

Two replicates from each of the experimental treatments were evaluated for changes in viscosity. Dilute solutions (0.10, 0.20, 0.30, 0.40 and 0.50 g/100ml) of nylon 6,6 dissolved in 90% formic acid were made. Viscosities of the dilute polymer solutions were determined at 25 $\pm$ 0.1^{o}C using a size 75 Cannon-Fenske capillary viscometer. Flow time measurements were repeated for each solution and for the pure solvent until three consecutive readings within 0.2 seconds or 0.1% of the mean were obtained (24). The average of the three consecutive flow times was used to determine the relative viscosity for each solution.

From the relative viscosity, the reduced and inherent viscosities were calculated and plotted against concentration. The line of best fit through each set of points was extrapolated to zero to determine the intrinsic viscosity $[\eta]$. The molecular weight (M) of the nylon was then calculated according to the Mark-Houwink equation, $[\eta] = KM^{a}$, where $K = 3.5 \times 10^{-4}$ and $a = 0.786$.

Amino end group concentration of the experimental fabrics was determined using the ninhydrin method (25). Three samples from each of the two replicates for each experimental condition were analyzed.

Surface characteristics of the nylon 6,6 fibers exposed to acid rain conditions were examined by scanning electron microscopy (SEM). Nylon fabric samples were attached to an SEM mounting stub using conductive silver paint. The mounted samples were sprayed with an anti-static spray and allowed to dry thoroughly. The samples were then sputter coated with gold, after which the anti-static spray was reapplied and allowed to dry. Micrographs were taken using a JSM-U3 SEM at 10 KV at magnifications of 1000 and 3000 with a Polaroid camera.

Statistical Analyses. Analysis of variance in a nested factorial design (BMDP) was used to determine the influence of the type of acid treatment and level of exposure on the measured variables. T-tests were also used to determine if differences existed between light and dark exposed samples at a given exposure level and acid treatment. All statistical evaluations were carried out at $P = 0.05$.

Results And Discussion

Breaking Strength. The breaking strength of the nylon was not significantly affected by exposure to light or by the combination of light and acid treatments until after 80 AFU (Figure 1). The light exposed control fabric showed a loss in breaking strength of 2% at 80 AFU, 8% at 160 AFU and 13% at 640 AFU. Thus, breaking strength loss for the control fabric was not as severe as expected from previous studies (9, 21). Breaking strength losses between 39-75% have resulted from light exposure at 40% humidity with and without water sprays (12). The higher humidity of 65% in this study and the padding of water on the fabric after each 40 AFU of exposure should have resulted in larger strength losses according to available literature. The high humidity and extra water may have been responsible for observed increases in fabric compactness due to fiber swelling and therefore lower strength losses. Light exposure and treatment of the nylon with pH 4.4 and 3.0 H_2SO_4 also produced relatively small losses in breaking strength. The strength losses for the pH 4.4 and 3.0 acid treatments at 640 AFU were 14 and 6%, respectively. Statistical analysis of fabric strength losses after 640 AFU indicated no significant differences among the control, pH 3.0 and pH 4.4 acid treatments.

After 80 AFU of exposure, the loss in breaking strength of the pH 2.0 treated fabric was greatly accelerated. After 80 AFU of exposure, the pH 2.0 H_2SO_4 treatment significantly accelerated the loss in breaking strength. The breaking strength losses of 16% at 160 AFU, 35% at 320 AFU and 75% at 640 AFU resulted in an almost linear relationship between amount of exposure and loss of strength. These strength losses were significantly greater than those observed for all other treatments. The relatively steep slope of the pH 2.0 strength loss curve is indicative of continued degradation beyond 640 AFU.

The combined effect of acid, heat, and humidity without light on the breaking strength of nylon fabric is presented in Figure 2. The control and the acid treated fabrics exposed to darkness with heat and humidity showed strength increases. The strength of the pH 4.4, 3.0, 2.0 and control fabrics increased 0, 3, 7 and 15%, respectively, after 640 hours of exposure.

The increase, rather than a decrease, in breaking strength over 640 AFU of exposure may have been due to shrinkage brought about by the presence of moisture resulting in bulkier, thus stronger yarns. The shrinkage of the yarns would result in a more compact fabric and thus a gain in breaking strength. The extra moisture which would contribute to fiber swelling and hence shrinkage of the fabric structure would not be present to the same degree in the light exposed fabric due to higher fabric surface temperature which would drive off some of the moisture. The effect of the humidity on the breaking strength of the fabrics exposed to dark conditions was observed primarily between 40 and 160 units of exposure.

The most dramatic differences in strength loss between the light and dark exposed samples occurred in the pH 2.0 samples and the controls (Figure 3). From this data, it is evident that the acid treatment, even at pH 2.0, had little effect under darkness conditions on the breaking strength of nylon fabric. The gain in strength in the darkness conditions may be due to swelling of the fiber leading to an increase in bulk and compactness of the fabric which could result in the apparent strength increases. However, the synergistic effect of light on the hydrolytic action of acid on nylon is clearly indicated by the difference in strength loss between the light and dark exposed pH 2.0 treated fabrics after 640 AFU. The anticipated loss in breaking strength due to the additive effect of light and acid can be calculated by adding the loss due to light (light control) at 640 AFU and the loss due to acid alone (dark pH 2.0 - dark control) at 640 AFU. The anticipated additive

... and ... (2). The current ... electrode is polarized from ... it decreases at elevated temperatures ... the corrosion reaction is stimulated. One of the most important features of the system is the fact that the maximum current (or limiting current) obtained during anodic polarization never exceeds the OCV corrosion rate. This unusual behavior and the impact of elevated temperature on the faradaic efficiency of the cell are examined in this paper.

Experimental

Details of the electrochemical cell and support equipment have been given previously (1). In summary, the test cell housed a circular 11.4-cm^2 area anode support to which were bonded lithium test specimens. The cathode comprised a wire screen spotwelded to a ribbed iron back plate. Anode-cathode contact pressure was controlled by an air pressure cylinder linked to the anode pushrod. Elimination of the hydraulic pressure component was accomplished by a thrust balance cylinder connected to the inlet and outlet lines of the cell. Electrolyte hydraulic pressure could be varied by a control valve on top of the test cell. A needle point penetrometer was mounted on the cathode compartment such that on activation the

* Electrochemical Society Active Member.

Key words: lithium, lithium hydroxide, anodic polarization, corrosion, current efficiency, limiting current.

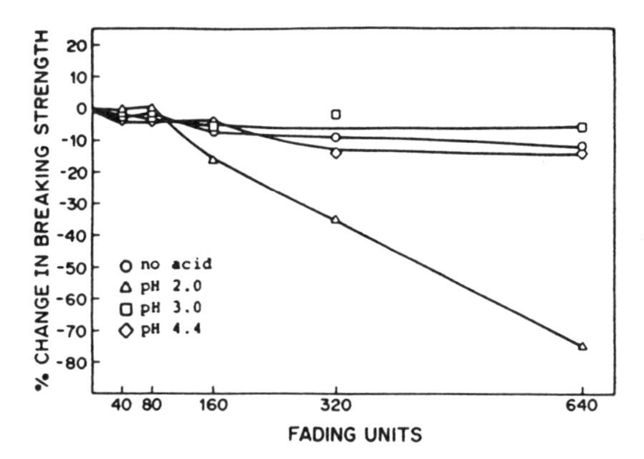

Figure 1. Percent change in breaking strength as a function of light exposure.

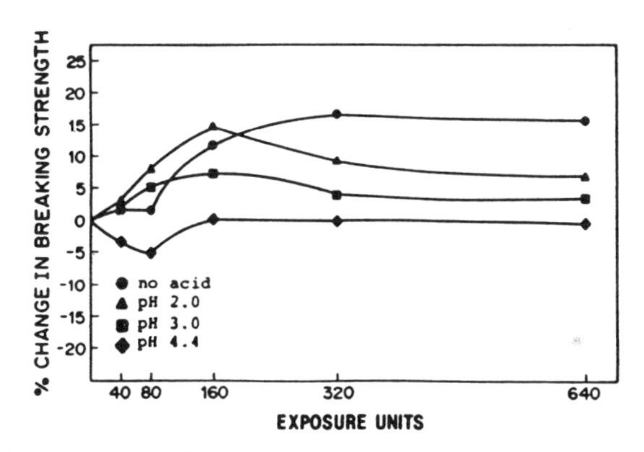

Figure 2. Percent change in breaking strength as a function of dark exposure.

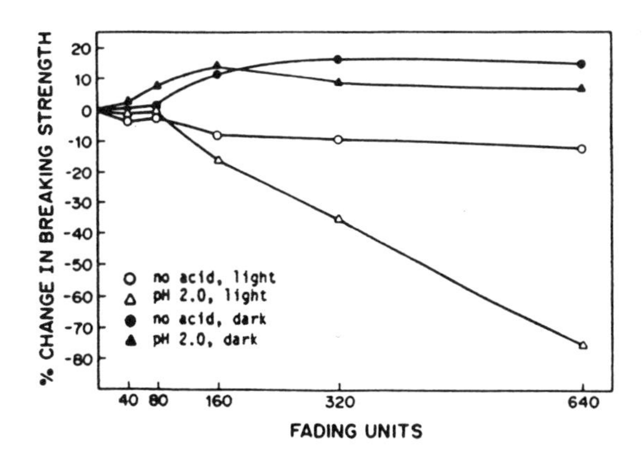

Figure 3. Percent change in breaking strength as a function of exposure.

loss in breaking strength for the pH 2.0 treated nylon at 640 AFU was 22%. The actual loss at 640 AFU was 75%. This indicated that 53% of the loss in breaking strength was due to the synergistic effect of light on the hydrolytic action of acid.

Molecular Weight. The percent change in molecular weight of the nylon as a function of light exposure is presented in Figure 4. All of the control and acid treated nylon fabrics exhibited a significant decrease in molecular weight, particularly after 80 AFU exposure. The rate of molecular weight loss up to 640 AFU was similar for the control, pH 3.0 and 4.4 fabrics. The 28 and 21% decreases in D.P. observed after 640 AFU for the pH 3.0 and 4.4 fabrics, respectively were not significantly different from one another.

The pH 2.0 treated sample exhibited a much higher rate of molecular weight loss as a function of light exposure, particularly after 80 AFU. As seen in Figure 4, the pH 2.0 sample had a 58% decrease in molecular weight after 640 AFU, which was 31% greater than that of the control.

For most polymers, breaking strength is a function of molecular weight. This relationship is clearly illustrated by comparison of Figures 1 (breaking strength loss) and 4 (molecular weight loss). The rate and relative extent of breaking strength and molecular weight losses due to light exposure are almost identical.

Molecular weight loss of the nylon was also determined following exposure to acid under darkness conditions (Figure 5). Little molecular weight loss occurred in the control, pH 3.0 and 4.4 treated fabrics up to 640 exposure units. Molecular weight loss of 19% for the pH 2.0 fabric at 640 exposure units was the only significant change observed in the dark exposed fabrics at 640 exposure units. The rate and relative extent of breaking strength and molecular weight losses due to darkness exposure were not highly correlated. The high humidity produced a physical change in the fabric which affected the breaking strength of the fabric but not the change in molecular weight due to chain scission.

The percent molecular weight losses observed for the control and pH 2.0 treated fabric as a function of light and dark exposure are presented in Figure 6. The change in molecular weight of the control after 640 AFU exposure was 1% in the dark and 27% in the light. Under the same 640 units of dark and light conditions, the pH 2.0 treated samples had 19 and 58% decreases in molecular weight, respectively. If the effects of light and acid on nylon were purely additive, a 45% molecular weight loss would be expected. However, a 58% molecular weight loss was observed, leading to the conclusion that light acts synergistically with the acid to degrade the nylon and was responsible for the 13% difference between the actual and additive molecular weight changes.

Amino End Group Concentration. Previous work on the degradation of nylon has shown that the $[NH_2]$ of nylon decreased if the

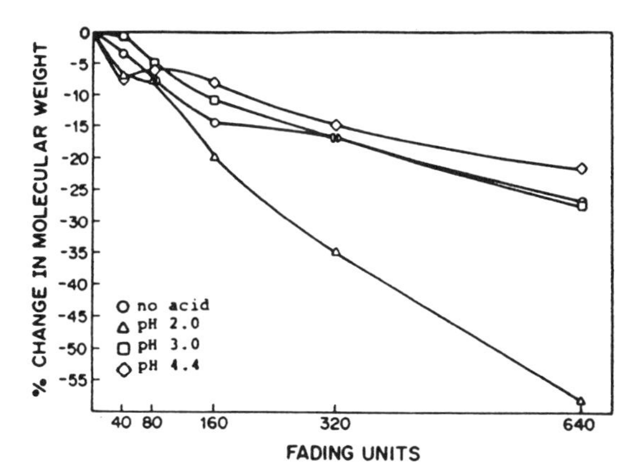

Figure 4. Percent change in molecular weight as a function of light exposure.

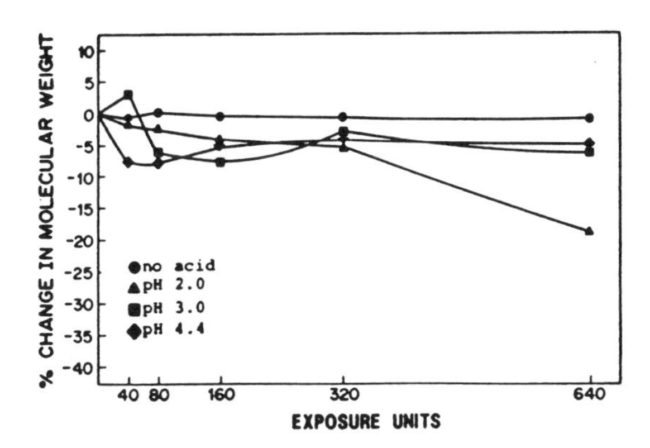

Figure 5. Percent change in molecular weight as a function of dark exposure.

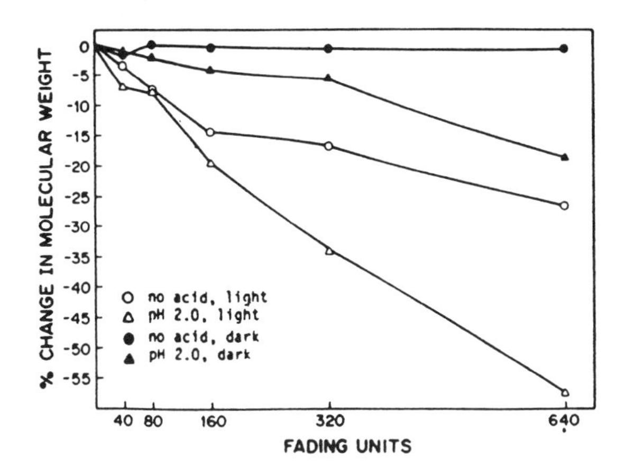

Figure 6. Percent change in molecular weight as a function of exposure.

degradation was due to light and increased if the degradation was due to acid hydrolysis (12, 13). Similar results were obtained in this study. As seen in Figure 7, the $[NH_2]$ of all of the control and all of the acid treated fabrics decreased through 160 AFU of light exposure. After 160 AFU the control and the pH 4.4 treated fabrics showed a continued decrease in $[NH_2]$ while the pH 3.0 and 2.0 treatments resulted in an increase in $[NH_2]$. After 640 AFU, the percent changes in amino end group concentration of the control, pH 4.4, 3.0 and 2.0 fabrics were -23, -17, -8 and + 26%, respectively. This gradual increase in $[NH_2]$ as pH of the acid treatment increased was indicative of acid hydrolysis, although the pH 2.0 treatment was the only acid treatment strong enough to cause an actual increase in the $[NH_2]$ when compared to an untreated nylon sample.

Amino end group concentration as a function of darkness exposure and acid treatment is presented in Figure 8. Trends similar to those for light exposure (Figure 7) were observed. After 640 AFU of exposure the changes in $[NH_2]$ for the control, pH 4.4 and pH 3.0 treatments were statistically equal, but were significantly lower than the $[NH_2]$ for an untreated nylon fabric. This indicated that the decreases in $[NH_2]$ for these fabrics were due to exposure to the heat and humidity in the darkness conditions and not to the pH 4.4 and 3.0 acid treatments. However, as seen in Figure 8, the pH 2.0 treatment produced a significant increase in $[NH_2]$ after 80 units of dark exposure, again indicating a change in mode of degradation as acid concentration increased.

A comparison of the effect of light and dark exposure on amino end group concentration of the control and pH 2.0 treated fabrics is presented in Figure 9. As expected from previous studies (12, 13), light and dark exposure of the control fabrics resulted in a net decrease in $[NH_2]$. The pH 2.0 acid treatment, however, indicated an increase in $[NH_2]$ under both light and dark conditions.

If acid and light worked together in an additive manner, a decrease in $[NH_2]$ of approximately 4% would have been expected for fabrics exposed to the combination of light and pH 2.0 acid. Figure 9 shows that exposure to pH 2.0 acid and light actually produced a 26% increase in $[NH_2]$. This data, like that of the breaking strength (Figure 3) and molecular weight measurements (Figure 6), indicated the synergistic effect of light and acid on degradation of the nylon. The ratio of acid hydrolysis to photodegradation steadily increased throughout the final 320 AFU of exposure, therefore shifting the predominant mode of degradation from that attributed to light to the acid mode of degradation.

Scanning Electron Microscopy. The surface of an untreated nylon sample appeared smooth and uniform (Figure 10(A)). Similar results were observed for the control nylon samples exposed to 640 AFU of darkness conditions (Figure 10(B)). However, following 640 AFU of light exposure, the fiber surface of the control fabric showed some pitting and cavities (Figure 10(C)).

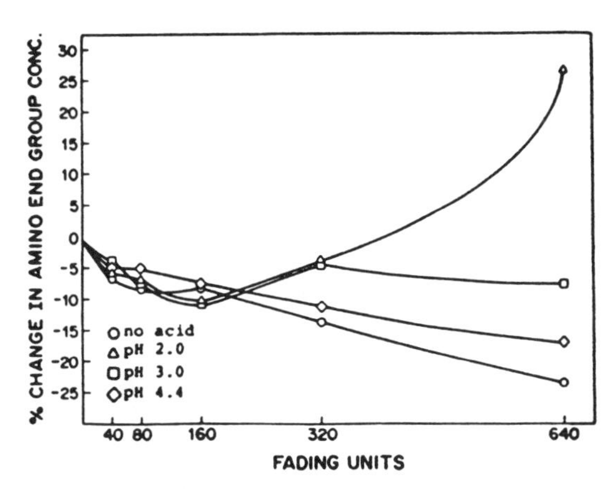

Figure 7. Percent change in amino end group concentration as a function of light exposure.

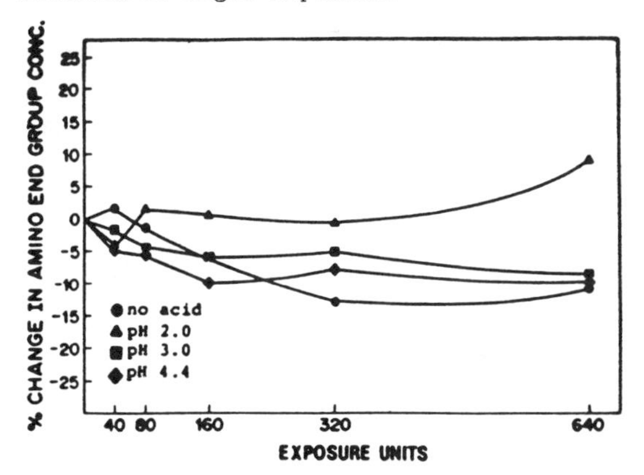

Figure 8. Percent change in amino end group concentration as a function of dark exposure.

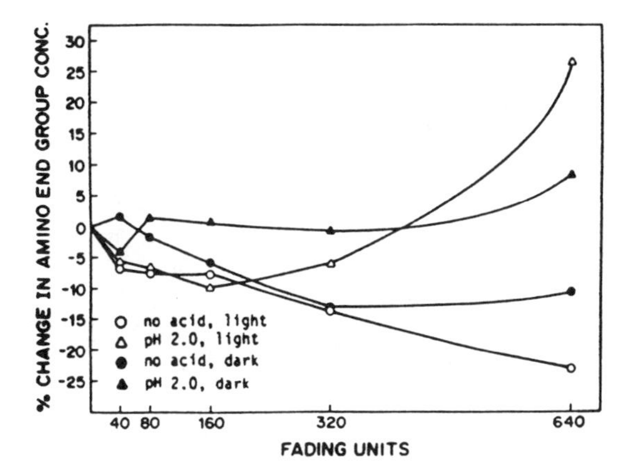

Figure 9. Percent change in amino end group concentration as a function of exposure.

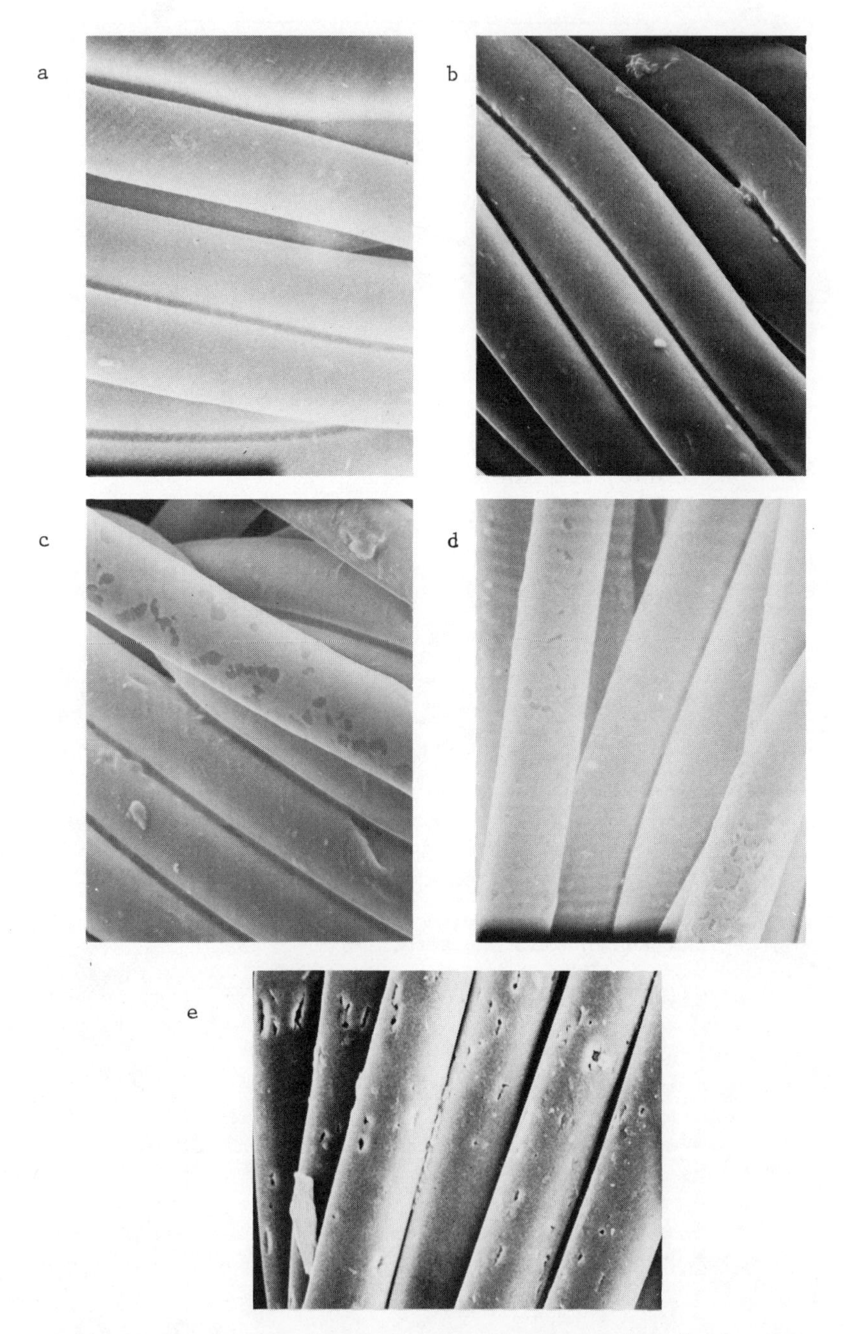

Figure 10. Scanning electron photomicrographs (480x) of nylon 6,6: (a) untreated fabric; (b) control, 640 FU dark; (c) control, 640 FU light; (d) pH 2.0, 640 FU dark; (e) pH 2.0, 640 FU light.

Dark exposure with pH 2.0 acid also produced some pitting and cavities as seen in Figure 10(D). The damage to the dark exposed fabric with pH 2.0 acid was similar in nature to that seen in the light exposed control fabric, which indicated that the damage was probably from heat and/or humidity rather than from the acid.

Exposure to 640 AFU light and the pH 2.0 acid treatment produced an increase in frequency of pitting in the fiber as seen in Figure 10(E). Although the pits do not appear to cover as much of the surface area as those seen in Figure 10(C), they do appear to be longer and deeper and some roughening of the surface is present. Immersion of nylon 6,6 in strong aqueous acid has been shown to cause a roughening of the fiber surface but no pitting of the fiber (12, 21). From this work, it appears that aqueous acid in combination with light and atmospheric oxygen has resulted in more physical damage to the fiber than exposure to light or to acid alone. This could account in part for the much greater loss in breaking strength when the nylon 6,6 was exposed to the light and H_2SO_4 treatment at pH 2.0 than when exposed to the light, no acid treatment. It also supports the theory that the synergistic effect of light and acid in simulated acid rain conditions is severely detrimental to the physical as well as the chemical properties of the fiber.

Conclusions

Exposure of nylon fabric to light, heat and humidity produced both physical and chemical damage in the nylon fiber. Losses in molecular weight and physical damage to the fiber surface contributed to losses in breaking strength. The decreases in $[NH_2]$ indicated that polymeric damage occurred at the amide linkages as well as at other sites along the polymer chain.

Exposure of nylon to dark conditions (heat and humidity) revealed that small amounts of damage detected by $[NH_2]$ were due to the combination of heat and humidity. As sulfuric acid was combined with light exposure to simulate acid rain, the severity and nature of the physical and chemical damage slowly increased as exposure time increased. Initially the effect of pH 4.4 and 3.0 acid was masked by the detrimental effect of light exposure. As the exposure level for the pH 3.0 acid increased to 320 and 640 AFU and as the acid strength increased to pH 2.0, the amount of degradation greatly increased. The combination of light and H_2SO_4 showed a synergistic effect which resulted in large and rapid decreases in breaking strength and molecular weight.

The nature of the degradation also changed as the exposure level and strength of the acid increased. The increase in $[NH_2]$ indicated that cleavage at the amide linkage due to acid hydrolysis was the predominant form of chain scission during light exposure with H_2SO_4 at pH 3.0 and 2.0 at 320 and 640 AFU. The roughening of the fiber surface shown in the SEM photographs supported this conclusion.

The increase in intensity of damage to nylon as the amount of exposure and strengh of acid increased is indicative of potentially shorter serviceable lifetimes for outdoor fabrics exposed to acid rain conditions. Although the current average pH of acid precipitation is above pH 3.0, the effect of continual exposure to sources of nonvolatile H_2SO_4 and the possibility of saturation for longer periods of time and more frequent exposure than experimentally simulated may result in damage equal to or exceeding that indicated in this study. Degradation might also be worse than indicated by this study in areas of more intense sunlight, higher relative humidity and more frequent and lower pH level acid precipitation events.

Literature Cited

1. Likens, G. E.; Wright, R. F., Galloway, J. N; Butler, J. Scientific American 1979, 4, 43.
2. Ember, L. R. Chemical and Engineering News 1981, 37, 20.
3. Wolff, G. F. Journal of the Air Pollution Control Association, 1979, 1, 26.
4. Likens, G. E.; Bormann, F. H. Science 1974, 4142, 1176.
5. Hershaft, A. Environmental Science and Technology 1976, 10, 992.
6. Gajendragdkar, S. K. Chemical Age of India 1977, 8, 673.
7. Fisher, T. Progressive Architecture 1983, LXIV, 99.
8. Egerton, G. S.; Shah, K. M. Textile Research Journal 1968, 2, 130.
9. Little, A. H.; Parsons, H. L. Journal of the Textile Institute 967, 10, 449.
10. Shah, C. D; Srinivasan, R. Journal of the Textile Institute 1978, 5, 151.
11. Singleton, R. W.; Kunkel, R. K.; Sprague, B. S. Textile Research Journal 1965, 35, 228.
12. Zeronion, S. H.; Alger, K. E; Omaye, S. T. Textile Research Journal 1973, 4, 228.
13. Zeronion, S. H.; Alger, K. W.; Omaye, S. T. Proceedings of the Second International Clean Air Congress 1971, p468.
14. Little, A. H. Journal of the Society of Dyers and Colourists 1964, 43, 527.
15. Lock, M; Frank, G. C. Textile Research Journal 1973, 43, 502.
16. Moore, R. F. Polymer 1963, 4, 493.
17. Stowe, B. S.; Fornes, R. E; Gilbert, R. D. Polymer-Plastic Technology Engineering 1974, 2, 159.
18. Clibbens, D. A. Textile Institute and Industry 1968, 1, 20.
19. Deland, M. R. Environmental Science and Technology 1980, 6, 657.
20. Beloin, N. J. Textile Research Journal 1968, 7, 130.
21. Zeronion, S. H. Textile Research Journal 1971, 2, 184.
22. Tweedie, A. S.; Milton, M. T.; Sturgeon, P. Z. Textile Chemist and Colorist 1971, 2, 22.
23. Zeronion, S. H. Textile Research Journal 1970, 8, 695.
24. "Annual Book of ASTM Standards," Parts 23 and 32, American Society for Testing and Materials, Philadelphia, PA, 1981.

25. Knott, J; Rossbach, V. Die Angewandte Makromolekulare Chemie 1980, 1307, 203.

RECEIVED January 2, 1986

ECONOMIC EFFECTS

25

Materials Damage and the Law

Roger C. Dower and Sarah E. Ball

Environmental Law Institute, 1346 Connecticut Avenue NW, Washington, DC 20036

Public debate concerning legislative and regulatory strategies for addressing acid rain damages has raised significant questions about the regulation of materials damage caused by air pollutants in general. Most environmental statutes (as well as public concern) have focused on environmental health effects, leaving us relatively unprepared to cope with non-health damages. Nowhere is this better illustrated than in the lingering controversy over acid rain, yet new concerns such as forest damages suggest that acid rain may not be an isolated case. The purpose of this paper is to highlight the current legal and regulatory structure for addressing materials damage, particularly the economic and technological dimensions of the system. The strengths and weaknesses of the present framework will be discussed and illustrated by the acid rain case. The focus will be on what can be learned from our ability (or inability) to handle acid rain and the implications for regulatory treatment of other air pollution-caused materials damage.

In recent years, the now much delayed re-authorization of the Clean Air Act (CAA) has been inextricably tied to resolution of what is referred to as the acid rain debate. Although toxic air pollutants have moved in recently to take the front seat, acid rain remains an issue of bitter dispute. This debate has brought to the public's attention the non-health related effects of air pollutants, including materials damages. The record suggests that this is a class of pollution impacts that we have a very difficult time resolving within our current statutory and regulatory framework. Yet materials damages and other non-health air pollution effects increasingly appear on the regulatory agenda. The purpose of this paper is to review some of the more obvious reasons for

0097–6156/86/0318–0360$06.00/0

our problems in coping with materials damage effects, particularly those that emerge from the acid rain debate, and to suggest some very modest proposals for beginning to regulate those effects.

Some General Observations

As a prelude to the main body of this paper, it might be useful to offer some broad observations concerning materials damages and environmental rule-making. Taken together, these form at least one perspective from which to judge alternative structures for materials damage control. They provide a framework for testing and considering certain basic hypotheses concerning the applicability of our current structures for developing rational and effective policies towards acid rain and other types of air pollution problems where materials damages and other environmental effects are dominant.

1) The statutory and regulatory structure contained in the CAA practically ensures that materials damage effects and other non-health related effects will not and cannot play a major role in rulemaking concerning air pollutants. The Act emphasizes public health concerns and provides little in the way of a substantive part for materials damages. Of course, this situation is not unique to air pollution control. EPA has been characterized as a public health agency and purely environmental concerns appear to have often taken a back seat. Where health effects predominate, this focus is sensible. However, where health effects are small or outweighed by materials damage or other environmental effects, the bias may be towards no regulation. This may be so even if the materials damage effects are quite large.

2) The two characteristics of the acid rain problem that have made the regulatory response so difficult are the uncertainties associated with the causes and effects and the interstate nature of the causes and effects. The first, of course, is hard to deal with in any rulemaking, but there seems to be more reticence in deciding one way or the other when it comes to environmental effects. The second comes up hard against the federalism of the CAA, where the goals of the Act are carried out, in large part, by independent state implementation plans. There is little to force one state to consider the environment of another. Both of these characteristics are common, in varying degrees, to other air pollutants with materials damage effects.

3) There appears to be a willingness on the part of EPA, Congress and presumably the public, to acknowledge the economic trade-offs associated with materials damage effects. This is in rather sharp contrast to the unwillingness to acknowledge economics in setting or reviewing health-based standards. While this is in part a function of the statutory language of the Clean Air Act, it also reflects the interstate and more regional nature of materials damage effects. Both tend to highlight the key questions of who pays for acid rain control, who benefits and by how much. We certainly do not ascribe the same ethical dimension to materials damage effects as we do to health effects.

This willingness to consider economics more fully tends to polarize the decision making process if there is no formal decision rule to guide the regulator. It is precisely the economic trade-off debate that has stalled acid rain legislation for so long. There is little question but that regulators are able to act more quickly, if not more rationally, when health effects are at stake.

4) The difficulty of dealing with the economic trade-offs of controlling materials damages from air pollutants is exacerbated by the obvious and direct costs of control and the uncertain economic benefits. This is not a particularly creative notion and affects to one degree or another most environmental rulemakings. It seems many more times pronounced when considering materials damage effects. It is important, though, because it suggests the importance of merging economic and technical information to provide the other half of the equation. The alternative is, as in the acid rain debate, no action or regulatory programs with uncertain outcomes. If, in fact, we are as a society willing to consider the economic trade-offs of regulatory pollutants on the basis of materials damage, and if we do not provide some formal framework within which these effects can be considered and balanced, there seems to be little hope for effective and rational regulatory outcomes.

The Clean Air Act and Materials Damage Effects

Having stated the general bias of the paper, it is necessary to provide the background that leads to these observations. The obvious starting point is the statutory language of the CAA as it relates to materials damage. In an earlier paper for the National Academy of Sciences 8th Symposium on Statistics, Law and the Environment (1), Dower has discussed in some detail the major section of the Act that relates to materials damages. Little purpose is served here by repeating that discussion except in summary. In addition, other sections of the Act offer some potential for regulating pollutants with materials damage effects, and these deserve some greater attention.

Ambient Air Quality Standards. The primary mechanism for regulating air pollutants on the basis of material damage effects is outlined in Sections 108 and 109 of the Act. These sections lay out the process for setting primary (health-based) and secondary (welfare-based) national ambient air quality standards for criteria pollutants. Materials damages fall under the secondary standard setting process, where welfare effects are defined as "effects on soils, water, crops, vegetation, man-made materials, animals, wildlife, property,... transportation, as well as effects on economic values and personal comfort and well being."

Congress distinguished health effects and welfare effects quite explicitly. EPA would set primary standards that allow for an "adequate margin of safety". Welfare standards are required "to protect the public welfare from any known or anticipated adverse effects." The functional difference between these two definitions and whether one suggests a more stringent standard than another is the matter of some debate. Nevertheless, literally interpreted, secondary standards are intended to be more difficult to achieve than primary standards and, theoretically, could be set at a zero level (since most materials damages dose/response functions are linear without thresholds).

Perhaps more important than the basis for standard setting is the fact that the CAA did not set deadlines for attainment for the secondary standards while establishing a formal set of compliance deadlines for the primary standards. Where deadlines are one method for demonstrating Congressional resolve with respect to a regulatory requirement, their absence is particularly noteworthy.

The legislative history of the CAA clearly prohibits EPA from taking into account the costs and benefits of primary NAAQS. These are strictly health-based standards. While many students of the CAA argue that the same is said for the secondary standards, the case isn't quite as clear. The Act itself is silent on the issue and the legislative history is quite vague. However, the definition of welfare effects might be interpreted to allow for trading off of one type of welfare damage (say materials damage) against another (for example, personal comfort and well-being). The issue has never been directly addressed by the courts, and although EPA appears to consider in a general sense the costs and benefits of certain regulatory effects in the setting of secondary standards, the question remains unanswered.

The NAAQS process does not provide the most reliable or useful mechanism for regulatory accommodation of materials damage effects. Lacking an attainment schedule, there is little incentive for EPA or the states to take the requirements seriously. Where there are secondary standards that differ from the primary standards (this is the case only for Total Suspended Particulates and Sulfur Dioxide), the attainment record is bleak. Second, the requirements for national ambient standards seems somewhat at odds with the fact that, at least to date, welfare effects in general appear to be more regional in nature and may not be easily handled by a process that requires standards to be geographically uniform. (Some states do have secondary standards in place that are either stricter than Federal standards or that regulate a pollutant in the absence of a Federal standard. Wyoming, for example, has a secondary standard for sulfates, and Maine's standard for ozone is stricter than the Federal standard.) Finally, most observers, until very recently, have assumed that EPA could not set a secondary standard without also setting a primary standard. If a pollutant were associated with materials damage effects but had no known health effects, the Agency would have to go elsewhere for statutory authority to regulate. This conventional view is being challenged by the Natural Resources Defense Council, which argues that EPA has a non-discretionary duty to regulate certain sulfate particle sizes under the secondary standards for their contribution to reduced visibility and acid rain. Even if it were possible to regulate sulfates under Section 108 and 109, the Act still lacks attainment deadlines to put teeth in the requirements.

Alternative Approaches. In 1977, Congress explicitly recognized the problems of dealing with interstate pollution problems, of which acid rain was most often considered, and added Section 126 to the Act. Section 126 allows states on the receiving end of transported air pollution to petition EPA and require polluting states to revise their state implementation plans to reduce emissions (2). The drawback of Section 126, besides the fact that it relies on the ability of existing air quality modeling to demonstrate the source of pollution problems and that EPA has placed the burden of proof of the petitioning state, is that it only applies to pollutants that have been addressed through the NAAQS process, which may miss the real concerns. In addition, Section 126 does not attempt to resolve a major impediment to acid rain control--the fact that states that benefit from the regulations do not generally bear the costs (3). Lacking a formula for allocation of costs, it is little wonder that there has been virtually no activity under this section of the Act.

Section 111 of the CAA involving the establishment of new source performance standards has also been identified as a mechanism for addressing the control of acid rain. On the surface, this section of the Act would appear to have the most flexibility to deal with welfare effects. Section 111 directs EPA to set technology-based "standards of performance" for new and modified sources of pollution that "cause or significantly contribute to air pollution which may reasonably be anticipated to endanger public health and welfare" (4). While these standards are intended to be nationally uniform, EPA is given rather broad authority to distinguish between "cause or significantly contribute to air pollution which may reasonably be of sources and to take into account the age of the source facility. Further, EPA is instructed to ensure that a reasonable balance is struck between economic, environmental and energy considerations. Although written in terms of new sources of pollution, Section 111(d) provides for the application of NSPS to existing sources where the NSPS apply to pollutants not regulated elsewhere under the Act (5). The history of applying NSPS to existing sources, as a possible means to overcome the limitations of setting NAAQS for welfare-only type pollutants, is limited. One somewhat relevant rulemaking, involving coal fired utilities, did deal with new and existing sources, for the performance standard was also combined with a percentage reduction requirement. The nationally uniform nature of the NSPS option may have inhibited its application to acid rain or other pollutants. Moreover, the NSPS for coal fired utilities may have worsened the acid rain problem by biasing investment decisions towards old or existing plants rather than new, cleaner plants. Its flexibility may also be limited by its technology basis which doesn't allow wide range of possible regulatory responses.

The final section of the CAA to be discussed here is Section 169A. Originally designed to protect national parks and wilderness areas from regional haze and sulfates, it has recently been suggested that it could also serve as a vehicle for controlling acid rain. Section 169A provides for a regional approach to the regulation of powerplants and other sources of visibility-impairing pollutants. To date, regulations under 169A have not advanced beyond a set of initial regulations aimed at powerplants near national parks in the West and an ambitious second phase proposal for large reductions in eastern SO2 emissions. While 169A does come close to dealing with some chacteristics of materials damages, it currently lacks much in the way of enforcement potential and may be just another case of using the wrong tool just because it is handy (6).

Lessons from Acid Rain

In attempting to come to grips with acid rain control, EPA, industry, states and environmentalists have had to seek out creative (or at least novel) approaches for overcoming the limitations of the CAA to address the unique characteristics of the problem. In doing so, they have constructed something of a track record for those concerned with regulating air pollutants associated with materials damages and other welfare effects. While there are differences, of course, between the problems posed by acid rain and those related to other pollutants with welfare effects, such as ozone (for example, the political/economic constraints posed by the coal switching problem), there are several critical similarities. Without attempting to review each and every acid rain proposal or to review the history of the acid rain legislation, there are broad characteristics of the alternative proposals which bear on materials damages in general.

1) Excess Emission Reduction Targets -- Most of the serious proposals for acid rain control (in fact, both the House and Senate bills of 1984) involve some formula for allocating reductions excess emissions of SO_2 over a defined population of emitters (in almost all cases, electric utilities). The principal focus of these efforts is to reduce total regional emissions, thus avoiding the nationally uniform, ambient requirements of the NAAQS. While the proposals differ as to who must come into compliance with what, when, and how, they share the common feature of requiring proportional reductions, in emissions above some threshold level (7). The political and economic implications of this approach have been discussed extensively elsewhere (8).

2) Cost Sharing -- Several of the most recent proposals provide for some formula for distributing the costs of acid rain control across a wider audience than those who would normally bear the direct burden of the control requirements. For example, the House bill would have imposed a tax of 1 mill per kilowatt hour on all electric utility consumers in the contiguous U.S. The intent of such cost sharing arrangements is to redress the distributional consequences of acid rain control benefits and costs falling on different regions in differing levels. Lacking an accurate basis for estimating who gains and who loses, the cost and benefit allocation problem is among the hardest to resolve as we have seen over the last several years.

3) Treatment of Uncertainty -- The basic approach of most acid rain proposals towards the uncertainties between emissions rates, ambient environmental concentrations and environmental impact is to ignore them. That is, an assumption is made that reductions in SO2 will result in improved environmental quality and that the exact nature of the relationship (and therefore the actual increase in environmental quality resulting from a given percentage rollback) is problematic. Perhaps this language is too strong. The acid rain debate is certainly being held in the name of environmental improvement--expected or actual. Yet we seem to know much less, or at least have greater uncertainty, about the expected environmental effects of current control proposals than we do for most other in-place regulatory programs. This may be a feasible political or legislative response, but not particularly appropriate or possible within the administrative requirements of environmental rulemaking. The administrative response to acid rain of doing very little is, in part, a function of the barriers set up by scientific, technical and economic uncertainty. Lacking even partial information on possible effects thresholds, regulators are understandably loathe to act. The burden of acting falls to Congress as its decision-making is not as constrained by uncertainty. Even Congress, however, has to balance competing interests, and because it lacked any firm or compelling scientific evidence, the issue was bound to stall.

One somewhat subtle effect of this uncertainty relating to materials damage effects is to complicate the public's ability to understand or value (in a general not economic sense) these kinds of environmental effects. For example, there is tremendous uncertainty regarding the health outcomes of some of our current regulatory programs. The same can be said of some programs affecting recreational values. This has not stopped EPA or Congress from regulating, perhaps because the public has a mental framework within which they can judge and understand regulatory outcomes with which they have some actual or perceived experience. It is more difficult to galvanize a broad base of public support for a costly regulatory program to reduce materials degradation. Besides the fact that one does not seem as important as the other, the general public has had little experience with materials damages. It may not be able to relate changes in air quality to these kinds of damages. This does not mean there is no impact: both health and environmental effects impose real economic costs.

This brief picture of some of the common elements in current proposals to curb acid rain damages does not offer much optimism concerning our ability to address other forms of welfare or materials damage effects. Even setting aside the distributional problems associated with long-range transport of pollutants, scientific and technical uncertainty, regional and local impacts, and a lack of public understanding concerning materials damage effects, the CAA remains a poor vehicle for effective control of materials damage effects. The current approach to acid rain control is to add a new program into the act directed specifically at acid rain. The desirability or political feasibility of this tact is suspect and is certainly not a functional model for treatment of materials degradation impacts in general. The polarization of the debate between fast regulatory action and additional research has led to a stalemate. It is not difficult to imagine the same outcome for other pollution problems with the similar characteristics.

Conclusions

The picture painted in this paper is not very revealing; it is always easier to say what is wrong with a program than to design solutions. Nevertheless, certain themes emerge from the review of the CAA and the current acid rain debate that need to be incorporated into any broad legislative or regulatory attempt to cope with materials damages from air pollution. First, it should be clear that none of the current sections of the CAA or the current proposals for acid rain control offers a functional and consistent basis for regulating air pollutants where materials damages or other environmental effects predominate. A re-thinking of the underlying philosophy of the Act (in terms of nationally uniform standards driven by health concerns) needs to take place. The simple (actually not so simple) addition of new programs to the current statute may be effective in specific cases but does not address the basic limitations of the Act. As public concern over environmental as well as health effects increases, these limitations can only become more pronounced. There have been suggestions made to incorporate the primary and secondary NAAQS into one process with varying dates for attainment or to allow for regional secondary NAAQS where appropriate. Both are reasonable possibilities, but need to be coupled with wider discussion.

Second, there is a tremendous need to develop a more solid information base on the scientific, technical and economic underpinnings of material damages. Further, such data needs to be more closely merged if it is to play an important part in promoting effective protection programs. This will not only to ensure efficient decision making, but will also aid in determining appropriate allocation formulas for costs and benefits and to help develop a balanced public awareness of materials damages and other environmental effects. Economic benefit analysis, by providing a framework for bringing together technical, scientific and economic information, can be a critical asset in this process. By providing some notion of the dollar value of damages associated with materials damages, a baseline for comparison with other types of impacts (even if they are expressed in non-dollar units) can be presented.

Finally, EPA must be granted under whatever resolution is ultimately chosen a high degree of flexibility to cope with the wide variations in the extent of knowledge concerning environmental effects and the characteristics of the control problem. All or nothing approaches will lead to acid rain-type impasses. The benefits of tailoring responses to the data at hand rather than imposing broad-based controls seems more befitting the characteristics of materials damages and social perceptions surrounding the issue. This does not mean there should be no regulation without 100 percent certainty, an unlikely situation. It does suggest we need not apply health effects ethics to environmental damages control. Allowing for explicit recognition of the costs and benefits (however expressed) of alternative regulatory responses is one way to provide such flexibility. Others exist as well (such as variable deadlines) and should be considered as we evaluate how best to deal with the environmental effects of air pollution.

Literature Cited

1. Dower, Roger, "Materials Damage: Economics and the Law", forthcoming, The American Statistician.
2. "National Emissions Standards for Hazardous Air Pollutants," Air and Water Pollution Control Law: 1982, Environmental Law Institute: Washington, D.C., 1982; p. 288.
3. Crandall, Robert W. "An Acid Test for Congress," Regulation, September/December 1984, Vol. 8., No. 5/6.
4. "New Source Performance Standards," Air and Water Pollution Control Law: 1982, Environmental Law Institute: Washington, D.C., 1982; p. 159.
5. Ibid., p. 170.
6. The discussion of Section 169A is drawn in large part from: Trisko, E.M., "Alternative Approaches to Acid Rain Control," The Environmental Forum, April 1985, Vol. 3, No. 12.
7. Crandall, supra p. 26.
8. For further discussion of this topic see: Acid Rain and Transported Air Pollutants: Implications for Public Policy, U.S. Congress, Office of Technology Assessment, OTA-0-204 June 1984; "An Analysis of Issues Concerning Acid Rain," U.S. Congress, General Accounting Office, Report to the Congress, GAO/RECD-85-13, December 11, 1984; Acid Rain Legislation and the Future of the Eastern Low - Sulfur Coal Industry, U.S. Congress, Congressional Research Service, Report No. 84-89-ENR, April 6, 1984.

RECEIVED January 2, 1986

Economic Features of Materials Degradation

Thomas D. Crocker

Department of Economics, University of Wyoming, Laramie, WY 82071

This paper provides an overview of the economic problem of valuing materials degradation. Differences in engineering, physical science, and economics perspectives are discussed, and a general economic framework in which engineering studies of materials damages are a necessary component is set forth. Particular emphasis is placed upon the role that materials play in "making markets." It is argued that conventional economic assessment techniques neglect this role and therefore existing estimates of materials damages from acid deposition likely embody a substantial downward bias.

All parts of society must regularly confront the problem of "optimally" managing abiotic materials. A factory manager's choice of equipment maintenance activities, a homeowner's choice of exterior trim for his house, and an architect's choice of artwork for a public square can all be cast as problems of materials management. Materials can then be regarded as components of processes that produce desired outputs and services which may be enhanced by some human activities (e.g., devoting labor to cleaning) and hindered by others (e.g., atmospheric pollution). Just as a farmer requires a description of his crop production process in order to allocate his resources efficiently, the manager of a collection of inanimate materials must have a description of the process by which the outputs he desires are produced as well as the effects on these outputs of any factors which he may control. A model of the internal workings of the production process, including the roles that materials play, must be a fundamental part of any economic version of its optimal management. In this paper, I try to explain and present examples of the economic approach, with special emphasis upon the information required from other disciplines. [See Johansen(1) and Marsden, et al.(2) for treatments which emphasize the connections between engineering information and the economic notion of a production function.]

0097-6156/86/0318-0369$06.00/0

In the next section, I discuss differences in the engineering and the economic perspectives on the problem of designing production processes. A general economic framework in which natural science studies of materials damage are a necessary component is set forth in the third section. Although this treatment attempts to elucidate rather than resolve, I try to provide an idea of the significance of dynamic elements in complete studies of materials damage. The approaches described in this section are well known. A fourth section adopts the perspective of the consumer to argue that materials damages have plausibly important features that are unconventional in economic assessments. In the interest of communication rather than the maximization of disciplinary integrity, the discussion throughout is thoroughly heuristic.

The Engineering and the Economic Perspectives

While drawing upon known physical principles, the engineer focuses upon the design and operation of processes capable of producing desired outputs with minimum expenditure of valued material inputs. The input prices that determine costs are taken as given. The engineer thus seeks to solve a constrained optimization problem involving detailed knowledge of the expected costs of material inputs. If all nonmaterial inputs, such as labor were free, these engineering solutions would alone, in economic terms, decide the optimal process design. However, nonmaterial inputs are also costly, and both material and nonmaterial costs can change. Engineering results thus identify the technically efficient set of materials combinations within the larger set of ultimate physical and biological limits for producing a given level and type of desired output.

Economists, such as Hildenbrand(3), who study physical production processes also fix their gazes upon a cost minimization objective but they assume that the engineer has already solved his constrained optimization problem. The economic theory of cost and production describes the effects of variable input prices upon cost-minimizing combinations of material and nonmaterial inputs. From the set of cost-minimizing combinations, the theory provides rules for identifying that combination having the lowest material and nonmaterial costs, and it specifies how this "minimum of the minima" will be altered as relative input prices change. It identifies economically efficient input combinations. In this fundamental sense, then, economics portrays the results of engineering reoptimization in terms of the effects of changes in technological and biological possibilities and the relative prices of inputs on the cost-minimizing combination of inputs. With respect to pollution impacts, it describes the effects of *differences* in these impacts upon the costs of the alternative physical and biological ways an economic agent has to meet his objectives.

The engineer considers all input combinations consistent with known physical laws since his objective is to develop and ultimately implement a detailed plan for a particular production process. Subsequent efforts to improve upon this plan will center upon any input or subset of inputs appearing to provide substantial opportunities for cost reduction. In contrast, most basic scientific studies of materials damage from air pollution (e.g., Nriagu(4) have concentrated on the pollution impacts for a single input combination in some production process. Many have studied the impact independently of any cost-minimizing production process. One need not depend only upon economic arguments to recognize how misleading the failure to consider the place of the affected material in a production process can be. Natural science findings will often serve the purpose. Sereda(5) points out, for example, that sulphur compounds, by reacting with carbonates, cause blistering, scaling, and loss of surface cohesion in stone. These reactions induce similar effects in neighboring materials which would otherwise be immune to attack. Hence, at best, the focus has been upon a small set of the feasible input combinations. These highly detailed studies typically assume that other input types and quantities remain constant. They therefore frequently fail to provide the design information the economist requires if he is to estimate the economic consequences of the changes in cost-minimizing input combinations that pollution induces.

The basic science, separable, piece-by-piece approach to the study of pollution impacts upon materials makes truly awesome the task of covering and synthesizing the technically efficient or even the feasible input combinations. In manufacturing alone, thousands of different types of material inputs exist. Each input type is, in turn, embodied in one or more production processes or outputs which may appear in a variety of forms and which can be put to a number of distinctive uses. Moreover, there may be environmental cofactors such as temperature and moisture that act in concert with pollution to aggravate or to soften its impact. Basic science studies of pollution impacts upon materials have received little guidance about which of these embodiments, varieties, and uses take on economic significance.

One key problem, then, is to reconcile the basic science and the economic approaches to materials damages by developing criteria for deciding how much materials science detail must be retained in any particular circumstance. When one of the several reasonable objectives of materials science pollution impact studies is to provide information useful for estimating economic consequences, the basic question is whether more or less materials detail will alter the economic estimates in a nontrivial way. Both the materials scientist and the economist must refine their knowledge such that the relations between pollution and materials damages are defined in dimensions corresponding to those in which economic agents choose to define them. Because the practicing engineer is responsive to economic phenomena, his insights can be highly instructive in prospective attempts to capture these dimensions.

The Economic Assessment Problem

The objective of efforts to assess the economic consequences of pollution-induced materials damages is to estimate differences in the sums of consumer surpluses and producer quasi-rents over two or more policy-relevant pollution levels. Consumer surplus portrays the difference between the maximum a representative consumer would be willing to commit himself to pay for a given quantity of a commodity and what he in fact has to pay. Similarly, a producer quasi-rent is the difference between what the owner of the inputs to a particular activity receives for supplying a particular output quantity and the minimum he must receive in order to be willing to commit to that supply. The sum of consumer surplus and producer quasi-rent is thus a measure of the net benefits associated with the consumption and the production of a commodity. The observable unit prices of other commodities that provide him equal satisfaction set an upper bound to the consumer's maximum willingness-to-pay; the observable earnings his inputs could obtain in other activities set a lower bound on the minimum reward the producer must receive. Maximum willingness-to-pay represents demand; the minimum necessary reward defines supply.

Figure 1 traces the set of factors that must be accounted for if a complete assessment of the economic consequences of pollution-induced materials damages is to be performed. Assume that a manufacturer is trying to decide what kind and how much of a product to produce. The materials used in some production processes for these products are susceptible to air pollution damages. For a given level of expected air pollution exposures, the portion of the figure lying to the left of the leftmost dotted line represents the problem of determining the least costly way of producing a given product. It answers the question: given the alternative ways I have to produce any particular level of output for this product, which ways allow me to employ the minimum combinations of those costly inputs physically necessary for production. Variations in pollution levels will alter these minimum necessary combinations. Identification of these combinations for each pollution level enables the manufacturer to maximize the output of the product that he obtains from a given expenditure on inputs. However, solution of this problem cannot tell him which of the alternative products to produce: it can only show the least costly plan for producing a particular product, and the manner in which this plan will vary as air pollution varies.

The portion of Figure 1 lying between the dotted lines depicts the problem of choosing among the alternative products, given prior identification of the least costly time and means with which to produce each good. In this portion, the producer is allowed to adapt by substituting products that are more or less prone to quantity and/or quality reductions from expected air pollution levels. The problem of this middle portion can be contrasted with the problem of the first portion, where the producer could only adapt by manipulating the time and/or the process used to produce a particular product. Basically, the producer has a set of alternative products dictated to him by biological and physical science knowledge of the laws of nature, as well as by institutional constraints (the laws of man) limiting allowable processes for producing these goods. For each of these known and allowable processes for each good, he estimates the least cost time and process; for given output prices and pollution levels, he then selects the combination of goods that he expects to generate the maximum quasi-rent.

In the rightmost portion of Figure 1 resides the consumer. There are two routes whereby pollution-induced materials damages can influence his behavior, thus altering the consumer surplus he obtains from the output in question. First, an increase (decrease) in materials damage that is registered in service flow reductions will increase (decrease) costs. Consequently, the minimum price the producer must receive in order to be willing to commit himself to supply any given quantity of the output will increase (decrease). In addition, altered levels of air pollution may impact the attributes of the output, thus changing the consumer's willingness-to-pay and the consumer surplus he acquires from any quantity of the output. The change in cost implies a shift in the producer's supply function, while the change in willingness-to-pay is represented by a shift in the consumer's demand function. Both result in a change in the market price of the output.

Given that producers and consumers have already adapted so as to minimize their prospective losses, or to maximize their prospective gains, Figure 2 depicts for a predetermined time interval one example of the changes an air pollution increase can have upon consumer surplus and producer quasi-rent. The air pollution increase reduces the desirable properties of the output, making smaller the consumer's willingness-to-pay and causing his demand function to shift from D^0 to D^1. It simultaneously increases the least-cost of producing any particular output quantity, thereby causing an upward shift in the supply function from S^0 to S^1. Market price for the output drops from P^0 to P^1, partly because of the greater relative magnitude of the shift in the demand function, and partly because of its lesser relative slope. Consumer surplus was the area aP^0b; it is now the area dP'e. Producer quasi-rent was the area fP^0b; it is now the area gP'e. Total economic surplus from the production and use of the output in question is thus reduced by the area fghb plus the area adeh.

Of course, alternative relative shifts in demand and supply relations, reflecting different impacts in the production and consumption sectors, will yield results of a modified character. For example, if the demand curve shifts to D" rather than D', market price will rise to P". If the supply shift from S^0 to S' is small enough, producers could thus actually see an increase in their quasi-rents. Qualitative results are unchanged, however: alterations in producer quasi-rents and consumer surplus result from the pollution-caused changes in the two sectors. These two examples serve to illustrate the issues of concern here: consumers and producers can bear very different economic gains or losses depending on the relative shifts of the demand and supply functions. Moreover, the distribution of these economic consequences can differ drastically with the slope of the demand function relative to the slope of the supply function.

The above observations set out in the most general of terms the substance of the problem of calculating the economic benefits of controlling pollution-related materials damages. Dynamic issues complicate this rather straightforward substance. [See Crocker and Cummings(6) for a formal analytical treatment.] Consider, for example, the output and investment decision problem confronted by a firm whose existing capital stock does not suffer from air pollution exposures. Increased current output, because it increases wear and tear and reduces the effectiveness of a maintenance program, draws down the capital stock available for all future periods. In any single future period, production costs will

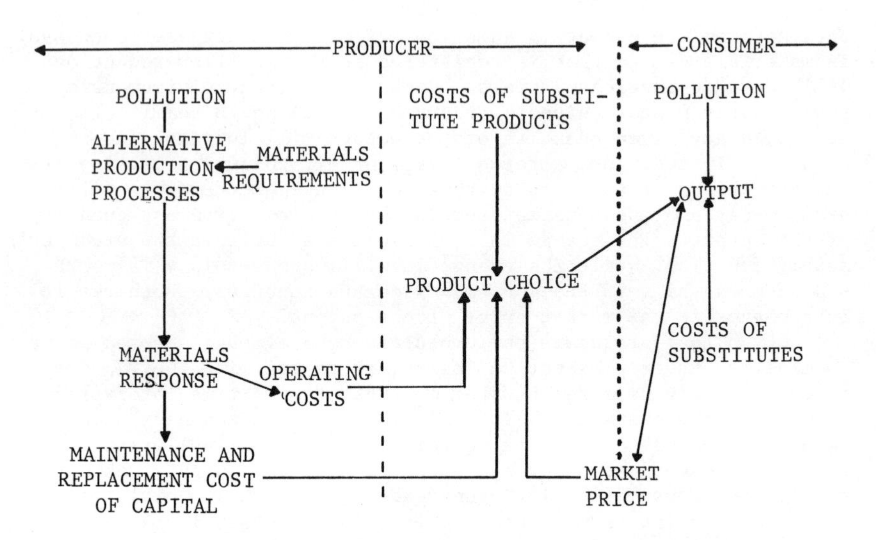

Figure 1. The sources of economic consequences

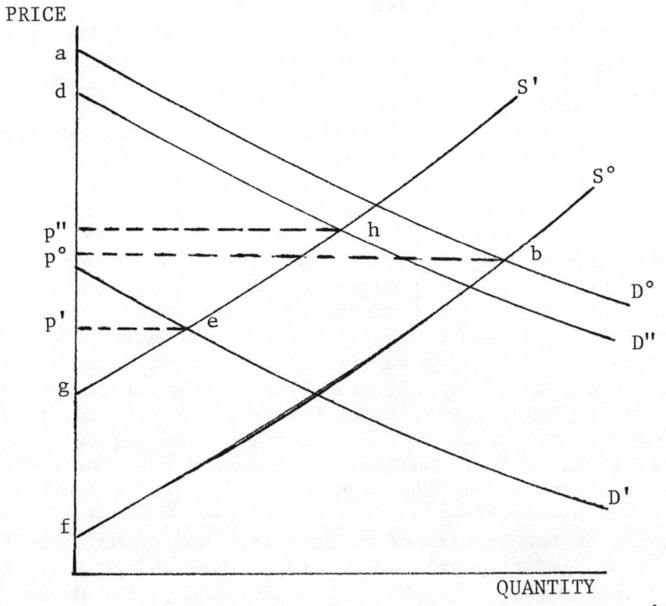

Figure 2. Changes in producer's quasi-rent and consumer's surplus.

be increased because less capital with which to combine other inputs will be available. Moreover, since the existing capital stock will be depleted more rapidly, some options for producing outputs in all future periods are foreclosed. The opportunity losses associated with the foreclosed options are user costs. They can be viewed as foregone future revenues.

Alternatively, the firm can increase the present value of its net revenues in all future periods by reducing current period production and/or by investing, perhaps via better maintenance, in current period capital stock which will, at the expense of current output, reduce future production costs and user costs. More generally, since the firm's output and investment decisions of each period affect the capital stock availabilities and the output possibilities of future periods, a sequential problem exists in which the impact that current decisions have upon the accumulated future consequences of past decisions must be weighed.

Suppose now that air pollution increases. All else equal this is equivalent to a reduction in current period maintenance activities. Its consequence is a reduction in future capital stocks. Production costs and users costs increase. These higher costs result in a leftward shift from S^0 to S^1 of the supply function in Figure 2. If demand is unchanged, higher future prices and lower future outputs result. The discounted value of the stream of changes in quasi-rents represent the economic consequences of the air pollution increase for the firm.

Implicit in the discussion immediately above is the assumption that the firm is forever locked into a particular production process and output bundle. However, the air pollution increase may differentially affect the production and the user costs of alternative processes and bundles. The firm will substitute toward the least costly alternative. Brick rather than painted wood may now, for example, enclose the firm's offices. A complete assessment of the quasi-rent consequences of the air pollution increase for the firm requires that these induced technical changes as well as production and user cost changes be taken into account. As Figure 2 demonstrates, these changes for the firm will also have consumer surplus consequences.

All capital stocks are not owned by firms; many are owned by consumers. The building of the previous paragraph could be a residence rather than an office. Fundamentally, the analytical problem is no different since direct analogs of changes in technique, production costs, and user costs also exist for consumers. [See Crocker and Cummings(6)]. In at least one respect however, there is a big difference between the two sets of problems: the aesthetic features of materials are likely to be significant for consumers. To the extent that soiling and corrosion affect these features, the psychophysical properties of human perceptions and perception thresholds will influence economic consequences. In particular, there is unlikely to be even an approximate one-to-one correspondence between physical soiling and corrosion measures and aesthetic perceptions. Numerous psychophysical experiments involving human perceptions of sensory events such as light and color changes have been codified in the form of Fechner's law, which states that the strength of a just noticeable increment in a sensation is proportional to the logarithm of its stimulus [Baird and Noma(7)]. These psychophysical results are consistent with that class of arguments in economics which posits a "threshold of sensitivity" [Georgescu-Roegan(8),(9) or a "zone of indifference" [Luce(10); March(11)] within which choices are randomly made or an "inertia of choice" prevails [Devletoglou(12)]. The likely absence of a known isomorphism between physical measures and human perceptions of materials damage implies that assessment studies must develop nonarbitrary mapping funtions. The presence of perception thresholds means that unperceived changes in physical material damages will have no economic consequences for the consumer [Watson and Jaksch(13)]; once the threshold is crossed, it can also mean that the marginal benefits of reduced damages will be increasing and that the marginal costs of increased damages are decreasing [Crocker(14)]. Accounting for these discontinuities and nonconvex economic damage functions requires a substantially more sophisticated analytical and empirical treatment of the consumer's problem than is conventional in most assessments of the economic benefits of pollution control. [See, for example, Arrow, et al.(15) and Hausman(16)].

Existence Values

Recently, the environmental economics literature [e.g., Schulze, et al.(17)] has distinguished between "existence value" and "use value," where the former refers to something for which an individual is willing to sacrifice some of his wealth even though he does not and will not use the good in question. Generally, the examples provided are unique natural, biological, or cultural assets such as the Grand Canyon, whooping cranes, and the Statue of Liberty. With very few exceptions, however, the literature fails to supply any formal framework for analyzing the value implications of existence value. Miller and Menz(18) is the first exception of which this author is aware. They present a model in which *stocks* as well as flows of services from wildlife are arguments in the objective function of the consumer's constrained utility maximization problem. When stocks appear thusly, they show that activities, such as current service flows or air pollution, which deplete the stock entail a user cost. For the consumer as well as the producer, it follows that damages to the stock must be examined not only in terms of foregone current service flows but also in terms of foreclosed future options. The latter include foregone future service flows and the utility value of the stock.

The Miller and Menz(18) argument is appealing; however, there are additional ways of analytically motivating existence values. In particular, the individual may regard assets that he does *not* own but to which he has potential access as part of his wealth. His unowned assets thus enter his constraint system rather than his objective function. Though for him they are completely illiquid, they may nevertheless be combined as inputs with other assets that the individual does own so as to produce a current period cash flow. Plausible examples range from the mundane such as toll bridges which reduce the costs of transporting goods, to the exotic such as a monument which inspires belief in and discourages deviant behavior from the cultural and economic norms which a society professes. The unowned assets enhance the individual's ability to discover, create, and exploit differences in wants and advantages among people, thus giving rise to gains from trade with others and with nature. Consider the following example from Crocker and Cummings(6). The structures and the statuary of an urban public square provide an aesthetically pleasing backdrop for the site-specific production of private musical performances and artwork. The backdrop reduces the artists' costs of attracting an audience that might purchase their work. Away from the square, the unit costs of attracting an audience exceed any artist's reservation price for producing art. In short, the art would not be supplied in the absence of the square. In essence, these assets reduce the individual's costs of transacting and, by making markets less thin, thereby expand his trading opportunities.

The unit cost of transacting varies directly with the thinness of markets. Baumol(19) points out that putting any coordination system like a market in place involves lumpy set-up costs that imply the presence of scale economies. Moreover, thin markets reduce demands for goods. Consequently, prices fall, production is discouraged, and the expected waiting time required to sell at any particular price is increased. Finally, as Clower and Howitt(20) show, the owned assets increase that the individual must store in order to achieve his optimal lifetime consumption path, thus raising his holding costs. With thin markets, he is less able to enjoy the fine arts when and where he wants to and he has to worry more about from where his next meal is coming.

Conventional benefits assessment techniques disregard transactions costs. They therefore implicitly take the size of the market, as defined by the number of potential trading partners, as given. Some of the implications for benefits assessments of allowing the size of the market to vary can be captured by a simple model adapted from a framework originally set forth by Brunner and Meltzer(21).

Let the individual be represented by a twice differentiable, quasi-concave utility function

$$U = U(Y_1, \ldots, Y_n) \qquad i = 1, \ldots, n \tag{1}$$

where any Y_i is a good, and U is expected utility. Assume a transformation function

$$t_{ij} = t_{ij}(Q, P_{ij}), \; i,j = 1,.., n, \; i \neq j \tag{2}$$

where t_{ij} is the quantity of good in transformed into a unit of good j, P_{ij} is the price ratio, p_i/p_j, and Q is a vector of assets, whether owned or unowned, which enhance the individuals access to trading partners. Some elements of Q could also appear in (1). For simplicity, P_{ij} is considered to be exogenous, and $t_{ij}(\cdot)$ is homogeneous of degree one.

Let the unit cost of t_{ij} to the individual be c_{ij}. A c_{ij} need not be equal to c_{ji}. The unit cost of exchanging clean air for building materials will likely differ from the unit cost of exchanging building materials for clean air, for example. The c_{ij} represent the individual's costs of gaining access to or engaging Q. They could represent the toll on a bridge or highway or even the opportunity cost of the time the individual expends promoting social cohesion.

The individual's problem is to maximize his expected utility in (1) subject to the budget constraint

$$\sum_n p_n(W_n - Y_n) - \sum_i \sum_j c_{ij}t_{ij} = 0 \quad (3)$$

where W_n is a measure of the individual's endowment at the beginning of the period. $(W_n - Y_n)$ is therefore that part of his original endowment which the individual retains at the end of the period. Expression (3) says that the value of what he retains at the end of the period most equal the net value of each good he buys and sells plus the cost of exchange. From the perspective of the market, this net value must be zero; that is, the market or accounting value of the goods the individual sells must equal the market or accounting value of those he buys.

The Lagrangian for this problem is

$$L = U(\ ) + \sum_n p_n\lambda_n [(W_n - Y_n) - \sum_i \sum_j c_{ij}t_{ij}] \quad (4)$$

Note that in contrast to the traditional statement of the consumer's problem, each good has its own multiplier, λ_n, which reflects differences among goods in the ease of converting them into money and thence into the expected marginal utility of money income.

One of the first-order conditions for (4) is

$$\frac{\partial U}{\partial Y_n} = \lambda_n p_n X_n \quad (5)$$

which, as usual, says that in equilibrium the marginal utility of a particular good will be equal to the marginal value of the expenditures on it. However, the presence of λ_n implies that differences in transactions costs among goods cause, from the individual's perspective, a given change in the dollar value of an endowment to generate different shadow prices for different goods. This result enables one to develop a simple diagrammatic insight into the role that Q plays in determining the value of a good to the individual consumer.

In Figure 3, the individual's initial endowment of Y_1 and Y_2 is at X. If the exchange act is costly, an initial endowment of Q implies a budget constraint of VXV, whereas if the exchange act is costless, the budget constraint is MM. The slope of MM corresponds to P_{12} in (2). When the individual has exhausted all gains from trade, he will finish the period with Y_1^0 and Y_2^0 if MM is operative. If VXV is the operative budget constraint, he will select Y_1^2 and Y_2^1. The kink in VXV implies that the transaction costs of exchanging Y_1 for Y_2 differ from the transactions costs of exchanging Y_2 for Y_1. If some point on MM other than X constitutes the initial endowment, costly acts of exchange will mean that a budget constraint different from either VXV or MM may be operative because the costs of exchange acts may differ by the relative quantities of goods in the initial endowment as well as by types of goods. Thus the individual's budget constraint may vary according to the form

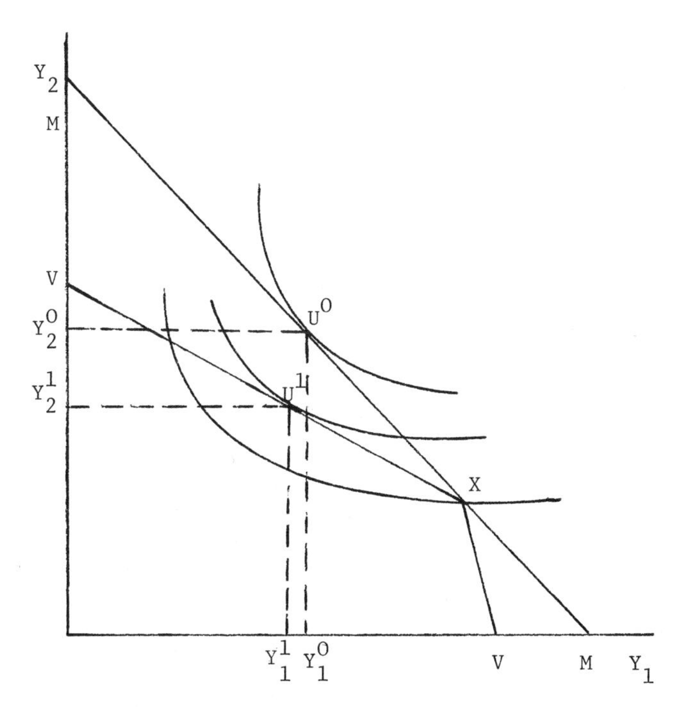

Figure 3. Welfare effects of costly exchange.

in which his initial endowment was accumulated, although the market value of this endowment as determined by P_{12} may be identical for many combinations of Y_1 and Y_2. Since costs of exchange differ according to the original (Y_1, Y_2) combination, each combination will result in a different, generally nonlinear budget constraint. It follows that, from the individual's perspective, a dollar is not an invariant pecuniary measure. Instead, the subjective value of an additional dollar depends on the form of the income change, i.e., on the good in which the increment is embodied. Moreover, it appears that realized market behaviors and the individual's subjective gains from these behaviors are dependent not only on money incomes and relative market prices of goods, but also upon the combination of goods in the individual's endowment. That is, rather than the customary

$$Y_n = Y_n(p_1, \ldots, p_n, W), \qquad i = 1, \ldots, n \qquad (6)$$

the demand function for a good is

$$Y_n = Y_n(p, \ldots, p_n, W_1, \ldots, W_n) \qquad (7)$$

[See Samuelson and Sato(22) for a development of the implications for consumer theory of this formulation].

Now consider changes in the Q vector of (2) which tend to reduce costs of exchange. In immediate terms, consider an acid deposition reduction which noticeably slows the corrosion and discoloration rates of building, monument, and public infrastructure surfaces in an urban area. The slowing of building decay increases the expected returns from building investments and plausibly increases the number of potential traders available for the individual. Improved public infrastructure enhances communication, therefore reducing the cost of trade. Similarly, if the appearance of the urban area is improved, more potential traders may be motivated to reside in it and the behavior of these traders after an exchange has been consummated may be made more predictable. Each of these features reduces exchange costs, and, for the original endowment at X in Figure 3, causes VXV to rotate outward toward MM. If the acid deposition reduction were adequate to wipe out all exchange costs, the value of this reduction would be the money equivalent of the difference between the expected utility levels U^0 and U^1 in Figure 3. A complete assessment of the economic consequences of the acid deposition reduction requires that this gain be added to those discussed earlier. The gain might reasonably be labelled existence value. The obvious problem with this formulation is that the budget line MM may be totally foreign to the individual; that is, it may lie totally outside his experience. He may be ignorant of the utility gains he could acquire if it were made operative. In these circumstances, it is indeed questionable whether the individual's revealed preferences should be used to inform the decision about supplying the Q. For example, an art consumer who would evaluate ex ante the provision of the public square must make calculations involving two kinds of conjectures about the future, conditional on his current action. First, he must state a belief about future world states and their respective probabilities; second, he must estimate the extent to which his tastes will be satisfied. This art consumer will find it exceedingly difficult to evaluate ex ante the provision of the public square if he is unable to acquire and array information on the art that will be supplied. Because his information about the art is distorted or diffuse, the preferences which he reveals may not reflect his actual tastes. Of course, he could be supplied with more information. Alternatively, in accordance with Hammond (23), the utility levels in Figure 3 could be interpreted as ex post utility levels and the Q-goods of expression (2) might, in accordance with Musgrave(24), be termed "merit goods" which have existence value.

It seems reasonable that planners in North American democracies need not respect individual beliefs about future possibilities in the same manner as they respect individual tastes. Tastes may be refined or vulgar; still, it may be held that a person's tastes must be respected. Beliefs, however, can be wrong and it may be very costly to correct them. In these circumstances, a planner might make decisions about Q-goods on this basis of his expectations of the individual's ex post utility level. Existence value would then be the planner's perception of differences in the individual's ex post utility level, where the planner's ex ante beliefs have been substituted for the individual's. The usual economic efficiency (Pareto optimality) conditions would still apply. The danger of paternalism in which the planner substitutes his own tastes for those of the individual is acknowledged. In addition, account would have to be taken of the accuracy and precision of the planner's knowledge of world states and the individual's tastes. This notion is not as odd as it perhaps seems at first glance. In fact, one might characterize an expert as someone whom others allow to make decisions for them about the provision of particular goods and services. The codes of ethics of most societies of experts ask them to respect the individual's sovereignty but to do so in an ex post rather than ex ante fashion. Perhaps a common phrase used by parents is apt; "You may not be grateful now, but you will be later." The higher education and the medical professions are transparent examples. People who know how to use materials to make markets less thin and more hospitable might also qualify.

Conclusions

In Crocker(25),(26), I stated my belief that the abiotic resources commonly known as materials suffer more from acid deposition than any single category of biotic resources. The statement achieved a fair degree of notoriety, not all of which was laudatory. The analytical basis of the statement, however poorly it was articulated at the time, was the existence value argument set forth here in Section IV. Although empirical support for the argument is notably weak, be aware that the $2.6 trillion U.S. 1980 gross national product (GNP) is a measure of the annual exchange value of final output only. The $2 billion that I assigned to acid deposition-induced materials damages is less than one-tenth of one percent of 1980 GNP.

Literature Cited

1. Johansen, L., Production Functions, Amsterdam: North-Holland Publishing Co. (1972).
2. Marsden, J., D. Pingrey, and A. Whinston, "Engineering Foundations of Production Functions," J. of Economic Theory 9 (1974), 124-140.
3. Hildenbrand, W., "Short-Run Production Functions Based on Microdata," Econometrica 49 (1981), 1095-1125.
4. Nriagu, J.O., Sulfur in the Environment, Part II, New York: John Wiley & Sons (1978).
5. Sereda, P.J., "Effects of Sulfur on Building Materials," in NRC Associate Committee on Scientific Criteria for Environmental Quality, Sulfur and Its Organic Derivatives in the Canadian Environment, Ottawa, Ontario: National Research Council of Canada (1977), 359-426.
6. Crocker, T.D., and R.G. Cummings, "On Valuing Acid Deposition-Induced Materials Damages: A Methodological Inquiry," in D.D. Adams, ed., Acid Deposition: Environmental, Economic, and Policy Issues, New York: Plenum Press (forthcoming).
7. Baird, J.C., and E. Noma, Fundamentals of Scaling and Psychophysics, New York: John Wiley & Sons (1978).
8. Georgescu-Roegan, N., "The Pure Theory of Consumer's Behavior, The Quarterly J. of Economics 50 Feb. 1936), 18-33.
9. Georgescu-Roegan, N., "Threshold in Choice and the Theory of Demand, Econometrica 26 (Feb. 1958), 157-168.
10. Luce, R.D., "Semiorders and a Theory of Utility Discrimination," Econometrica 24 (1956), 178-191.
11. March, J.G., "Bounded Rationality, Ambiguity, and the Engineering of Choice," The Bell J. of Economics 9 (Autumn 1978), 587-608.
12. Devletoglou, N.E., "Thresholds and Transactions Costs," The Quarterly J. of Economics 85 (Feb. 1971), 163-170.
13. Watson, W.D., and J.A. Jaksch, Production of Household Cleanliness in Polluted Environments, Working paper, Reston, VA: U.S. Geological Survey (undated).
14. Crocker, T.D., "On the Value of the Condition of a Forest Stock," Land Economics 61 (Aug. 1985), 244-254.
15. Arrow, K.J., L. Hurwicz, and H. Uzawa, "Constraint Qualification in Maximization Problems," Naval Research Logistics Quarterly 8 (June 1961), 175-190.
16. Hausman, J.A., The Econometrics of Non-linear Budget Sets, Fisher-Schultz Lecture for the European Econometric Society Meetings, Dublin, Ireland, Sept. 1982, Revised draft (April 1984).
17. Schulze, W.D., D.S. Brookshire, et al., "The Value of Visibility in the National Parklands of the Southwest," Natural Resources J. 23 1983), 149-165.
18. Miller, J.R., and F.C. Menz, "Some Economic Considerations for Wildlife Preservation," Southern Economic J. 45 (Jan. 1979), 718-729.

19. Baumol, W.J., "The Transactions Demand for Cash: An Inventory Theoretic Approach," The Quarterly J. of Economics, 66 (Nov. 1952). 545-556.
20. Clower, R.W., and P.W. Howitt, "The Transactions Theory of the Demand for Money: A Reconsideration," J. of Political Economy 86 (June 1978), 449-446.
21. Brunner, K., and A.H. Meltzer, "The Uses of Money: Money in the Theory of an Exchange Economy," The American Economic Review 61 (Dec. 1971), 784-805.
22. Samuelson, P.A., and R. Sato, "Unattainability of Integrability and Definiteness Conditions in the General Case of Demand for Money and Goods," The American Economic Review 74 (Sept. 1984), 588-604.
23. Hammond, P.J., "Ex-ante and Ex-post Welfare Optimality under Uncertainty, Economica 48 (1981), 235-250.
24. Musgrave, R.A., The Theory of Public Finance, New York: McGraw-Hill Book Co. (1959).
25. Crocker, T.D., "What Economics Can Currently Say About the Benefits of Acid Deposition Control," in H.M. Trebing, ed., Adjusting to Regulatory, Pricing, and Marketing Realities, East Lansing, MI: Institute of Public Utilities, Michigan State Univ. (1983), 724-744.
26. Crocker, T.D., "Statement," Select Committee on Small Business and Committee on Environment and Public Works, Economic Impact of Acid Rain, U.S. Senate, 96th Cong., 2nd Sess. (Sept. 23, 1980), 100-111.

RECEIVED January 2, 1986

27

Economic Effects of Materials Degradation

E. Passaglia

Institute for Materials Science and Engineering, National Bureau of Standards, Gaithersburg, MD 20899

When materials are placed into service they are subject to a number of degradative processes, examples of which are corrosion, wear, fatigue, fracture, U-V degradation, mildew and rot. These degradative processes cause the producers and users of durable goods to incur costs for special materials, for maintenance, repair and early replacement. On a national level, these costs have an economic effect in that they represent resources in the form of materials, capital, energy and labor that in the absence of these degradative processes could be used for other purposes. Using corrosion and fracture as examples, a method for accounting for these effects, and their magnitude, will be discussed, as well as the relevance to the economic effects of acid rain on materials.

When materials are formed into goods and put into use as structures, machinery, equipment, appliances, art objects, etc., they are acted upon by a number of degradative processes. Foremost among these are corrosion, wear, fatigue and fracture, oxidation, ultra-violet degradation, mildew, rot, and a number of other processes caused by the action of living organisms such as microbes and insects. All these processes have in common the fact that their action requires specialized materials and processes during manufacture, constant maintenance and repair that would not otherwise be necessary, and causes earlier replacement than would be necessary in their absence. Thus, both in manufacture and while in service, these degradative processes require the use of resources in the form of materials, energy, and labor, that would not otherwise be necessary and hence could be used for other purposes. These extra resources associated with manufacture, maintenance and repair, and early replacement, may be termed the "economic effects" associated with the degradative processes. And since these economic effects may be accounted for in terms of money, they are usually called the <u>cost</u> of the degradative process, e.g., the cost of corrosion, the cost of wear, etc.

Aside from the fact that it is difficult if not impossible to place a value on a human life, while it is easy to do so on a durable good, there is an apt analogy between the national costs of these degradative processes and the national cost of health care. Indeed, these degradative processes may be thought of as the illnesses of durable goods. In fact, certain art objects such as famous paintings, sculptures, buildings, and documents like our Declaration of Independence and Constitution, are essentially irreplaceable and the analogy becomes even more apt.

A number of studies in various nations have determined the cots of these degradative process (1-17). Perhaps the most complete are those carried out for corrosion and fracture by the U.S. National Bureau of Standards (NBS) in collaboration with Battelle Columbus Laboratories (BCL). In addition to these referenced works there are a large number of studies, particularly in corrosion, of the costs in limited, more specialized circumstances.

It is the intent of this paper to review some of the methodology and results of the NBS-BCL studies, and their relevance to the economic effects of acid rain. Since the effect of acid rain on materials would appear to be a form of corrosion (although affecting all materials, not only metals), the NBS corrosion study (1,2) will be the principal one discussed.

The Concept of Avoidable Costs and "Worlds"

Not all the costs associated with a given degradative process are unavoidable. A portion of them is caused by the fact the best preventive and control measures are not always used throughout the economy. This "best" does not mean the use of the most resistant materials, e.g., the use of gold or platinum in place of steel, in the case of corrosion, but means "best" in an economic sense of lowered life-cycle costs. Often the use of more resistant materials or processes during manufacture, while increasing the purchase price of an item, may decrease the maintenance costs of the item and prolong its useful life sufficiently to more than overcome the added initial costs. This latter situation may be termed "economic best practice", and to the extent that it is not used throughout the economy, the national costs incurred because of a given degradative process will be higher than they need to be. Thus, two types of costs may be considered: the total costs incurred in the economy, and the reducible costs incurred because economic best practice is not universally used. The difference between these two are presently irreducible costs, which cannot be lowered with the present state of technology, but might be lowered as control and prevention technology develops.

These considerations bring into sharp focus the issue of the baseline from which costs are calculated. What is generally intuitively meant when costs associated with a given degradative process are discussed is that they are costs above and beyond those that would be incurred in a world in which that process did not exist. This intuitive definition of costs associated with corrosion and wear seems to have been used in all studies prior to the NBS corrosion study (1,2), where it was specifically defined. In that study, three "worlds" were defined as follows:

1. The world as it exists is called World I.
2. The baseline world in respect to which total costs are counted in a hypothetical (and unattainable) corrosion-free world. This is called World II.
3. A world in which everyone uses best practice is called III. This world is in principle attainable, although this may be difficult.

In addition to these, in the NBS fracture study, a future world, World IV, was defined, in which "best practice" would have been improved and hence give the possibility of reducing the unavoidable costs. This situation is depicted in Figure 1.

While corrosion-free and wear-free worlds are relatively easy to envisage, because the form of objects would not change significantly, a world without fracture could lead to the concept of glass automobiles and other unlikely changes. The resolution of this was a more complicated definition. In the base world for the fracture study (World II) it was recognized that design would still be load or stiffness limited, but it was taken to be world of perfect knowledge and perfect materials. Thus, there would be no material variability, and properties would be four standard deviations above the present mean; there would be a safety factor of 1.0 on load or stiffness; there would be no time dependent fracture such as fatigue or stress corrosion; and no inspection would be required. However, fracture would still be permitted but only when design limits are exceeded. This more complex definition results from the fact that all engineering design has as its principal goal to insure that fracture does not occur.

In the case of acid rain, the baseline world (World II) is clearly a world in which acid rain does not exist. Presumably, this means acid rain caused by man. It is thus easy to define what this baseline world is, but it might be difficult to determine the difference in costs in the present world (World I) and in the hypothetical acid rain-free world. This will be discussed more fully later.

The Elements of Costs

The elements of costs are categories of expenditures that go to make up the costs. For the corrosion study they are as follows:

Capital Costs. These include costs associated with shortened lifetime, any excess capacity that may be required because of corrosion, and the costs of any redundant equipment that corrosion may make necessary. These costs are borne by all sectors of the economy, both the producing sectors and final demand. Hence they apply to consumer items and equipment owned by government as well as commercial and industrial equipment.

Control Costs. These include costs for maintenance and repair and for corrosion control (cathodic protection, water treatment, etc.). They are borne by all sectors of the economy.

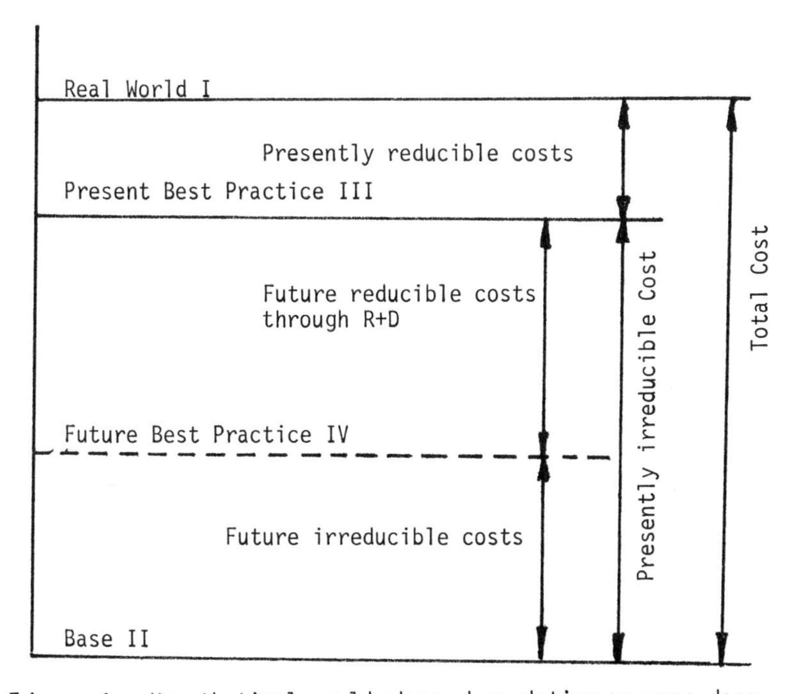

Figure 1. Hypothetical world where degradative process does not exist.

Design Costs. These include the costs of special materials (stainless steel, copper alloys, etc.) when used because of corrosion, costs of extra materials (heavier sections on beams, etc.) and special processing. These are incurred during manufacture.

Associated Costs. These include such items as loss of product, research and development, insurance, and the cost of maintaining the inventory of parts and equipment needed for corrosion control. They do not include the cost of items in inventory since this is accounted for under control and repair costs.

For the fracture study a somewhat different breakdown was used. This breakdown was occasioned by the fact that fracture leads to a specific identifiable event which is often catastrophic. Thus, costs are estimated in two main categories. Direct costs, or costs of occurrence, and indirect costs, or cost of prevention. The specific items included in these two categories are as follows:

Direct Costs. These include the loss of capital equipment, loss of product or production, and the cost of injury and death.

Indirect Costs. These are costs incurred to prevent fracture, and include overdesign in the form of excess materials, energy and labor; redundant equipment; and the costs of maintenance, repair, and inspection. As will be seen later, the indirect costs far exceed the direct.

It should be noted that costs due to stress corrosion were included in both the corrosion and fracture studies. There is thus some double counting of the costs of this process. The magnitude of this is unknown since costs were not accounted for on the basis of types of corrosion and fracture process, but is not expected to affect the results substantially.

Methodology

The intent here is to give only a brief summary of the methodology by which the studies were carried out. Briefly, input-output analysis was the basic tool used. The economy was modeled as a steady state, full employment economy for 1975 and 1978 for the corrosion and fracture studies respectively. The economy was broken down into 130 sectors for the corrosion study and 150 sectors for fracture study. In both cases, capital equipment was treated as an input into production rather than a part of final demand as normally done. Having established the steady state for the chosen year for the world as it is (World I), steady state World II (corrosion or fracture free world) and World III (best practical world) were established. Final demand and the coefficients in the transactions matrix and the flow and stock capital matrices were changed as appropriate. In the case of the flow matrix, changes in the coefficients by column were collected in a special "social savings" row. This precluded the necessity to renormalize the coefficients and gave a convenient way for calculating the corrosion costs associated with the producing sectors (intermediate output). References 1 and 2 give details of the method.

In this methodology, the national costs of corrosion and fracture become the difference in final demand in two steady state worlds - World I and World II for overall costs and World I and World III for avoidable costs. It should be noted, however that in each of these worlds, the same number of useful activities are carried out (i.e., not associated with the prevention of corrosion or fracture), and the same number of useful goods are produced, and in that sense the useful GNP is the same in all worlds.

No attempt was made to redistribute the savings incurred because of the absence of corrosion or fracture; they were simply collected.

Results

Before discussing the results of the NBS-BCL corrosion and fracture studies, it is well to look at the results for the costs of corrosion of other and less complete studies for the U.S. and other countries. The results, taken from Refs. 3-14 and summarized in Ref. 1 are shown in Table 1. The important point to be noted from these studies is that the estimate of corrosion costs amount to 2-3% of the GNP, with 20-25% of that being avoidable by the use of best corrosion control practice. It is felt that these costs are too low in that, to varying degrees, one or several of the elements of corrosion costs as described above were omitted from the accounting. In particular, costs associated with the shortened lifetime of capital equipment were not handled as thoroughly in any of the studies as they were in the NBS-BCL studies. Nevertheless, the results of these earlier studies show that the costs of corrosion amount at least to the substantial sum of 2-3% of the GNP, and this amount, within the validity of the results, is the same for all the nations which have been studied.

The results for corrosion and fracture from the NBS-BCL studies are shown in Table 2. Costs are shown by the sector of the economy where the costs are incurred: Intermediate output (IO, the producing sectors of the economy); Personal consumption (PCE, the consumer part of final demand); Federal government and state and local government (FG and S/LG); and Producer Capital (PFCF). The elements of the costs included in these sectors are as follows: IO includes all elements except capital costs, which are in PFCF; FG and S/LG include all elements except design costs; and PCE includes capital costs and control costs.

For each of these sectors, total costs and avoidable costs are shown, as well as the range of uncertainty. The latter was done in considerable detail for the corrosion study, and in less detail for the fracture study, the results for which are felt to be of less uncertainty partly as a result of the experience gained in the corrosion study and partly because the costs associated with fracture are easier to identify, since fracture control is the most important aim of engineering design.

Table I. Cost of Corrosion, Various Nations

Nation	Year	Cost	Avoidable	% GNP	Ref.
USSR	1969	6 B Rubles $6.7 B			2
West Germany	1969	19 B DM $6 B	4.4 DM $1.5 B	3 (0.75)[1]	3
Finland	1965	150-200 M Markaa $47-62 M	--	--	5,6
UK	1969-1970	1.365 B Pounds $3.2 B	0.31 B $0.74 B	3 (0.69)[1]	8,9
Sweden	1964	0.03-0.4 B Crown $58-77 M	25%		7
India	1961	1.54 B Rupee $320 M			10
Australia	1973	$470 M $550 M	--	1.5(3)[2]	11
USA	1947	$5.5 B	--	2.3	13
	1965	$15 B[3]	--	2.2	
	1975	$9.7 B[4]	--	--	14
Japan	1977	2500 B Yen $9.2 B	--	1.8	12

[1] Avoidable Costs
[2] Authors estimate that total costs are approximately twice those accounted for
[3] Unpublished NBS Results
[4] Incomplete Study

Table 2. Cost of Corrosion and Fracture

	Corrosion, 1975				Fracture, 1978	
	Total Costs		Avoidable Costs		Total Cost,	Avoidable $B
Sector	$B	Range	$B	Range	$B	$B
IO	24.5	23.5-25.0	2-Y*	-Y-2	72.3	21.1
Final Demand						
PCE	15.8	10.3-21.3	4.9	3.8-15.9	4.6	1.6
FG	7.9	6.2-9.6	1.7	0.8-2.5	1.3	0.23
S/LG	2.4	1.2-3.6	0.9	0.15-1.4	0.9	0.07
PFCF	19.1	11.5-26.7	6.2	3.0-19.1	8.4	2.94
Total Final Demand	45.2		13.7		15.2	4.8
Total	69.2	52.7-86.2	15.7-Y	(8.1-Y)-40.9	87.5 (78.8-96.3)	25.8
%GNP	4.2	3.2-5.2	-0.9		3.9 (3.5-4.3)	1.2

*The value of Y is a matter of speculation, but assuming it costs between 10 and 70 percent of the expected final demand gain of $13.7 B for best practice (extra coatings, etc.), Y would be between $1.4 and $9.6 B, and the total avoidable costs would be between $6.1 and $14.3 B, or about 10 and 20 percent of the total cost. Note that these values of Y could make the avoidable IO contribution negative. This would mean an increased cost to manufacturers in a best practice world, to achieve a net savings to manufacturers plus final demand (life-cycle costs).

There are striking similarities as well as differences in the results. First is the total cost. Within the range of uncertainty, total fracture costs and corrosion costs are both 4% of the GNP, which, considering the greater completeness of these studies, correlates well with previous studies. However, the similarity of the corrosion and fracture results must be somewhat fortuitous since the corrosion study was limited to metals, while the fracture study included all materials. Similarly, the avoidable costs of approximately 20% of the total for corrosion (for technical reasons, this number has a large uncertainty (1)) and 29% of the total for fracture also correlate well, and correlate with previous studies (Table I).

There is, however, a striking difference in the sectors in which the results are incurred. Thus, for corrosion, 65% of the corrosion costs are incurred in final demand, whereas for fracture, only 17.4% are in final demand. The bulk of this difference is in design costs, i.e., the cost associated with prevention at the design-manufacture stage, and is caused in large measure by the imposition of large safety factors in fracture control. Except for stress corrosion, corrosion does not generally result in catastrophic collapse, whereas fracture does, and is therefore the primary item to be avoided. This leads to the large difference in the costs.

Overall Costs of Degradative Processes

With these results for corrosion and fracture we are in a position to summarize the overall costs of degradative processes. Corrosion (of metals) and fracture (of all materials) each amounts to approximately 4% of the GNP in the U.S., and probably in other industrialized nations. Wear, the other degradative process of capital equipment, has not been discussed as yet. However, from the results presented in Ref. 15 and 16, it can be concluded that the application of tribology could save 1.1 to 1.5% of the GNP in the U.S. in 1971. From the definition used (16) this represents costs somewhere between avoidable costs and total costs, the reference "World" being essentially the World IV of the NBS fracture study. While the subject merits further study, the results indicate that the costs of wear are roughly comparable to the costs of corrosion and fracture, but perhaps somewhat less. It can thus be concluded that the overall costs of these three degradative processes amount to somewhat more than 10% of the GNP. Costs associated with the other degradative process mentioned earlier are not known to this author, but clearly will raise this total.

Acid Rain

It can be asked at this point what these results have to do with the economic effects of acid rain on materials. First, the baseline or world from which the acid-rain accounting is made is examined. In parallel with definitions used in the corrosion study, this is clearly a world in which man-made acid rain does not exist. However, unlike the case in the corrosion study, where a corrosion-free world can be imagined but cannot be realized given the thermodynamic state of metals, a man-made acid-rain free world can be both imagined and, at least in principle, be realized One could therefore consider the cost of preventing acid rain to be the cost of acid rain on materials. This would overstate the costs of acid rain on materials, for it would include all other effects of acid rain, not only that on materials. Hence a more accurate accounting would appear to be achieved by the methodology described above for corrosion. The costs of acid rain on materials would, of course, be lower than the total costs of corrosion (on all materials, not only metals), but could be estimated by the same methodology.

Second, an attempt may be made to arrive at the actual costs of acid rain. Since acid-rain is a form of atmospheric corrosion, the total costs of corrosion for metals of approximately 4% of the GNP would certainly be a high upper limit for the costs of acid rain on metals. Even the inclusion of irreplaceable artistic and architectural items in the cost of acid rain would appear to leave the total costs well short of the total corrosion costs.

Some idea of the costs that might be associated with acid rain can be obtained from the study of the corrosion costs on metals associated with air pollution carried out by Fink, Buttner and Boyd (18). This was a study of extra corrosion costs caused by air pollution throughout the U.S. Basically the methodology was similar to that described above. The baseline world was taken to be a rural area with minimal or no air pollution. The principal costs were associated either with extra maintenance (mostly painting) over the economic life of structures, or with early replacement in cases where maintenance was not feasible. By an analysis of all structures with surfaces exposed to the environment throughout the nation, and from an analysis of the effect of atmospheric corrosion on various metals, the steel and galvanized structures listed in Table 4 were chosen as the important ones in which added costs due to air pollution were expected. Then, knowing from a number of studies the extra corrosion and maintenance costs per unit area and per unit time, and knowing the total exposed area from the integrated value of shipments over the economic life of the structures, the overall yearly costs could be evaluated. This is shown in Table 3, and amounts to a total of $1.45B in 1970, which is 0.15% of the GNP. This represents approximately 4% of the overall costs of corrosion. Again, of course, this includes only metals. This figure may give some order of magnitude estimate for the costs associated with acid rain. There are, of course, costs associated with the effects of acid rain on materials that are not easily included in economic analyses. Among these are the conservation costs of cultural artifacts whose loss has few readily identifiable economic benefits. Some of these issues have been addressed in a recent study (19).

Table 3. Summation of Annual Extra Losses Due to Corrosion Damage by Air Pollution to External Metal Structures for 1970[1]

Steel System of Structure	Basis for Calculation	Annual Loss in $1000
Steel Storage Tanks	Extra Cost of Maintenance	$ 46,310
Highway and Rail Bridges	Extra Cost of Maintenance	30,400
Power Transformers	Extra Cost of Maintenance	7,450
Street Lighting Fixtures	Extra Cost of Maintenance	11,910
Outdoor Metal Work	Extra Cost of Maintenance	914,015
Pole-Line Hardware	Extra Cost of Maintenance	161,000
Chain Line Fencing	Extra Cost of Maintenance and Cost of Replacement	165,800
Galvanized Wire and Rope	Extra Cost of Maintenance	111,800
Transmission Towers	Extra Cost of Maintenance	1,480
		$1,450,165

[1]From Ref. 18

Conclusions

The overall national costs of corrosion of metal and fracture of all materials are each about 4% of the GNP, 20-30% of which could be saved with complete use of economic best practice. The costs of wear, which are not as well known, raise the total costs of these three degradative processes to somewhat over 10% of the GNP. The costs are expected to be comparable in other industrialized nations. From the results in Ref. 18, the extra corrosion costs of metals associated with air pollution are about 0.15% of the GNP, or 4% of the total corrosion costs.

Literature Cited

1. NBS Special Publication 511-2. "Economic Effects of Metallic Corrosion in the United States. Appendix B. A Report to NBS by Battelle Columbus Laboratories." SD Stock No. SN-003-01927-5. (1978). NBS GCR78-122. Battelle Columbus Laboratories Input/Output Tables. Appendix C of "Economic Effects of Metallic Corrosion in the United States." PB-279-430, (1978).
2. NBS Special Publication 511-1. "Economic Effects of Metallic Corrosion in the United States. A Report to Congress by the National Bureau of Standards." (Including Appendix A, Estimate of Uncertainty). SD Stock No. SN-003-003-0192607. (1978).
3. Y. Kolotyrkin, quoted in Sov. Life, 9, 1968, (1970).
4. D. Behrens, Br. Corros. J. 10 (3), 122 (1975).
5. V. Vlasaari, Talouselama (Economy), No. 14/15, 351 (1965) (Quoted by Linderborg, Ref. 6).
6. S. Linderborg, Kemian Teollusius (Finland), 24, (3), 234 (1967).
7. F. K. Tradgaidh, Tekn. Tedskrift (Sweden), 95 (43), 1191 (1965) (Quoted by Linderborg, Ref. 6).
8. T. P. Hoar (Chairman), "Report of the Committee on Corrosion and Protection," Dept. of Trade and Industry, H.M.S.O., London (1971). An independent corrosion survey by P. Elliot, supplement to Chem. Engr. No. 265, Sept. (1973), substantiates the findings of the Hoar report.
9. T. P. Hoar, Information Conference "Corrosion and Protection" presented at the Instn. Mech. Engrs., April 20-21, 1971, to discuss the Hoar Committee results.
10. K. S. Rajagopalan, Report on Metallic Corrosion and Its Prevention in India, CSIR. Summary published in the "The Hindu," English language newspaper (Madras), Nov. 12, 1973.
11. R. W. Rene and H. H. Uhlig, J. Inst. Engr. Austral. 46 (3-4), 3 (1974).
12. Boshoku Gijutse (Corrosion Engineering Journal), 26 (7), 401 (1977).
13. H. H. Uhlig, Corrosion 6 (1), 29 (1950).
14. NACE Committee Survey Report, "Corrosion," October 1975.
15. Lubrication (Tribology) --- A Report on the Present Position and Industry Needs, Department of Education and Science, (1976), H. M. Stationery Office, London.

16. H. Peter Jost, "Economic Impact of Tribology," NBS Special Publication 423, pp. 117-139 (1976).
17. "The Economic Effects of Fracture in the U.S.," NBS Special Publication 647-1, 647-2, 1983.
18. F. W. Fink, F. A. Butner, and W. K. Boyd, "Technical-Economic Evaluation of Air-Pollution Corrosion Costs on Metals in the U.S.," Report to Air Pollution Control Office, Environmental Protection Agency, Feb. 19, 1971, NTIS, PB198-453.
19. "Conservation of Historic Stone Buildings and Monuments," National Academy Press, 1982, 82-082101.

RECEIVED January 2, 1986

28

Model for Economic Assessment of Acid Damage to Building Materials

Thomas J. Lareau[1], Robert L. Horst, Jr.[2], Ernest H. Manuel, Jr.[2], and Frederick W. Lipfert[3,4]

[1]U.S. Environmental Protection Agency, Washington, DC 20460
[2]Mathtech Inc., 210 Carnegie Center, Princeton, NJ 08540
[3]Brookhaven National Laboratory, Upton, NY 11973

Using a "damage function" approach, material damages to building components associated with SO_2 and wet acidic deposition are estimated for four case-study cities. The damage function method links the physical change (corrosion or erosion) associated with changes in pollutant levels with material inventory and economic data. The valuation of damages in monetary units is based on the increased maintenance and replacement costs associated with the reduced lifetimes of material components, e.g. painted exterior walls. The damage estimates are better than those heretofore available, given improvements in the quality and resolution of the material inventory and air quality data that have been developed as part of the National Acid Precipitation Assessment Program.

Air pollution sources in the United States and Canada currently emit more than 25 million tons of sulfur dioxide each year. SO_2 and wet acidic deposition are believed to cause damage to aquatic life, crops, forests, and materials. The effects on materials include damages to common construction materials including galvanized steel (zinc), paint, copper, building stones and mortar, as well as damages to cultural or historic objects and buildings.

The response of both manufacturers and households to pollutant-induced damage is to increase maintenance and to find ways to circumvent the deterioration of materials through development of more resistant materials and the use of substitutes. The economic cost associated with these activities is potentially large, given the widespread distribution of exposed buildings, infrastructure components such as bridges and transmission towers, and cultural resources. Prior economic studies indicate dollar losses due to sulfur and sulfate ambient concentrations could amount to as much as 2 billion dollars annually (1). However, the poor quality of the data available for these earlier studies clouds the confidence one can place on these estimates.

[4]Current address: 707 Continental Circle, Mountain View, CA 94040

0097-6156/86/0318-0397$06.00/0

Accurate quantification of damages to materials is difficult. Most of the previous studies utilized available dose-response functions, which predict physical damage as a function of environmental variables. Translating the physical damage rates to dollar denominated losses required additional data on the spatial distribution of pollutants and materials at risk and the mitigative behavior of consumers and businesses. Significant compromises were necessary in previous efforts, due to the incompleteness of these data. For example, detailed geographic information on the amount of materials at risk has generally not been available. In the absence of such detailed data, inventory estimates have sometimes been based on national production data. Since new, high resolution data has been acquired in the last two years through the National Acid Precipitation Assessment Program (NAPAP), an assessment of the costs associated with increased maintenance can now be made which avoids compromises of this kind. In this study, the spatial resolution of damages is 5 km., permitting a more accurate appraisal of the interactions between the distribution of air quality and material inventories than was previously possible.

The new, higher quality data do not resolve all the estimation difficulties, however. Because quantified damage functions are unavailable for some materials at risk, comprehensive coverage is still not possible. In addition, estimating the aesthetic losses from the deterioration of cultural resources requires survey data, which are not available. Furthermore, major uncertainties in the physical damage functions are yet to be resolved, and these uncertainties are directly translated to the final economic estimates. Thus, at this point in time, the analysis cannot definitively establish the magnitude of all adverse effects; the analysis can, however, provide a better indication than heretofore possible of whether or not the material damages from acid deposition are economically important.

The objective of this paper is to present economic estimates of damage to common construction materials. The damage calculations focus directly on the damages to material-building component combinations that can be attributed to exposure to SO_2 and wet acidic deposition. The estimates presented here capture <u>total</u> damages associated with current loadings (relative to natural background concentrations and pH) in four urban areas in the Northeastern quadrant of the United States. Given that SO_2 emissions can be reduced by no more than 50 percent at a cost acceptable to society in this century, less than half of the estimated damages are potentially recoverable as "benefits" of a control effort.

This paper is divided into three sections. In the first section, we outline the conceptual basis for valuing material damages. The second section describes the data, the computations, and the economic damage estimates for four case-study Metropolitan Statistical Areas (MSAs): New Haven, Pittsburgh, Cincinnati, and Portland, Maine. The last section summarizes the results and provides some perspective on the uncertainties in the analysis.

The General Problem of Estimating Material Damages

In broad terms, material damages arise when maintenance or repair costs increase, when there is aesthetic degradation because maintenance is postponed or not undertaken at all, or when producers or consumers substitute more expensive materials or processes to offset the effects of corrosion or erosion on materials. The conceptual economic framework offers a variety of insights into damage measurement issues and indicates the difficulty of comprehensive measurement of these damages, given available economic data. Only a brief discussion of the insights and limitations derived from economic theory is presented in this section. A more detailed theoretical treatment of this subject can be found in Horst et al. (2).

Economists emphasize the concept of willingness to pay in the theory of value. This reflects the judgment that the sum of the dollar votes of individuals best represents society's valuation of any good or service, whether that good or service is private or public. A crucial distinction is that willingness to pay is not equal to what one has to pay, which is an observable market result. Rather, willingness to pay can be inferred from the information embedded in the demand function for a good or service. A market demand curve, D, shown in price-quantity space in Figure 1, represents marginal willingness to pay as a function of quantity. That is, a consumer would be willing to pay at most p_0 for the q_0th unit of the good. Total willingness to pay is represented by the area under the demand curve up to a specified quantity. Frequently, only a portion of total willingness to pay is relevant. When consumers pay for the good in an established market, the focus is on consumer surplus, which nets purchase expenditures from total willingness to pay. Consumer surplus, shown in Figure 1 as the area enclosed by Bp_0C, thus represents what individuals are willing to pay over and above what they do pay.

Consumer surplus is only a part of measured societal well-being from which a change attributed to pollution can be computed. To describe the additional surplus value, it is necessary to introduce a supply relationship. The supply curve, S, in Figure 1 represents the incremental cost of producing one more unit of the good or service. At the market price, p_0, some providers of the good will be able to supply it profitably at a unit cost less than p_0. This gives rise to producer surplus, shown as the area enclosed by Ap_0B in Figure 1. This surplus represents the aggregate difference between price and marginal cost. Since producer surplus depicts the gain to the owners of productive inputs, it is added to consumer surplus to determine the total economic surplus that results from market activity. This total amount is shown by area ABC in Figure 1.

To examine the effect of acid deposition on economic surplus, it is convenient to focus on the market for "building services." (This argument is valid for user-related damages; it does not account for the disutility to nonusers that may result from aesthetic degradation of properties with cultural or historical significance.) Building services, for both household and commercial properties, can be defined to include both quantity and quality dimensions. The price of building services would be the annualized life-cycle cost of a maintained building. This price would include the costs of periodic maintenance, cleaning, and general upkeep. Starting from an environmental state, e_0, associated with zero damages, increased acidic pollution, e_1, would increase the cost of building services. This is shown in Figure 2 by the shift in supply from $S(e_0)$ to $S(e_1)$. As can be seen, the economic surplus decreases by the area ACDE.

In this application, a damage function methodology, rather than

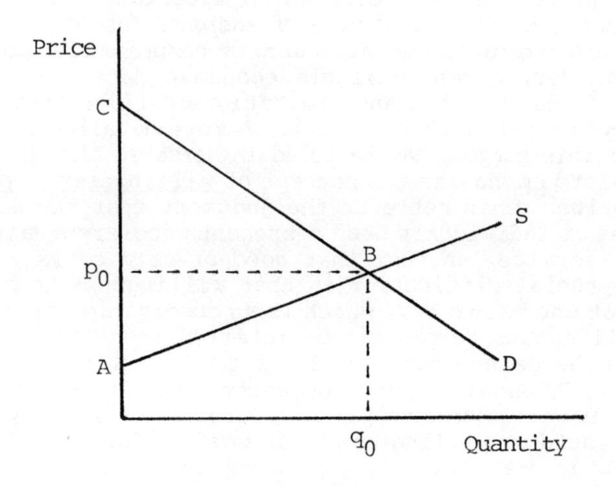

Figure 1. Consumers' and Producers' Surplus

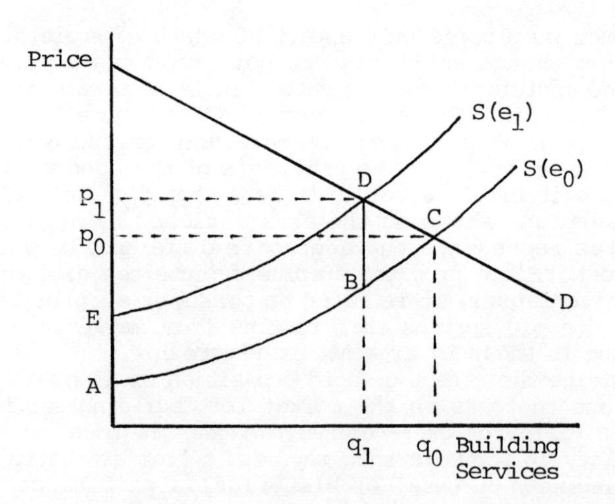

Figure 2. Surplus Change due to Pollution

a willingness to pay measure, is used to estimate the monetary damages of increased maintenance or replacement of building materials. It is important, therefore, to examine how closely measured maintenance or repair cost increases match the generalized building service cost increase described above (area ACDE). Even under fairly general conditions, the two are not the same, since the generalized cost change is predicated on a broad definition of costs and behavioral flexibility. The damage function method, as generally defined, does not include the costs of substituting pollutant-resistant materials to mitigate damages, nor does it account for loss of serviceability or aesthetic value. If the increase in maintenance cost is based on computing the change in maintenance frequency, little behavioral flexibility exists. The maintenance frequency depends on the time it takes to reach the critical damage level, which is determined by engineering standards and damage relationships assuming regular maintenance practice. The problem with this approach is that individuals may choose to perform maintenance on a different schedule. One might expect the optimal level of maintenance to depend on a building's current usage and age, for example.

The biases resulting from physical and economic data limitations lead to underestimates of economic damages in some cases and overestimates in other cases. Maintenance cost increases will overestimate actual damages if behavioral flexibility is ignored, since property owners would consider alternatives to maintenance if the alternatives were less costly. On the other hand, the omissions, principally those associated with aesthetic values and the higher costs associated with substituting pollution-resistant materials either at the point of manufacture or at the time of construction, lead to underestimates of total damages. For example, galvanized gutters are more commonly used where acidic deposition is low, while more pollutant-resistant vinyl-coated gutters are more prevalent in high deposition regions (3). The cost associated with the pollution-resistant gutters is not counted in a calculation based solely on more frequent maintenance. Of greater concern is the noninclusion of potentially important materials at risk, e.g. reinforced concrete structures, automobile paints, and infrastructure materials. In sum, the limited coverage of affected resources probably dominates other biases, though inaccuracies in the measurement of physical damages and inventories of materials at risk are not inconsequential.

Computing Construction Material Damages

Using physical damage functions to value increased maintenance or more frequent replacement of building components in the presence of pollution requires the joint application of the following data:

- o Distribution of the pollutants, SO_2 and wet deposition of H^+, and other factors, such as time of wetness, that enter into dose-response functions.
- o Distribution of resources at risk.
- o Cost of maintenance, repair, or replacement along with other economic data that indicate how consumers and producers respond to the more rapid deterioration of building components.

The problem of estimating material damages using disaggregated data is difficult. There are many different types of buildings, reflecting regional construction patterns and changes in architectural styles over time. Each of these buildings contains a mixture of materials with varied sensitivities to different pollutants. Thus, building materials and the damages to them are not distributed uniformly, either within urban areas or across urban areas. For example, there is more steel in central business districts, and brick exteriors are more prevalent in cities in the Midwest than those on either coast. Variations in material selection also seem to reflect ambient pollution concentrations. There is, for example, anecdotal evidence that use of aluminum siding is greater in areas more susceptible to pollution-induced paint damage (3). Given this variability in material usage, an accurate estimate of damages requires detailed surveys to estimate the distribution of materials.

Four MSAs in the Midwest and Northeast regions--Cincinnati, Pittsburgh, New Haven, and Portland--were chosen for this analysis. While material inventories will be estimated for over 100 metropolitan areas in the forthcoming NAPAP Assessment, highly detailed ground surveys were conducted in 1984 by the Corps of Engineers only in these four cities (4). The resulting tabulations of building components (painted surfaces, gutters, etc.) by building type (residential, commercial, etc.) will be used to extrapolate materials usage in the nonsampled MSAs. However, only results in the four case-study cities, where greater accuracy is possible, are reported in this analysis.

The basic procedure is outlined in Figure 3. The first step, Inventory Accounting, utilizes randomly chosen 100 x 100 ft. to 400 x 400 ft. "footprints" of sampled data on building components. This detailed inventory included approximately 1100 buildings in the four cities. Surface area or linear footage was recorded by material type for exterior walls, roofs, gutters and downspouts, and fencing for all structures within a given footprint. Window trim area was indirectly inferred from the recorded area of glass. We used these data to compute average material usage and probabilities of occurrence for each material-building type combination. There were four building categories: single-family residences, multiple-family residences, commercial and industrial structures, and tax-exempt structures. Census and property tax records are used to obtain accurate counts of the numbers of buildings in each category by census tract for each of the four case-study cities. Applying the probabilities and areas of material usage by building type to the census count of buildings allowed us to extrapolate from the sampled data to an estimate of the amount of building material in an entire metropolitan area (3).

In the second step, Damage Calculations, shown in Figure 3, physical damage functions for paint (two types), zinc galvanized material, stone (marble and limestone), and mortar are used to compute damage rates for each material. These damage functions have been developed from field and laboratory data as well as stoichiometric theory (see Lipfert et al. (5) for summary report on metals; Haynie (5) for paint; and Reddy et al. (7) for stone). The damage coefficients are summarized in Table I, with lower- and upper-bound estimates provided for paint. These coefficients are used to develop a range estimate for monetary damages in the third stage of the analysis.

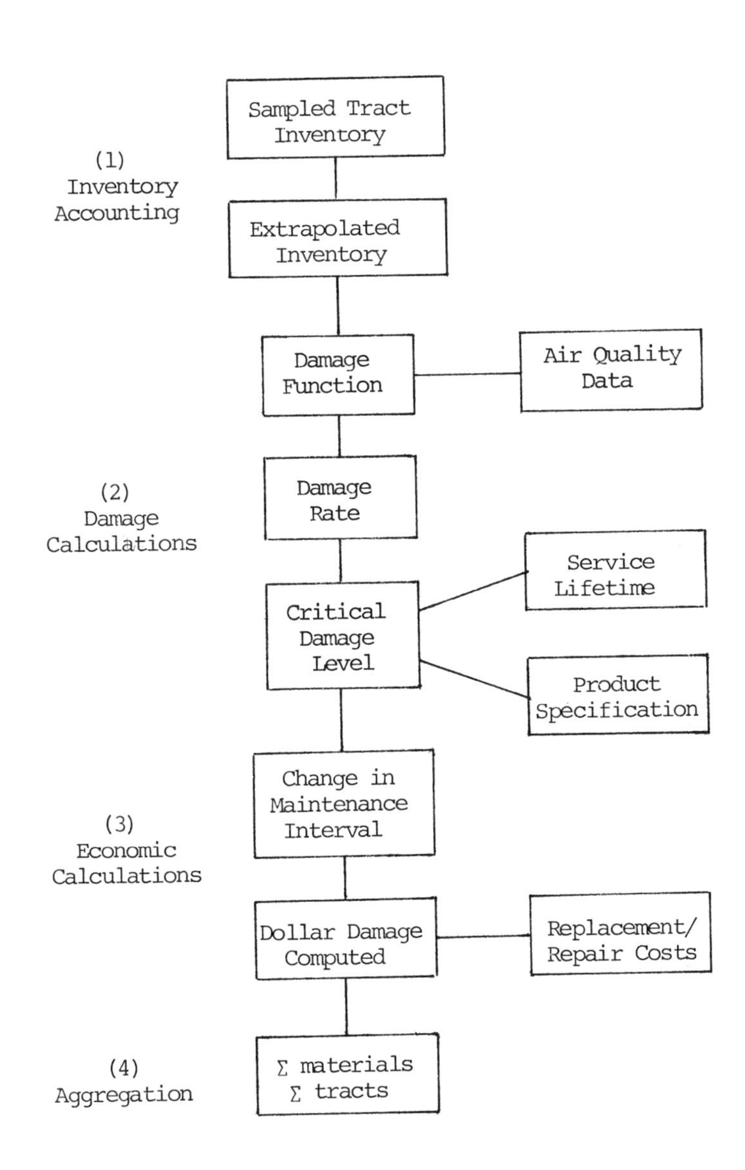

Figure 3. Computation Flowchart

Table I: Estimation Assumptions

	Low	Point	High
A. Damage Coefficients			
1. Paint --Carbonate SO_2	.0400	.1200	.1966
--Carbonate H^{+2}	.0212	.0625	.1040
--Silicate SO_2	.0098	.0194	.0261
--Silicate H^{+2}	.0050	.0098	.0132
2. Zinc --SO_2	--	.5470	--
--H^{+2}	--	.1070	--
3. Mortar--SO_2	--	.3330	--
--H^{+2}	--	.2655	--
4. Stone --SO_2	--	.1110	--
--H^{+2}	--	.0885	--
B. Critical Loss Level			
Painted walls	--	33.0um	--
Painted gutters/downspouts	--	19.1um	--
Painted trim	--	36.8um	--
Mortar	--	.375in	--
Stone	--	.250in	--
Zinc gutters/downspouts	--	229g/m^3	--
Zinc fencing	--	229g/m^3	--
C. Repair/Replacement Costs (\$/ft^2)			
Painted walls	0.53	0.99	1.97
Painted gutters/downspouts	0.54	0.77	1.75
Painted trim	0.73	1.54	3.96
Mortar	2.58	6.73	10.81
Stone	0.00	0.17	20.07
Zinc gutters/downspouts	0.54	0.77	0.77
Zinc wire-mesh fencing	1.35	1.35	1.35
Zinc chain-link fencing	1.57	1.57	1.57
D. Carbonate-Silicate Paint Split	.05	.15	.25

a. Units: SO_2, ug m^{-3}; H^+, ueq m^{-2} yr^{-1}; paint erosion, ug; zinc, g/m^2; mortar/stone, um/yr.

b. Reported costs are national averages; costs for specific MSAs are adjusted for regional labor rates and other local factors.

SOURCE: Mathtech (12), see text

Although there are many varieties of paint in use, Haynie (6) has chosen to simplify them into two types according to the composition of the extender, or thickener. Of the two paint types, the more pollutant-resistant contains silicate extenders while older, carbonate extender paints are less pollutant-resistant. Silicate extender paints have become dominant in the exterior paint market only in the last decade, accounting for most (probably over 80 percent) of current sales. The proportion of silicate paint on buildings is unknown, however, but is conservatively assumed to be 85 percent. The switch to silicate paints may have occurred partially as a result of concern with the environmental susceptibility of the carbonate extender paints. These substitution costs are not factored into the damage estimates provided in this analysis.

Air quality data for SO_2, wet deposition of H^+, and other general environmental data are inputs to the damage functions. For the resolution level required for reasonable overall accuracy, monitored data are inadequate so that modeled data are developed using dispersion algorithms. Damages are computed for 5 km. grids within the defined boundaries of each metropolitan area. "Local" SO_2 concentrations were calculated using emission sources within an envelope defined by a 50 km. boundary around the metropolitan area (8). The "nonlocal" contribution of SO_2 was determined from ASTRAP model runs (provided by J. Shannon at Argonne National Laboratory). Finally, H^+ deposition was measured from regionally monitored data.

The damage rate, in weight or linear loss units, is used in the Damage Calculation stage of Figure 3 to compute the number of years until the critical damage level (defined as the level of damage at which maintenance or replacement is required) is reached. As pollutant loadings increase, the time interval between required maintenance or replacement decreases, and maintenance costs go up. The critical damage level is determined in one of two ways. One, the critical loss of material thickness is specified by industry standards or by materials specialists. Two, the critical damage level can be computed from service lifetime information and knowledge of current pollution levels. The critical damage levels used in this stage of the analysis are summarized in Table I.

Economic damages are computed in the third stage shown in Figure 3. The basis for economic damages is the increased maintenance, repair, or replacement of building components under current pollution loadings relative to pristine conditions (SO_2 at .5ug/m^3 and a pH of 5.2). Annual monetary damages are computed for each material component as the product of three factors: the cost per unit area ($/$ft^2$), the exposed area in the inventory (ft^2), and the maintenance time interval difference predicted from the damage function and the critical damage level (yr^{-1}). The third factor effectively distributes the additional cost of maintenance uniformly over the reduced maintenance interval. These annual costs represent damages in the 1980's, but because long-term life-cycle considerations have not been taken fully into account, the estimates may not be representative of future annual damages. All calculations are made for the existing stock of materials, assuming no net growth in this stock. Furthermore, all buildings are assumed to be optimally maintained in the future.

The underlying maintenance cost data are summarized in Table I.

Maintenance and replacement cost data were obtained from industry manuals, such as Robert Snow Means Co. (9), and from a NBS study (10). Lower- and upper-bound estimates of unit costs are provided to facilitate computation of range estimates of the economic damages.

For repainting, the low unit cost assumption reflects the application of a primer and one surface coat of paint. The midpoint unit cost assumption includes minimal surface preparation and two surface coats, while the high estimate includes additional scraping and sanding preparation costs. The repainting estimates for single-family units have been adjusted to account for "do-it-yourself" labor, which had the effect of lowering homeowner painting costs by 18 percent. Galvanized gutters/downspouts are assumed to be painted once the zinc coating fails. Galvanized fencing is assumed to be replaced when the critical damage level is reached.

The cost of repointing mortar was based on detailed case studies of rehabilitation of several buildings in New York City (11). Unlike paint or galvanized materials, mortar and stone material failures occur over long time periods. Thus, annual maintenance costs may need to be adjusted by a discount factor, even given an implied assumption of constant historical and future emissions. With repointing of mortar occurring at roughly 50 year intervals and many older buildings exposed to comparable (or higher) pollutant loadings, pollution-associated mortar repair is already a reality. Thus, unadjusted annual cost estimates are appropriate. For stone the situation is quite different. The replacement period for facade stone, under current pollutant conditions, is hundreds of years. Buildings using this material, e.g. skyscrapers, are all less than 100 years old. Thus, damages will not be realized for a long time. For this reason, the low and midpoint unit cost estimates are discounted. All the cost estimates are adjusted to 1984 dollars and to the city construction price index for the appropriate MSA.

The final step in the process illustrated in Figure 3 is the aggregation over building components, building types, and the 5 km. grids, to estimate total damages in each of the four cities.

Case-Study Damage Estimates and Findings

The results are summarized in Table II. Monetary estimates of annual damages are shown by city as a function of material, building type and pollutant source. The estimates are presented in range form using the assumptions listed in Table I. Essentially, three main sources of uncertainty are captured in the range estimates: uncertain physical damages, an uncertain split between carbonate and silicate paint extenders, and uncertain maintenance and replacement costs.

It is clear from Table II that while the relative damages among the cities vary, the potential dollar amounts are large. In just the four case-study MSAs, the total annual damages amount to approximately \$130 million. On a per capita basis, these damages range from approximately \$8 to \$43 in the four case-study cities, with Pittsburgh the highest and Portland the lowest. Much, but not all, of the variation among the cities can be explained by differences in the average size of buildings, by regional material usage, especially for walls, and by the relative size of the commercial/industrial sector. In addition, the pollutant levels vary among the case-study cities. Thus, it is not surprising that Pittsburgh with many large

Table II: Summary of Damage Estimates for Four Case-Study MSAs

ANNUAL DAMAGE IN MILLIONS OF 1984 DOLLARS BY SECTOR, MATERIAL, AND POLLUTANT SOURCE. One-year damages arise from increased costs associated with reduced maintenance/replacement intervals at ambient relative to "pristine" pollutant levels. Low estimate reflects minimal maintenance, low damage coefficient, and 95% silicate assumptions. High estimate reflects extensive maintenance, high damage coefficient, and 75% silicate assumptions. Variation in damages is explained by differences in average building size, local material usage, size of industrial/commercial sector, and pollutant levels.

	Annual Damage ($\$10^6$)		
	Low	Point	High
Cincinnati			
Paint	3.2	18.6	77.5
Mortar & Stone	3.0	8.3	16.4
Zinc	2.1	2.7	2.7
Single-family	4.1	15.0	50.8
Multi-family	0.9	4.0	13.6
Commercial/Industrial	2.7	9.0	27.7
Tax-exempt	0.6	1.7	4.4
Local SO_2	1.6	5.9	19.8
Nonlocal SO_2	2.4	9.6	31.2
H^+	4.0	14.1	45.1
Total	8.3	29.6	96.6
Per Capita ($)	6.	22.	69.
New Haven			
Paint	0.7	3.7	15.3
Mortar & Stone	0.3	0.7	1.5
Zinc	0.2	0.3	0.3
Single-family	0.6	2.8	10.9
Multi-family	0.2	1.0	3.5
Commercial/Industrial	0.1	0.4	1.5
Tax-exempt	0.2	0.5	1.3
Local SO_2	0.1	0.5	1.8
Nonlocal SO_2	0.4	1.8	6.7
H^+	0.6	2.3	8.7
Total	1.1	4.8	17.2
Per Capita ($)	2.	11.	34.

(Table II continued on following page)

Table II. Continued

	Annual Damage ($\$10^6$)		
	Low	Point	High
Pittsburgh			
Paint	10.8	61.2	248.6
Mortar & Stone	9.0	24.9	49.0
Zinc	5.9	9.8	9.8
Single-family	6.4	24.6	77.6
Multi-family	2.8	10.6	33.0
Commercial/Industrial	15.5	56.9	185.3
Tax-exempt	1.0	3.8	11.4
Local SO_2	5.8	22.6	74.5
Nonlocal SO_2	9.0	35.2	116.8
H^+	9.7	36.1	114.0
Total	25.7	95.9	307.3
Per Capita ($)	12.	43.	138.
Portland			
Paint	0.2	1.2	5.0
Mortar & Stone	0.1	0.3	0.5
Zinc	0.1	0.1	0.1
Single-family	0.2	0.9	3.4
Multi-family	0.1	0.3	1.1
Commercial/Industrial	0.1	0.3	0.8
Tax-Exempt	0.0	0.1	0.3
Local SO_2	0.0	0.2	0.7
Nonlocal SO_2	0.1	0.5	1.7
H^+	0.2	0.9	3.3
Total	0.4	1.6	5.7
Per Capita ($)	2.	8.	29.

SOURCE: Mathtech (12)

buildings and high SO_2 levels has higher damages, both total and per capita, than Portland.

Several other findings are illustrated in Table II. Most obvious is the importance of paint in this analysis, accounting for well over half of total damages in each of the four case-study cities. Paint damages are less dominant only in the unlikely situation (i.e., with a low subjective probability) that the lower-bound estimate with a low damage coefficient, no surface preparation, and a 95-5 percent silicate-carbonate paint split is actually correct. Mortar damages are the second most significant category, increasing in importance for cities, like Cincinnati, that are further west. In the larger cities, e.g. Pittsburgh, the commercial/industrial sector accounts for the largest proportion of damage; in smaller cities, the greatest proportion of damage occurs in the single-family residence building category. Damages to single-family residences may be more certain than damages to other types of buildings, given the presumption of "ownership pride" and concern of appearances on the part of home owners. Thus, the assumption of regular maintenance is probably close to the mark so that the damages are relatively certain. In contrast, changes in local business conditions can sometimes lead to unexpectedly shorter economic lifetimes for commercial/industrial buildings, and for these buildings maintenance on a continuous "optimal" schedule is less likely, and damages are consequently less certain.

Finally, as indicated in Table II, "local" SO_2 sources account for a fairly small fraction of total damages--less than 25 percent in each of the four cities. Apparently, the Clean Air Act has effectively reduced SO_2 levels in major cities to the point where nonlocal SO_2 damages and damages associated with wet deposition are far more important than damages attributed to local sources.

Two general observations on how these results should be interpreted are worthy of emphasis. First, the current total damage estimates reported here represent damages associated with ambient pollution loadings; thus, any practical reductions of emissions will only buy a partial reduction of these damages. These damages represent losses at this point in time; future damages could be lower if, for example, the silicate extender share of the paint market increases.

Second, as is apparent from the order of magnitude span between the low and high estimates, considerable uncertainty is present in the analysis. Much of the uncertainty associated with the increased maintenance and replacement costs of the materials for which dose-response functions were available has been captured in the range estimates. Nonetheless, there are additional unquantified uncertainties. None of the damage functions has been developed from, or checked against, actual deterioration rates on real buildings. In addition, most of the damage is to paint, for which the uncertainty is particularly large, especially with respect to the H^+ portion of the dose-response function. Finally, the estimates presented here do not fully capture society's willingness to pay to avoid material damages. Only repair and replacement costs are estimated; aesthetic losses are ignored. Further, the estimates are not comprehensive. Materials such as automobile paints and concrete have not been included in this analysis. Material usage in some sectors, e.g. structures such as bridges and rural areas, have also been excluded.

While the caveats listed above should not be underestimated, we believe that this analysis provides more convincing evidence than was heretofore available that damages to materials from acidic air pollutants are economically important. This outcome is attributed to improvements in the data describing the joint distribution of materials and air pollutants and in the damage functions. The non-comprehensive coverage of sectors and materials and the substantial uncertainty reported are strong arguments for further research.

Acknowledgments

This research was conducted as part of the National Acid Precipitation Assessment Program (NAPAP) from which most of the data were obtained. The views expressed in this paper are those of the authors and do not necessarily reflect those of NAPAP or its supporting agencies.

Literature Cited

1. Freeman, M. A.; "Air and Water Pollution Control"; John Wiley & Sons, New York, NY, 1982; pp. 91-100.
2. Horst, R. L.; Manuel, E. H.; Bentley, J. T.; "Economic Benefits of Reduced Acidic Deposition on Common Building Materials: Methods Assessment"; Mathtech Inc., Report prepared for Office of Policy Analysis, U.S. Environmental Protection Agency, 1984.
3. Novak, K. M.; Coveney, E. A.; Torpey, M. R.; Lipfert, F. W.; "Data Bases on Residential Construction Practice"; Brookhaven National Laboratory, 1984.
4 Merry, C.; LePotin, P.; "A Description of New Haven, Conn. Building Data Base"; U.S. Army Corps of Engineers Report, 1985.
5. Lipfert, F.; Benarie, M.; Daum, M.; "Derivation of Metallic Corrosion Damage Functions for Use in Environmental Assessments"; Brookhaven National Laboratory, 1985.
6. Haynie, F.; "Atmospheric Damage to Paints"; EPA Environmental Research Brief, EPA/600/M-85/019, 1985.
7. Reddy, M.; Sherwood, S.; Doe, B.; "Limestone and Marble Dissolution by Acid Rain"; Proceedings of 5th International Congress on Deterioration and Conservation of Stone, 1985.
8. Lipfert F.; Dupuis, L.; Schaedler, J.; "Methods for Mesoscale Modeling for Materials Damage Assessment"; Brookhaven National Laboratory, 1985.
9. Robert Snow Means Co.; "Repair and Remodeling Cost Data: Commercial/Residential 1983"; Kingston, MA.
10. Weber, S.; Lippiatt B.; Wiener, M.; "A Life-Cycle Cost Data Base for Assessing Acid Deposition Damage to Common Building Materials"; National Bureau of Standards, Department of Commerce, 1985.
11. Ottavino, K.; Prudon, T.; "Facade Repair: New York City Case Studies"; Brookhaven National Laboratory, Report prepared for National Park Service, 1985.
12. Mathtech Inc.; "A Damage Function Assessment of Building Materials: The Impact of Acidic Deposition"; Report prepared for Office of Policy Analysis, U.S. Environmental Protection Agency, 1985.

RECEIVED January 13, 1986

Application of a Theory for Economic Assessment of Corrosion Damage

Frederick W. Lipfert[1,3] and Ronald E. Wyzga[2]

[1]Brookhaven National Laboratory, Upton, NY 11973
[2]Electric Power Research Institute, Palo Alto, CA 94303

Deterioration of materials in the built environment is one of the considerations with regard to the justification for more stringent controls on anthropogenic pollution sources. There are both economic and aesthetic concerns involved: reduced service life of common construction materials is primarily an economic consideration, while irreversible damage to art objects and historic buildings has a strong emotional context as well.

Damage to monuments and carved building stone has been observed and documented, both in this country and more extensively in Europe. Sorting out cause and effect, i.e., separating "natural" weathering from pollution-induced damage involves use of planned experiments to develop damage functions, either in the field or under laboratory conditions. Most of our knowledge of the deterioration of building materials comes from such experiments, which have usually been performed on standard test configurations rather than on actual building components.

The current capability for economic assessment of materials damage then depends heavily on the extrapolation of test conditions to the real world. The real world of today's built environment differs from both current and past testing conditions in many important ways. It is the purpose of this paper to explore these differences on a theoretical basis and to consider whether the concomitant uncertainties in the prediction of reduced service lives are tractable and whether important biases are involved.

The general problems of forecasting materials damage from both air pollution and acid precipitation have been considered before in previous papers (1-2). This previous work focussed on zinc as a paradigm for all materials at risk, and considered uncertainties in damage functions, variability in atmospheric conditions over space and time, and the general problem of inventorying the materials at risk. An important topic in these previous papers was whether the interaction between SO_2 ambient concentrations and relative humidity (RH) would require the consideration of time scales

[3]Current address: 707 Continental Circle, Mountain View, CA 94040

0097-6156/86/0318-0411$06.00/0

shorter than annual averages, in order to accurately predict corrosion losses, since SO_2 deposits more readily on wet surfaces (high humidity). Wyzga and Lipfert (2) concluded that SO_2-RH interactions did in fact exist and that they were variable in magnitude among the locations considered. Their results show that the error incurred in neglecting these interactions is generally less than 10%, and therefore annual averages are likely to be acceptable when other sources of uncertainty are considered.

The research being performed to support the National Acid Precipitation Assessment Program (NAPAP) has also led to considerable progress with regard to the other problem areas mentioned above. Damage functions have been developed by analyzing experimental data which delineate the separate effects of dry deposition of SO_2 from those of wet deposition of acidity (H^+), not only for metals (3) but for paints (4) and calcareous stones (5). The spatial variability of relative humidity and thus time-of-wetness has been analyzed (6) and prediction algorithms developed which incorporate urban heat island effects, based on the areal density of buildings. Thus it is now possible to predict relative humidity spatial gradients within a metropolitan area. Such gradients (lower humidity and thus fraction of wet time in city centers compared to outlying areas) have the effect of reducing the gradients in the corrosion effects of air pollution (primarily SO_2), when air concentration patterns are converted to deposition patterns.

Better methods for materials inventorying have been developed based on stratified random sampling (7) and have been carried out in four Northeastern cities (8). Analysis of these data, for about 1100 buildings in total, has shown predictable patterns in material usage for the major building materials, and extrapolation methods have been developed based on the 1980 Census of Housing and building counts (9). Since the overwhelming number of structures in the U.S. are residential, use of a building count inventory basis insures a realistic basis for estimating the total exposed surface area of all material types, with the principal remaining uncertainties stemming from the probabilities of use of specific materials, especially those used in small quantities such as marble, galvanized steel, or copper.

This paper then returns to the issue of using damage functions to predict reduced service lives of common building materials or the probability of irreversible damage to cultural resources. The primary issue is the applicability of the results of controlled laboratory and field experiments to actual conditions in the real world. The factors to be considered are:

- atmospheric variability
- effects of scale
- effects of configuration
- interactions between pollutants and atmospheric conditions.

These factors all relate to boundary layer theory and behavior, either the atmospheric boundary layer which governs meteorological behavior, or the boundary layers on the surfaces of the objects at risk, which control the delivery of corrosive materials to those surfaces. We will use boundary layer theory as a tool to try to gain a more detailed understanding of atmospheric effects on corrosion.

We intend to explore the concept of a damage function as a theoretical expression incorporating both atmospheric behavior and chemical reactions on the surface, rather than just an empirical correlation of test results. Previous considerations of the aerodynamics of gaseous deposition include Livingston's analysis as applied to stone (10), which outlined many of the concepts presented here, and Haynie's analysis (11) of zinc, which emphasized potential flow effects and the stoichiometry of the zinc-sulfur reaction. This analysis is intended to be general and to draw on the theoretical methods that are available for many practical situations, in order to examine possible biases in current assessment methods. The primary emphasis is on (dry) deposition of gaseous pollutants (SO_2); some thoughts are also given on corrosion due to (wet) acidic precipitation.

Boundary Layer Concepts

The concept of a "boundary layer" with respect to the motion of a fluid over a solid body was first expressed by Prandtl in 1904 (12), in which he established that the influence of fluid friction is limited to a very thin layer in the immediate vicinity of the body, outside of which fluid friction may be neglected. Subsequent developments have established the similarity between forced convective heat transfer and fluid friction, and between mass transfer and heat transfer. Deposition of (gaseous) pollutants is an example of mass transfer, and can be described by these same boundary layer concepts. The atmospheric (or planetary) boundary layer properties primarily reflect the effects of objects on the earth's surface in obstructing the wind flows set up by pressure gradients and other meteorological forces. We must therefore consider not only the details of wind flows around buildings, statues, monuments, etc., but also effects of these objects (usually in large agglomerations) on the atmospheric structure of the wind flow per se.

The Atmospheric Boundary Layer. The atmospheric boundary layer can be loosely defined as that portion of the lower atmosphere which manifests the effects of surface features in influencing wind flow. It often extends up to heights of the order of 1 km or to the height of the mixing layer, above which the thermal properties of the atmosphere may effectively insulate it from ground effects. The atmospheric boundary layer is the carrier for pollutants that affect corrosion. There are several properties of the atmospheric boundary layer of concern here:

- the velocity distribution within the layer
- the temperature distribution within the layer, which will govern the turbulence intensity and hence the dispersion of pollutants as they are released from sources
- the distribution of pollutants within the layer, which will be affected by their release heights as well as by the above two factors.

The velocity distribution is often given by the relation

$$u/u^* = (^1/_k) \ln (z/z_o + 1) \quad (1)$$

where u is the local velocity, u* is the friction velocity given by $\sqrt{\tau/\rho}$ (shear stress/density)$^{1/2}$, k is the von Karman constant, z is the height above ground, and z_o is the characteristic roughness height of the surface. For grassland, z_o may be of the order of 1 cm; for a suburban neighborhood, perhaps 1 m. Note that z_o is determined from velocity profile measurements and not from the physical size of objects on the ground. Standard National Weather Service (NWS) wind measurements are often referenced to a height of 10 m, in which case the equation above may be used to develop wind speed ratios (u/u @ 10m), cancelling out the u* factor. The turbulence intensity will be highest near the ground, and can be estimated from

$$Tu = k/\ln (z/z_o + 1) \quad (2)$$

There are two reference heights above ground of interest to this analysis: first, the height at which standard corrosion tests are usually made, about 1 m. Secondly, the appropriate average height for buildings or structures to which these test results may be applied:

- residential buildings, say 3 m
- fences 0.5-1 m
- non-residential buildings or structures 3-30 m

As an illustration of these wind speed variations, Table I presents sample calculations for the three classes of structures and two (extreme) values of z_o. We see that the urban-rural variations are the most extreme for smaller objects. Note also that in an urban area with regularly and closely spaced buildings, wind flow patterns will be highly irregular, depending on direction with respect to street orientation, for example (Figure 1). The dramatic increase in turbulence intensity in urban areas is also shown.

The distribution and dispersal of pollutants within the atmospheric boundary layer have been thoroughly discussed elsewhere, and will not be elaborated here.

Boundary Layers on Structures. All objects immersed in the atmospheric boundary layer perturb its flow in some way. A new boundary layer is formed on each structure by this flow, and the characteristics of this boundary layer govern the transfer of momentum, heat, and mass from the atmosphere to the structure (and vice versa). One of the important characteristics is the physical scale of the object or structure being considered, as well as its shape and surface roughness (texture). This is true not only for isolated objects but for agglomerations (cities, forests), which may in turn have a large influence on the scale of atmospheric turbulence as well as its magnitude.

A well developed theory (12) is available to deal with simple situations: flow along a flat plate, around a cylinder or sphere, over an airfoil, etc. Blunt objects such as buildings are generally handled empirically. Figure 2 depicts the perturbations created by boundary layer flows whose surface characteristics differ from those of the unperturbed atmosphere nearby or upstream:

> Momentum: the requirement of zero flow velocity at the surface creates a shear stress or drag on the object due to skin friction (C_f);
>
> Heat transfer: if the surface temperature differs from the stream temperature, heat will flow;
>
> Mass transfer: if the concentration of some component of the flow differs at the surface, either because of injection into the boundary layer or because of removal from the stream, mass will flow.

An analogy (due to Reynolds) has been postulated relating these three flux terms, which has been verified by numerous classical experiments, usually under conditions which are mathematically tractable. Our interests here are with removal from the air stream of pollutants, which in turn react with the surface and cause corrosion. This process is referred to as dry deposition, although the presence of a liquid (water) film on the surface is essential for rapid removal of soluble gases such as SO_2. The presence of such a film could require a two-layer analysis including phase changes, which is beyond the scope of this preliminary inquiry.

Reynolds' analogy allows estimates to be made of SO_2 deposition velocity (V_d) based on heat transfer or skin friction tests (or theory), of which the literature abounds. In so doing, one must realize that such a calculation deals only with the delivery of pollutant to the surface, through diffusion. If we assume that the concentration is zero at the surface (perfect absorption), we have tacitly assumed that the physical chemistry is not limiting, which will only be the case with reactive materials such as zinc or calcareous stones. For less reactive materials, the surface concentration in the pollutant profile may not be zero, leading to an interaction between physical and chemical processes. Such a situation may occur if the pH in the liquid film drops too low to permit additional SO_2 dissolution, as given by Henry's law. Buffering of the film with corrosion products can prevent this from

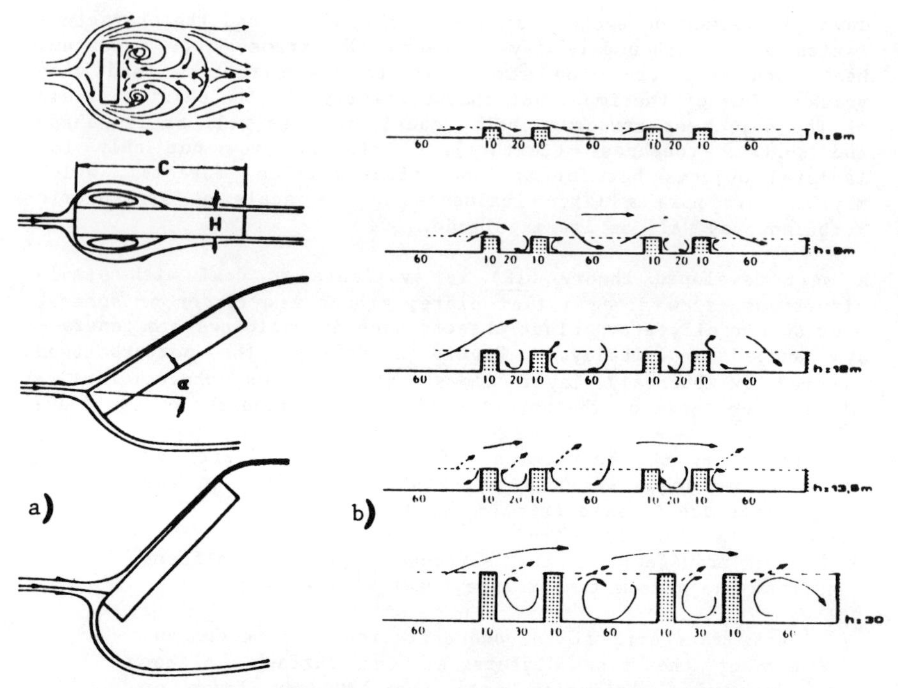

Figure 1. Flow patterns around a building (a) at various wind directions; (b) in relation to building heights.

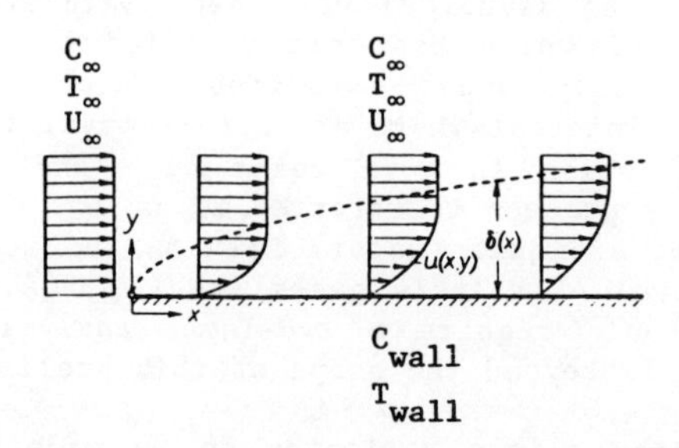

Skin friction shear stress: $\tau_o(x) = \mu\left(\frac{\partial u}{\partial y}\right)_{y=0}$

Heat transfer: $q(x) = -K\left(\frac{\partial T}{\partial y}\right)_{y=0}$

Mass transfer: $J(x) = D\left(\frac{\partial C}{\partial y}\right)_{y=0}$

Figure 2. Boundary layer profiles on a flat plat aligned with the air stream.

happening (at modest SO_2 gas concentrations). This could also be the case for materials such as copper or aluminum which tend to build-up protective surface layers over time. However, those materials of most interest are in fact the sensitive ones which either do not build up a protective layer or whose corrosion products are readily soluble at the pH values encountered in precipitation.

However, we should always expect the deposition velocity derived from boundary layer theory to be somewhat larger than observed in practice (neglecting measurement errors). Since the physical chemistry limitations should be independent of size and shape of the object in question, we may use boundary layer calculations to indicate the relative characteristics of different situations.

Boundary layer properties (governing mass transfer coefficients, for example) are mostly strongly influenced by the transition from laminar to turbulent flow (Figure 3). On a flat plate at low turbulence, such transition occurs naturally at Reynolds' numbers between 3.5×10^5 and 10^6 (12). For air flow at 20°C and 5 m/s, this corresponds to plate lengths between 1.1 and 3.1 m. Free-stream turbulence and roughness of the surface can reduce these values under certain conditions. These two parameters can also alter heat transfer and skin friction and thus mass transfer (according to Reynolds' analogy). However, according to stability theory (12), at Reynolds numbers below about 6×10^4 (about 3 cm. in length under the conditions above), transition to turbulent flow cannot begin, since turbulence disturbances in the boundary layer will die out.

Atmospheric turbulence near the earth's surface is generally much higher (Table I) than found in most wind tunnels (up to about 2%). Unfortunately, very few heat or mass transfer tests have been performed under natural outdoor conditions. Surface roughnesses of practical structures of interest may also deviate from laboratory conditions, although boundary layer theory may be used to compute critical roughness sizes and maximum permissible roughnesses, below

Table I.
Average Wind Velocities and Turbulence Intensities for Rural and Urban Conditions
u_∞ = 5 m/s (measured at 10 m)

	Local Wind Velocity (m/s)		Local Turbulence Intensity(%)	
	Rural z_o = 1cm.	Urban z_o = 1m.	Rural z_o = 1cm.	Urban z_o = 1m.
Residence	4.15	2.9	7.0	29.0
Fence	3.15	1.15	9.2	71.0
Large Building	5.45	6.15	5.3	13.6

which the surface is said to be "hydraulically smooth." Table II presents some of these values for surfaces of interest for construction materials sensitive to atmospheric corrosion or attack. Note that since such attack may enhance surface roughness, it may actually accelerate the process of corrosion.

Table II. Surface Roughness Data

A. Typical building material surface roughness (mm)

Material	Roughness (mm)
Smooth stone	0.005 - 0.01
Paint	0.1
Galvanized steel	0.15
Weathered stone	1.0
Corrugated siding	25
Carved stone	150

B. Admissible roughness, below which surface is "hydraulically smooth": 0.33mm

C. Roughness size creating transition from laminar turbulent (u_∞=5 m/s): 3mm

Note that the surface roughnesses associated with smooth stone, galvanized steel, and painted surfaces are "hydraulically smooth", that is they should have no substantial effects on boundary profiles and hence deposition velocity. Between 0.33 and 3mm, transition from laminar to turbulent flow may occur, depending on the Reynolds number.

Reynolds Analogy for Heat and Mass Transfer. For a flat plate, total frictional drag = $b \int_0^\ell \tau_o(x)dx$ where b = width of the plate, ℓ = length of the plate, and the total heat transferred $Q = b \int_0^\ell q(x)dx$.

$$\text{If P (Prandtl number)} = \frac{\mu g C_p}{k} = 1 \tag{3}$$

$$\text{then } \frac{\partial}{\partial y}\left(\frac{u}{u_\infty}\right) = \frac{\partial}{\partial y}\frac{t - t_w}{t_\infty - t_w} \tag{4}$$

$$\text{Define the Nusselt number } N = \frac{q\ell}{k(t_w - t_\infty)}$$

then $N_x = 1/2\ R_xC_f'$ for $P = 1$
for $P \neq 1$, $N_x = 1/2\ \sqrt[3]{P}\ Re_xC_f'$ (local heat transfer)
and $N_\ell = 1/2\ \sqrt[3]{P}\ Re_\ell C_f$ (total heat transfer).

Note: relationships exist for both laminar and turbulent flow defining the skin friction coefficient C_f as a function of Reynolds number (Re).

For mass transfer, the Schmidt number (Sc) replaces the Prandtl number, defined as $S_c = \gamma/D$, where D is the mass (molecular) diffusion coefficient. For SO_2 diffusing in air, $S_c = 1.18$. A mass transfer coefficient is defined as

$$j_D = \left(\frac{k}{u_\infty}\right) S_c^{2/3} \qquad (5)$$

where k is the ratio between mass flux ($\dot{m}$) and concentration gradient ($\partial C/\partial y$). k is thus seen to be defined identically to the "deposition velocity" V_d used in atmospheric science parlance. However, note that it is not expected to be a constant, but directly proportional to the free stream (for atmospheric boundary layers, local) velocity. According to Reynolds analogy, j_D for mass transfer is numerically equivalent to the Stanton number (St) in heat transfer:

$$j_D = StP^{2/3}\ ; \qquad St = \frac{Nu}{RePr} \qquad (6)$$

$$\text{Thus, } St = \frac{1}{2} C_f, \text{ and } j_D = \frac{V_d}{u_\infty} S_c^{2/3} = \frac{1}{2} Cf\ Pr^{2/3} \qquad (7)$$

$$\text{Simplifying, } V_d/u_\infty = \frac{1}{2} C_f\left(\frac{Pr}{S_c}\right)^{2/3} \qquad (8)$$

For SO_2 in air, $V_d/u_\infty = 0.36\ C_f$, providing a means to estimate deposition velocities from boundary layer theory or experimental results from first principles.

Application of Boundary Layer Theory to Corrosion Testing

Chamber Tests. Edney and his coworkers (14) have conducted some interesting tests using a rectangular flow channel through which various humidified polluted air mixtures are passed, with corrosion test samples mounted on the side walls and equipped with a chilling system to induce condensation and hence absorption of SO_2 on the test samples. Their test conditions are about 3 m/s air flow through a duct 13 x 13 cm. in cross-section. There is a smaller duct section upstream. According to pipe flow results, the flow should be turbulent under these conditions but not fully developed. Turbulent flow over a flat plate is probably an appropriate model for this situation.

The first sample in line along the wall should see a Reynolds number of about 2.7×10^4; the last one, about 3.2×10^5. At the last station, C_f should be about 0.0056, and thus the deposition velocity should be 0.63 cm/sec. The earlier stations should have higher values. Edney et al. report a measured V_d value of 0.9 cm/sec (14). This higher value could either be a result of increased turbulence in this particular flow system or the effects of lower Reynolds numbers at the more upstream test positions. In any event, this comparison indicates that, under these conditions, the flux of SO_2 to the surface appears to be controlled by the "atmospheric" resistance, and is apparently not limited by uptake on the surface.

Outdoor Tests. One standard protocol for atmospheric corrosion testing is to mount small rectangular plates (1.6 x 2.4 cm) on a test rack at about 30° from horizontal, about 3 feet off the ground, usually facing south (Figure 4). The test plates are held off the rack by porcelain insulators, and although strictly speaking a new boundary layer should form on each plate, there may be some positional differences in corrosion rate due to turbulence created by the plates first encountered by the wind flow.

For a classical laminar flow situation,

$$C_f = \frac{1.328}{\sqrt{Re_\ell}} \tag{9}$$

and thus $V_d/u_\infty = .478\ Re_\ell^{-1/2}$
Taking the average plate dimension at 2 cm.,
yields $V_d/u_\infty = 0.00244$. For $u_\infty = 5$ m/s, $V_d = 1.22$ cm/sec.

Experimental tests in the outdoor atmosphere (16) (simulated solar collectors) found substantially higher heat transfer coefficients than predicted by laminar theory or than measured in a wind tunnel, presumably because of the higher turbulence levels encountered. The relationship was approximately

$$St = \frac{j_D}{Pr^{2/3}} = \frac{2.0}{Re}, \text{ or } V_d/u_\infty = 0.00736. \text{ For } u_\infty = 5 \text{ m/s}, \tag{10}$$

$$V_d = 3.7 \text{ cm/sec.}$$

However, considering the atmospheric boundary layer profiles as discussed above, the local velocity at the rack may be considerably lower, especially for test sites in either forested or urban areas. This result emphasizes the need to measure local wind speeds at atmospheric corrosion test sites, at the test rack height. Turbulence intensity measurements might be useful as well.

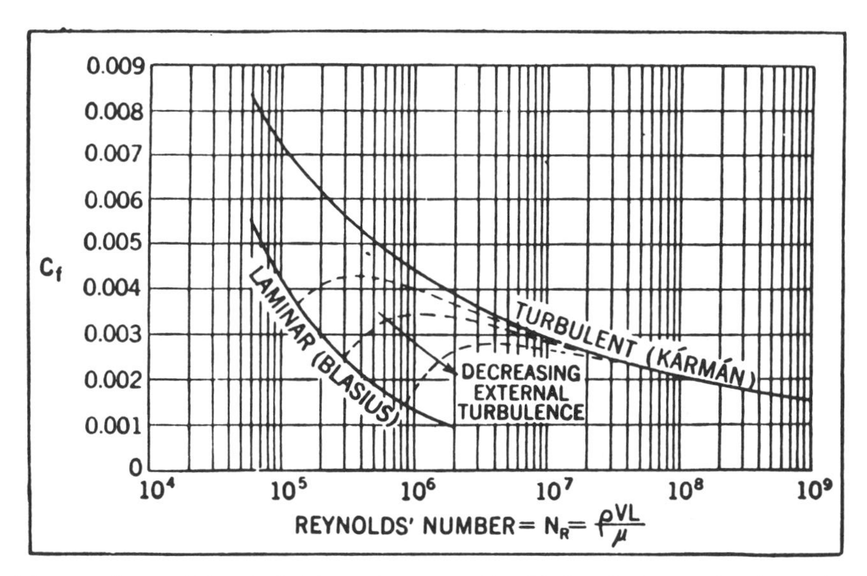

Figure 3. Skin friction for smooth flat plates, showing the transition from laminar to turbulent flow.

Figure 4. Typical outdoor corrosion test site.

Reference (3) reanalyzed data from a number of such outdoor test sites and derived an SO_2 deposition velocity of 1.55 - 1.75 cm./sec. for zinc, operable only during times of surface wetness. In that report, reference was made to SO_2 deposition velocities over water surfaces (1.6 cm/sec), which are likely to be low turbulence situations. It was also noted that the deposition velocity to water is dependent upon atmospheric stability. SO_2 deposition velocities for copper and aluminum were less straightforward (3), but appeared to be somewhat lower, perhaps reflecting less chemically active surfaces. The zinc result is reasonably consistent with the theoretical values developed above from boundary layer theory and tests, since it lies between these values (1.2 - 3.7 cm/sec). Incorporating actual test site wind speeds could obviously help reduce the scatter in these determinations.

In both the outdoor tests reanalyzed by Lipfert et al. (3) and the chamber tests of Edney et al., (13) it was deduced that zinc was removed stoichiometrically by the deposited SO_2, implying no chemical surface resistance. However, in the case of outdoor tests, this conclusion is conditional on the deposition velocity.

Application of Boundary Layer Theory to Buildings

The assessment question of concern here is the application of these test results to real buildings. Use of damage functions such as developed by Lipfert et al. (3) or Haynie (4) implies a direct 1:1 correspondence regardless of size or configuration, in addition to the assumption that the time-of-wetness (presence of liquid film) will be unaffected by size, shape, or surface orientation.

Rectangular Shapes. The flow around a building is highly dependent on its situation with respect to neighboring buildings. An isolated building presents a blunt obstacle to the wind: the front face will see stagnation point flow, which will separate around the front corners and reattach at the back corners. The rear face will be in a cavity zone (Figure 1). Flow around the building at a 45° angle will be less chaotic. Since during the course of a year, all of these situations may be expected along each of the building facades, one might try to deduce some sort of average conditions. However, surface maintenance actions may be triggered by the worst case location, such as a corner.

In contrast, buildings aligned sufficiently close together along streets will act as a quasi-continuous flat plate, and indeed may channel the wind flow in this way and cause local increases in wind speed. The trailing and leading edges of the block would tend to see somewhat different boundary layer conditions; persistent wind direction would be an important consideration.

We were unable to find in the literature any test data of blunt building-like shapes at sufficiently high Reynolds numbers to simulate real buildings. We did find tests of small square prisms in a wind tunnel (low turbulence), (17) and outdoor tests with natural turbulence for small spheres (18). Comparison of these two sets of tests may be useful.

The tests on prisms were carried out at various flow angles at Reynolds numbers up to 5.6 x 10^4. In contrast, real buildings would have Reynolds numbers greater than 10^6. The highest local heat transfer (and by analogy, mass transfer) occured at the rear corners. The average heat transfer was given by:

$$Nu_m = 0.27\ Re^{0.59},\ \text{or}\ St = 0.375\ Re^{-0.41} \qquad (11)$$

By analogy,
$$V_d/u_\infty = St \frac{P_r^{2/3}}{S_c^{2/3}} = 0.27\ Re^{0.41} \qquad (12)$$

For the tests, Re_{max} was 5.6 x 10^4, yielding $V_d/u_\infty = 0.00405$.

If extended to a full-scale building, say Re = 5 x 10^6, V_d/u_∞ = 0.00048.

However, this somewhat low average value should be tempered by the realization that local free stream velocities may be higher, and that the "hot spots" will be higher by about 70-90%.

Rounded Shapes. The tests of spheres outdoors yielded average heat transfer values up to 2.2 times higher than in a wind tunnel. This may be due to movement of the flow separation point around the sphere, which would not be appropriate for a less rounded building-like shape. The heat transfer enhancement was most pronounced near the ground, and correlated with turbulence intensity. The data at high turbulence may be represented by:

$$Nu_m = 0.8\ Re^{0.54} \qquad (13)$$

or about 70% higher than the average square prism heat transfer.

Similar results are given by Schlicting (12) for the heat transfer to a circular cylinder in cross-flow at varying degrees of turbulence. At the highest turbulence value,

$$Nu_m = 0.187\ Re^{0.659} \qquad (14)$$

which would indicate a factor of 4.8 reduction in V_d/u_∞ for 2 orders of magnitude change in Reynolds number. This would yield a full scale building estimate of V_d/u_∞ = 0.0006, or 0.30 cm./sec. for u_∞ = 5 m/s. However, as mentioned above, u_∞ might well be substantially higher in such a situation.

Flat Plate Model. Modelled as a smooth flat plate for, say, $Re_\ell = 10^7$, the average skin friction coefficent would be 0.003, and thus V_d/u_∞ = 0.00108. Turbulence created by window recesses, breaks in the walls, etc., might increase this value. For example, flow in a very rough pipe becomes independent of Reynolds number, but the heat transfer enhancement effects of free stream turbulence are considerably less along a flat plate than for flows around cylinders, etc., where flow separation plays a role. Pedisius et al. (19) showed a heat transfer enhancement on a flat plate of about 20% for a turbulence intensity of about 15%. Drizius et al. (20) showed a heat transfer enhancement of about a factor of 2 for roughness elements of 1.4 mm and Re_x up to about 3 x 10^6.

For a plate with regularly spaced ribs (simmilar to corrugation), Veski and Kruus (21) found local heat transfer enhancement to be greatest on the projections, given by the formula

$$Nu = 0.03\ Re^{0.8} \qquad (15)$$

which was also the formula for the average heat transfer for the closest rib spacing. For $Re_x = 10^7$, this gives V_d/u_∞ = 0.00096, in good agreement with the preceeding.

For a 5 m/s flow, then, we might expect SO_2 deposition velocities of about 0.56 - 0.9 cm./sec. on a rough or corrugated building with external freestream turbulence, but only about 0.56 cm./sec. on a smooth building. Note that these values are considerably lower than obtained from outdoor corrosion tests on small plates (discussed above).

Application of Boundary Layer Theory to Non-Buildings (Fences, Towers, Flagpoles, etc.)

We have seen how heat transfer and thus dry deposition of SO_2 is reduced on large surfaces, due to the buildup of boundary layer thickness (which reduces the local gradients). However, there are economically important structural objects composed of many elements of small dimension which show the opposite effect. These include fence wire and fittings, towers made of structural shapes (pipe, angle iron, etc.), flagpoles, columns and the like. Haynie (11) considered different damage functions for different structural elements such as these, but only from the standpoint of their effect on the potential flow in the atmospheric boundary layer. The influence of shape and size act in addition to these effects, and could also change the velocity coefficients developed by Haynie (11), which were for turbulent flow. Fence wire, for example, as shown below, is more likely to have a laminar boundary layer.

We will consider flow at right angles to an infinite circular cylinder, with varying diameters, as shown in Table III. A dramatic increase in deposition velocity is shown for small diameter objects. This would also apply to isolated portions of a statue, for example.

Schlicting (12) shows data on roughness effects on circular cylinders; for Re_D below about 2500, there is no effect. For

Table III.
Deposition Velocities to Circular Cylinders
(smooth surface, low turbulence)
(u_∞ = 4.5 m/s)

	Diameter (m)	Re_D	Nu_m	Average V_d (cm/sec)	Peak V_d (cm/sec)
Fence Wire	0.001	301	9	13.3	
Fence Post	0.025	7520	45	2.7	~4.5
Flag Pole	0.100	3×10^4	100	1.49	~2.5
Structural Steel	0.300	9×10^4	220	1.09	~1.9
Stone Column	1.	3×10^5	700	1.04	~1.8
Storage Tank	10.	3×10^6	4830	0.72	

smooth cylinders at low turbulence, there is virtually no Reynolds number effect (or drag) between Re_D = 1000 and $Re_D = 2 \times 10^5$. For large cylinders, with turbulent boundary layers over most of their surface, roughness effects should be similar to those on a flat plate.

Discussion of Boundary Calculations and Results

Where data for SO_2 deposition to zinc are available, good agreement is shown with the boundary layer calculations (Table IV). Unfortunately, experimental data are not directly available for other situations, so inferences must be made by comparing the calculations for similar flow situations.

Table IV.
Summary of Deposition Velocity Data
(cm/sec)

	Calculated	Measured
Flat Plates		
Chamber Tests	0.63	0.9[a]
Outdoor Test Racks[c]	1.2-3.7	1.55-1.75[b]
Large Buildings[c]	0.5-0.9	---
Circular Cylinders		
Fence wire	13.3	---
Posts--columns	1.0 - 2.7	---
Storage Tanks	0.72	---
Blunt Shapes		
(Entire Buildings)[c]	0.30	---

(a) Edney et al. (14) (b) Lipfert et al. (3) (c) u_∞ = 5 m/s

Larger structures have lower calculated deposition velocities as a result of their larger Reynolds numbers. This effect will be partially countered by higher free-stream velocities for taller structures. Blunt objects will tend to have lower average deposition as a result of their zones of separated flow. This may not pertain to local "hot spots", however.

Perhaps the most critical situation will occur for stone objects in the 0.3-1 m. diameter range, which could include either statues or columns. A roughened surface due to weathering will increase deposition and hence enhance further surface erosion.

Overlaid on all of these results is the tacit assumption that the surface offers no chemical resistance. This appears to be valid for zinc with a water film on the surface; it is less clear for stone, for example, for which the moisture may be trapped below the surface in the pores of the stone. The model is not valid for essentially inert surfaces.

In the event the surface does offer chemical resistance by virtue of slower reaction rates, resistance to acid attack, etc., the gas concentration will not be zero at the surface and the dry deposition rate will be reduced accordingly. Thus structures made or coated with such materials will show a different relationship between calculated deposition velocities, as presented here, and the actual operational values, which may be governed more by the material properties and less by aerodynamic resistance. This would be an important property to establish, by testing over a range of aerodynamic conditions.

Interactions Among Pollutants and Atmospheric Parameters

Boundary layer theory can only be used to estimate dry deposition coefficents when it is reasonable to assume complete and rapid uptake at the surface. The presence of other pollutants can affect the viability of this assumption, and the cases for which the assumption appears valid can be used to hypothesize the effects of wet deposition.

The solubility of SO_2 in water is strongly dependent on its pH, becoming limited below pH = 4. The presence of other pollutants can be important as they affect the pH of the liquid layer on the surface, which may also be buffered by corrosion products per se. Nitric acid deposits quite readily, for example, and could lower the pH and thus inhibit SO_2 uptake. On the other hand, many atmospheric particulates are basic, and the limited literature on dew chemistry (Cadle and Goblicki) (22) does not indicate acidic dew composition, (it should be noted that these data were all taken in low SO_2 environments).

The apparent rapid uptake of (dry) SO_2 on wet zinc surfaces with stochiometric removal of zinc seems to indicate rapid reactions. This is important in considering a damage function due to wet deposition of acidity (H^+). If the damage function employs total H^+ deposition (say, the annual sum) as the "driver," then the implicit assumption has been made that all of this precipitation-borne acidity is able to react with the surface before it runs off. Obviously the nature of both the precipitation event and the size and shape of the surface will modify the validity of this assumption. Further controlled experiments are badly needed.

We have shown the importance of local free stream velocity (u_∞) in controlling deposition velocity. There may be a relationship between u_∞ and time-of-wetness, either through variations in synoptic conditions or through heat transfer and evaporation. Such an interaction could modify the estimation of an effective annual average SO_2 deposition rate. As mentioned above, one must also account for surface orientation in predicting time-of-wetness.

Economic Assessment Ramifications

Economic assessments have usually been based on a common damage function for all structural elements, according to the material. Table IV indicates that loss rates to fencing may be underestimated and losses for large buildings will be overestimated, if damage functions based on small plate tests are used. Since the larger structures carry larger dollar losses as a rule, this may result in an overprediction bias to the assessment. The role of fencing in the economic assessment may be particularly important, especially since it tends to be replaced than painted as a maintenance action.

According to boundary layer theory, the effects of surface roughness and irregularities, including for example carved decorations, can create local increases in deposition velocity, or "hot spots." This is an important consideration for assessing potential damage to cultural resources, since the fine details of such objects that make them culturally interesting may be subject to much higher loss rates. In addition, the finely detailed portions of a carving may have a higher surface-to-volume ratio, subjecting the piece to a higher likelihood of mechanical failure for a given amount of surface recession.

In all cases, local wind conditions should be accounted for, especially since SO_2 concentrations can be correlated with wind speed (1). The correlation can be either positive or negative; positive for areas dominated by point sources and negative when dominated by area sources. This, of course, could lead to biases in either direction if such correlations are neglected.

We have seen how consideration of theoretical deposition velocities has identified potential biases in economic assessments. An additional consideration is the relative uncertainties in the determination of theoretical vs. experimental deposition velocities. The heat transfer data on which the theoretical deposition velocities are based are generally very precise, within a few percent. In contrast, the damage functions developed by Lipfert et al. (3) for metals from extant corrosion test data are only capable of predicting corrosion losses at a given time and place within a factor of two, although the individual regression coefficients are much better than that. Most of the uncertainty in the experimental approach is felt to be in test site characterization rather than errors in the determination of corrosion rates per se. Thus including the detailed aerodynamic conditions at each site would very likely reduce the overall uncertainties in predicting corrosion damage.

Concluding Remarks

This analysis has shown that there are pitfalls in applying corrosion rate data from small test plates to objects that are greatly different in size. To be sure, the original ASTM protocol for such tests was never intended for this purpose. We have used boundary layer theory to try to quantify the extent of possible biases (factors of 3 or more) and to suggest remedies through experiments under controlled aerodynamic conditions. In some cases, appropriate theory and experimental data have been lacking. These possible biases and uncertainties also suggest that building component service life predictions made in the course of performing economic assessments should be corroborated against real world experience.

Acknowledgments

This research was partially supported by the U.S. Department of Energy under contract DE-AC02-76CH00016. Members of NAPAP Task Group G and my colleagues at Brookhaven have provided valuable input and consultation. The manuscript was prepared by Donna Cange and Liz Seubert.

Nomenclature Used

C_f	Coefficient of skin friction $D/\frac{1}{2}\rho u_\infty^2$
C_p	Specific heat at constant pressure
D	Diffusion coefficient, drag, diameter
g	Acceleration due to gravity
j_D	Mass flux parameter
k	Thermal conductivity, mass transfer parameter, Von Karman constant
ℓ	Plate length
Nu	Nusselt number = $q\ell/K(T_w-T_\infty)$; Nu_m, average over the surface
Pr	Prandtl number = $\mu g C_p/k$
Re	Reynolds number = $u\,x/\gamma$
S_c	Schmidt number = $\gamma/\rho D$
St	Stanton number = $Nu/RePr = C_f/2$
T	Temperature
Tu	Turbulence intensity (fraction or percent)
u_∞	Undisturbed free stream velocity
V_d	Deposition velocity (cm/sec)
q	Heat flux
γ	Kinematic viscosity
μ	Dynamic viscosity
x	streamwise coordinate.

y width or height above surface coordinate
z height above the earth's surface
ρ air stream density

Literature Cited

1. Wyzga, R. E.; Lipfert, F. W. "Forecasting Materials Damage"; Presented at Eighth ASA Symposium on Statistics, Law, and the Environment: Washington, D.C., 1984.

2. Wyzga, R. E.; Lipfert, F. W. "Forecasting Materials Damage From Air Pollution"; Presented at the 78th Annual Meeting of the Air Pollution Control Association: Detroit, Michigan, 1985.

3. Lipfert, F. W.; Benarie, M.; Daum, M. "Derivation of Metallic Corrosion Damage Functions for Use in Environmental Assessments"; Brookhaven National Laboratory, Upton, NY 11973, 1985.

4. Haynie, F. H. "Atmospheric Damage to Paints", EPA Environmental Research Brief, EPA/600/M-85/019, June 1985.

5. Reddy, M.; Sherwood, S.; Doe, B. "Limestone and Marble Dissolution by Acid Rain"; to be sumitted to Proc. 5th Int. Cong. on Deterioration and Conservation of Stone, Lausanne, Switzerland, Sept. 1985.

6. Lipfert, F. W.; Cohen, S.; Dupuis, L. R.; Peters, J. "Predictor Equations for Relative Humidity from Relevant Environmental Factors"; Brookhaven National Laboratory, Upton, NY 11973, 1985.

7. Rosenfield, G. H. 1984. "Spatial Sample Design for Building Materials Inventory for Use With an Acid Rain Damage Survey"; U.S. Geological Survey, Reston, VA. in Schmitt R. R.; Smolin, H. J., Eds.; "The Changing Role of Computers in Public Agencies"; Presented at the Ann. Conf. of Urban and Regional Information Systems Assoc., Seattle, WA, 1984

8. Merry, C. J. and LaPotin, P. J. "A Description of the New Haven, Connecticut Building Material Data Base"; Report prepared for the U.S. Environmental Protection Agency, 1985. U.S. Army Corps of Engineers Cold Regions Research and Engineering Laboratory.

9. Lipfert, F. W.; Daum, M. L.; Cohen, S. "Methods for Estimating Surface Area Distributions of Common Building Materials"; Brookhaven National Laboratory, Upton, NY 11973, 1985.

10. Livingston, R. A.; Baer, N. S. "Mechanisms of Air Pollution-Induced Damage to Stone"; Presented at the VIth World Congress on Air Quality, May 1983, Paris, France.

11. Haynie, F. H. "Theoretical Air Pollution and Climate Effects on Materials Confirmed by Zinc Corrosion Data"; Durability of Building Materials and Components, ASTM STP No. 691, Sereda, P. J.; Litvan, G. G., Eds.; American Society for Testing and Materials: 1980, pp. 157-175.

12. Schlichting, H. Boundary Layer Theory; Fourth Ed., McGraw-Hill Book Company, Inc., 1960.

13. Binder, R. C. Fluid Mechanics; Second Ed., Prentice-Hall, Inc., 1953.

14. Edney, E. O.; Stiles, D. C.; Spence, J. W.; Haynie, F. H.; Wilson, W. E. "Laboratory Investigations of the Impact of Dry Deposition of SO_2 and Wet Deposition of Acidic Species on the Atmospheric Corrosion of Galvanized Steel"; Atmospheric Environment, (in press).

15. Flinn, D. R.; Cramer, S. D.; Carter, J. P.; Spence, J. W. "Field Exposure Study for Determining the Effects of Acid Deposition on the Corrosion and Deterioration of Materials: Description of Program and Preliminary Results." Durability of Building Materials 3, pp. 147-175, 1985.

16. Test, F. L.; Lessmann, R. C.; Johary, A. "Heat Transfer During Wind Flow Over Rectangular Bodies in the Natural Environment"; J. Heat Transfer, 103, 262-267, 1981.

17. Igarashi, T. "Heat Transfer from a Square Prism to an Air Stream"; Int. J. Heat Mass Transfer, 28, No. 1, 175-181, 1985.

18. Kowalski, G. J.; Mitchell, J. W. "Heat Transfer from Spheres in the Naturally Turbulent, Outdoor Environment"; J. Heat Transfer, p. 649-653, 1976.

19. Pedisius, A. A.; Kazimekas, P.-V. A.; Slanciauskas, A. A. "Heat Transfer from a Plate to a High-Turbulence Air Flow"; Heat Transfer-Soviet Research, 11, No. 5, 1979.

20. Drizius, M. R.; Bartkus, S.I.; Slanciauskas, A. A.; Zukauskas, A. A. "Drag and Heat Transfer on a Rough Plate at Various Pr Numbers"; Heat Transfer-Soviet Research, 10, No. 3, 1978.

21. Veski, A. Yu.; Kruus, R. A. "Local Heat Transfer from Plates with Regular Macroroughness (Ribbed Plates)"; Heat Transfer-Soviet Research, 9, No. 4, 1977.

22. Cadle, S. H.; Groblicki, P. J. The Composition of Dew in an Urban Area; in "The Meteorology of Acid Deposition"; APCA Specialty Conference, Hartford, CT, 1983. Edited by P.J. Samson, pp. 17-29.

RECEIVED January 2, 1986

INDEXES

Author Index

Ang, Carolina C., 92
Baboian, Robert, 200,213
Ball, Sarah E., 360
Cadle, Steven H., 92
Carter, J. P., 119
Charola, A. E., 250
Comeau, T. C., 200
Cramer, S. D., 119
Crocker, Thomas D., 369
Davidson, Cliff I., 42
Doe, B. R., 266
Doe, Bruce R., 226
Dolske, D. A., 259
Dower, Roger C., 360
Edney, Edward O., 172,194
Escalante, E., 152
Flinn, D. R., 119
Gamble, James S., 42
Haynes, Gardner, 213
Haynie, Fred H., 163,172,194
Hoffmann, Michael R., 64
Horst, Robert L., Jr., 397
Hurwitz, D. M., 119
Husar, R. B., 152
Kucera, Vladimir, 104
Kukacka, L. E., 239
Kyllo, Karen E., 343
Ladisch, Christine M., 343
Lareau, Thomas J., 397
Laughlin, J. S., 301
Lazzarini, L., 250
Linstrom, P. J., 119
Lipari, Frank, 92
Lipfert, Frederick W., 397,411
Manuel, Ernest H., Jr., 397
Mulawa, Patricia A., 92
Neal, K. M., 285
Newnam, S. H., 285
Ottar, B., 2
Paparozzi, Ellen T., 332
Passaglia, E., 384
Patterson, D. E., 152
Petuskey, W. T., 259
Pokorney, L. M., 285
Rands, D. G., 301
Reddy, Michael M., 226
Rosenow, J. A., 301
Rybarczyk, J. P., 285
Semonin, Richard G., 23
Sherwood, Susan I., 226
Sinclair, J. D., 216
Spence, John W., 172,194
Stiles, David C., 172,194
Turcotte, R. C., 200
Vandervennet, Rene T., 92
Webster, R. P., 239
Weschler, C. J., 216
Williams, R. Sam, 310
Wilson, William E., 172
Wyzga, Ronald E., 411
Youngdahl, C. A., 266

Subject Index

A

Abiotic materials, management, 369
Abiotic resources, acid deposition, 381
Acetaldehyde, deposition velocity on steel, 182
Acid deposition
  atmospheric corrosion of metals influenced, 104
  economic damages to building materials, 397-410
  effect on economic surplus, 399
  effect on portland cement concrete, 239-247
  effect on wood and cellulosic materials, 324-327
  effect on wood coatings, 318-324

Acid deposition--Continued
  need for concrete effects study, 246
  reduction, economic effect, 380-381
  See also Dry deposition
  See also Wet deposition
Acid dew, environmental concern, 100
Acid precipitation, 2-20
  economic effects, 359-428
  performance of building materials, 240
  pH, 343-344
  sources, 343
  sulfuric, nitric, and hydrochloric acid contribution, 241
  trends in the United States, 34-37
Acid rain
  cause and effect uncertainties, 361
  composition of synthetic samples used in coquina study, 304t

Acid rain--Continued
control, 365-366
cost, 393
definition, 310
effect of repeated wetting and drying on chemistry, 210t
effect on anodized aluminum automotive trim, 213-215
effect on mortars, 254-255
effect on painted wood surfaces, 310-328
effect on unfinished wood products, 338-340
effect on woody plant diseases and insects, 336-337
effect on woody plant growth, 337-338
emission uncertainties and environmental impact, 366
environmental factors in Indiana study, 286-287
focus of studies, 286
geographic distribution in the United States, 210f
indoor concentrations of associated species, 219t
leaf-litter interaction, 335
nylon degradation, 343-356
procedure used in Indiana study on limestone, marble, and bronze, 288-289
reagents and equipment used in Indiana study, 287
samples used in Indiana study, 287-288
sources, 332-333
upper limit for costs on metals, 393
See also Acid precipitation
Acidification
bodies of water effected, 15
drainage water, 17
rivers and lakes, 15-19
Aerochem-metrics bucket
dry deposition collection devices, 53
surrogate surface, 48
Aerometric monitoring, Anasazi civilization sandstone, 262-263
Aerosols
composition influenced, 13
condensation nuclei for cloud droplets, 15
Air flow, building, 422
Air pollutants, deposition processes, 105
Air pollution
effect on consumer surplus and producer quasi-rant, 373
increase, effect on capital stock of firm, 375
Aldehydes, presence in fog, 72
Alpha coefficients, weathering steel, 169
Altinelle
chemical analysis, 252t
compared with modern bricks, 254
Aluminum
ammonium to sulfate ratios, 222
anodizing process, 213
automotive trim use, 213
corrosion examined, 145-148
corrosion rates, 152-161
dry deposition effect on degradation, 221-222
indoor, acid rain effects, 216-223
nitrogen pollutants effect, 111
oxides as buffers, 221
pitting, 213
rain influence on corrosion, 113
SEM examination, 145
skyward side, 147f
sulfur pollutants effect, 108
weight loss as function of sulfur dioxide deposition, 107f,109f
Aluminum alloys, U.S. corrosion rates, 156
Aluminum hydroxide, effect on sulfate ions in soil, 17-19
Ambient air quality standards
CAA, 362
costs and benefits, 363
regulatory accommodation of materials damage, 363
Amino group
atmospheric conditions influencing nylon, 350-352
nylon, influence of light exposure, 353f
Ammonium
atmospheric ratio to nitrate ions, 13
concentration and deposition in the United States, 30,31f
dew deposition rates compared, 97
fog water concentration, 85
sources in precipitation, 28
surface accumulation rates due to anthropogenic influences, 220t
Anasazi civilization, aerometric and materials deterioration studies, 259-265
Anasazi ruins
air sampling system location, 263
atmospheric chemistry measurement, 262
erosion mechanism of greatest concern, 264
meteorological sensors location, 263
monitoring in deterioration study, 261
objectives of materials monitoring, 260

Anasazi ruins--Continued
rate of recession measured, 261
sandstone masonry, 260
Anthropogenically derived substances
estimating accumulation rates indoors, 217
indoor effects, 217
indoor-outdoor ratios for office buildings, 219t
interactions with typical indoor surfaces, 221-222
Aspen wood, acid rain effect on decomposition, 339
Associated costs, description, 388
Atmospheric boundary layer
definition and description, 413-414
properties, 413-414
Atmospheric corrosion
local nature, 113-115
solution, 115
See also Corrosion
Atmospheric deposition, measurement methods, 43
Authigenic quartz, characteristics in sandstone, 264
Auto-body steel
corrosion data in poultice test, 209t
range of appearance in cities, 208f
typical appearance after exposure in the northeast United States, 208f
Autocatalytic pitting, mechanism, 203f
Automotive corrosion
function of road salt usage, 201f
geographic distribution in the United States, 202,210f
Automotive trim, anodized aluminum, acid rain effect, 213-215
Average mass median diameter, elemental, 57t
Avoidable costs, concept, 385-386

B

Backscattered electron imaging microscopy (BSEI), chemical changes in sandstone studied, 261-262
Bark beatles, forest damage problem, 19
Benefits assessment, model for variable market size, 377
Beta coefficients, weathering steel, 169
Blush and bloom, anodized aluminum, 213-215
Boundary layer
application
buildings, 422
corrosion testing, 419-422
nonbuildings, 424-425
calculations and results, 425-426
concepts, 413-419
deposition velocity, 427
limitations, 426
turbulent flow, 417
profiles, flat plate aligned with stream, 416f
structures, 414-418
See also Atmospheric boundary layer
Breaking strength
effect of light on nylon, 348-350
function of light exposure, 349f
function of molecular weight, 350
nylon under atmospheric conditions, 347-350
Brick
composition and deterioration, 251-254
definition, 251
deterioration
due to acid rain,
due to soluble salt crystallization, 256f
history, 250-251
erosion rain water runoff, 253f
exposed to rain water runoff, 252
porosity during setting of mortar, 255-256
SEM examination, 252,253f
soluble salt occurrence, 255
sulfuric acid attack, 251-252
surface erosion, 252
Bronze
acid leaching in Indiana study, 296
acid rain stripping results, 291-293f
copper and zinc leaching rates, 297f
corrosion dependence on alloy, 160
corrosion rates, 152-161
elemental analysis of simulated acid rain stripping, 285-298
prepared in Indiana acid rain study, 288
total acid in Indian study, 298t
U.S. corrosion rates, 157-160
zinc leaching in Indian study, 296
Budget constraint, benefits assessment for variable market size, 378-380
Building materials
case-study damage estimates and findings, 406-410
computation flowchart of economic damage, 403f
damage estimation assumptions, 404t
economic damages exposed to acid deposition, 397-410

Building materials--Continued
  maintenance cost data, 405-406
  summary of damage
      estimates, 408-409t
Buildings, flow patterns, 416f

C

Calcite, solubility under varying
    conditions, 305t
Calcium
  concentration
    coquina, 302-305
    United States, 28,29f
  deposition
    effect on dew, 99
    United States, 28
  leaching dependence on pH, 293
  leaching rates for micrite, marble,
      and Salem limestone, 293
  plot versus incident rainfall
      pH, 236f
Calcium carbonate
  acid rain-induced surface
      recession, 235-237
  effect in oil-based house paint, 321
  ionic strength effect on
      degradation, 305
  materials
    environmental damage
        studies, 266-267
    surface material loss, 266
  reaction with acid rain, 226
  solubility in coquina, 302-305
CAPITA, Monte Carlo regional model of
    pollutant transmission, 154
Capital costs, definition, 386
Capital stocks, influence of air
    pollution, 375
Carbon dioxide, concrete affected, 243
Carbon steel
  atmospheric rust films, 126
  corrosion
    by sulfur dioxide, 245
    differences compared with
        weathering steel, 126
    rate, 126-134
    rate as function of distance from
        emission source, 114f
  groundward side, 135f
  relative sensitivity factor to
      corrosion, 124
  sulfur pollutants effect, 106
Carbonate rock, relation between
    environmental variables and
    surface recession, 230-234
Carbonic acid, effect on
    concrete, 243-244
Cellulose, definition, 330
Cellulosic material
  degradation caused by pollution, 327
  sulfur dioxide effect on tensile
      strength, 327
Cement paste, acid attacks, 242
Chloride
  accumulation on aluminum
      surfaces, 222
  concentration and deposition in the
      United States, 30,32f
  corrosion mechanism, 202
  corrosion rate of metals
      affected, 222
  relationship to coastal
      influences, 30
  surface accumulation rates due to
      anthropogenic influences, 220t
Chromium, presence in galvanized steel
    film surface, 142
Clean Air Act (CAA)
  addition of new programs, 367
  ambient air quality
      standards, 362-363
  effect on sulfur dioxide levels, 407
  EPA role, 362-363
  interstate pollution problems, 364
  linked to acid rain debate, 360
  materials damage effects, 361-365
  new source performance
      standards, 364
  philosophy, 367
Cloud water
  chemical constituents, 67f
  concentration, LWC, and solute
      loading for sequential
      samples, 81f
  deposition, 64-86
  histograph of pH frequency collected
      near Altadena, CA, 79f
  low pH values, 65
  nitrate-to-sulfate ratio, 80
  pH in Los Angeles, 76
  sulfate production rates, 70
Coal, consumption in United
    States, 4,6f
Coatings--See Wood finishes
Concrete
  acid effects, 241-243
  acid reactions, 241-242
  carbon dioxide effects, 243
  resistance to chemical
      attack, 241-246
Consumer surplus
  change due to pollution, 400f
  changes, 374f
  definition, 371-372,399
  effect of air pollution
      increase, 373
Contact materials, electronics,
    nitrogen pollutants effect, 111
Control costs, definition, 386-387

Cooperative program for Monitoring and Evaluation of the Long Range Transmission of Air Pollutants in Europe (EMEP), history, 3
Copper
  catalysts for sulfate formation, 72
  corrosion
    air containing sulfur dioxide and/or nitrogen oxide, 112f
    film examined, 143-145
  ISS analysis of corrosion films, 145
  nitrogen pollutants effect, 111
  rain influence on corrosion, 113
  relative sensitivity factor to corrosion, 124
  SEM micrographs of corrosion, 145
  skyward side, 144f
  sulfur pollutants effect, 108
  weight loss as function of sulfur dioxide deposition, 107f,109f
  XPS analysis, 145
Coquina
  deterioration by acid rain, 301-307
  plot
    calcium versus ionic strength, 306f
    calcium versus pH, 306f
  samples in deterioration study, 301-302
  sea spray contribution to degradation, 305
Cor-Ten A steel
  corrosion film contents, 128
  corrosion rates, 126-134
  groundward side, 130f
  mass of corrosion product retained, 123f
  X-ray diffraction analysis of corrosion film, 129
Corrosion
  acidification, subdivisions, 105
  amount, equation, 164
  annual extra losses to external metal structures, 394t
  application of boundary layer theory, 419
  automotive, 206
    acid deposition effect on poultice-induced, 200-211
    design factors, 200
    environmental factors, 200
  cost
    metals, 393
    relationship to fracture costs, 392
    various nations, 390t
  economic assessment, 411-428
  factors, weathering steel, 163
  film
    chemistry, 148-149
    composition for zinc, 134
Corrosion--Continued
    composition independent of environment, 125
    composition on steel described, 126-134
    corrosion-time function, 163-164
    growth on metal, 121-122,148
    mass balance results, 127t
    thickness, equation, 164
    weathering, 149
  fracture cost, 391t
  influence of rain and its composition, 113
  mass balance for film, 122
  metals, contributors, 153
  National Bureau of Standards study, 389-392
  outdoor test site, 421f
  outdoor tests, 420
  products
    environmental effects formed in short-term exposures, 119-149
    loss described, 125
  parameters in metal study, 153
  rates
    function of space and time, 152-161
    over time periods in United States, 160
  thermogravimetric analysis (TGA) of film on microanalysis samples, 128t
  time function, weathering steel, 197
  water effect, 153
  water supply pipelines, 20
  x-ray diffraction analysis film, 129
  See also Automotive corrosion
  See also Materials degradation
Corsican pine, acid treatment effects, 325
Cost sharing, acid rain control, 365
Critical damage level, determination in building damage study, 405
Cultural resource monitoring, concurrent aerometric and materials deterioration studies, 259-265

D

Damage calculations, damage functions used in economic damage study, 402-405
Damage function
  corrosion by sulfur dioxide, 245
  general form, 282
  method, 401
  sulfate attack on cement, 245

Damage function--Continued
uncertainty in building damage study, 407
Degradation
organics, 309-357
See also Materials degradation
Demand function, economic good, 380
Deposition
impacts of primary emissions, 154-155
See also Dry deposition
See also Wet deposition
Deposition rates
dew, 96-97
urban dew, 97t
Deposition velocities
acidic substances, 217-219
bias in economic assessments, 427
boundary layer theory, 417,419
calculation, weathering steel study, 166
chemicals contained in dew, 97-98,98t
circular cylinders, 425t
data for various shapes, 425t
estimated artificial dew, 99t
indoors, 217-219
MMD relationship, 59
plotted versus particle aerodynamic diameter for surrogate surfaces, 57-58f
shielded Teflon plates, 58f
sulfur dioxide estimation, 415
sulfur dioxide on zinc, 422
Design costs, description, 388
Dew
acidity controlled by basic particles, 99
analysis of ions in dry deposition study, 94-95
artificial
concentration of ions, 96
generation, 94
higher deposition rates, 96
pH, 96
calcium deposition effect, 99
chemical composition in propylene nitrogen oxide study, 185
collection in dry deposition study, 93
composition, 93
concentrations of ions, 95t
ions measured in dry deposition study, 94
pH, 95
pH range, 93
See also Acid dew
Dimensional stabilization, definition, 330
Direct costs, description, 388
Dose-response functions
sulfur pollutant effect on carbon steel, 106
Dose-response functions--Continued
sulfur pollutant effect on copper, 108
sulfur pollutant effect on zinc, 106-108
Douglas fir
germination affected by acid rain, 333
variation in individual tangential shrinkage values, 314f
Dry deposition
definition, 42-43,415
magnitude, 216
measurement
difficulties, 43
surrogate surfaces, 42-60
monitoring in the United States, 37-40
NADAP/NTN network sites, 39f
partitioning effects on metals, 194-198
techniques for assessment, 44-48t
urban dew influence, 92-100
See also Surrogate surfaces
Dry deposition velocities
calculation, 179-180
equation, 48,179
nitrogen dioxide and sulfur dioxide compared, 180-181
sulfates, function of increasing rim height, 60
Dry plate, concentrations of ions, 95t

E

Earlywood, definition, 330
Economics
definition of effects, materials degradation, 384
influence on materials damage legislation, 361-362
materials degradation, 384-395
See also Benefits assessment
See also Capital stocks
See also Welfare effects
Ectomycorrhizae, acid rain effect, 337
Energy dispersive analysis of X-rays (EDAX), sandstone surfaces examined, 262
Ettringite
formation, 242,245
production in portland cement concrete, 254
Excess emission reduction targets, acid rain control, 365
Existence values
analytically motivating, 376-377
definition, 381
distinction from use value, 376

F

Fechner's law, described, 375
Fish, decline in Scandinavian
 lakes, 18f
Flat-grained wood, definition, 330
Fog
 acid deposition, 43
 aldehydes, 86
 carboxylic acid presence, 86
 chemical constituents, 66f
 cloud droplet capture, 80
 deposition, 64-86
 evolution of fog water
 concentration, 68f
 low pH values, 65
 partial pressure of sulfur
 dioxide, 85
 pH in Los Angeles, 85
 secondary chemical constituents in
 southern California, 67f
Fog water
 acidification processes, 84
 acidity in southern San Joaquin
 Valley, 84
 collector comparison with cloud
 water, 83
 droplets
 high acidity sources, 76
 size, 70
 pH, 83
 sulfur(IV) equilibrium with sulfur
 dioxide, 84
Forest, damage due to pollution, 19
Formaldehyde
 deposition velocity on steel, 182
 fog concentration, 72,85
 hydroxymethanesulfonate (HMSA)
 formation, 75
Fossil fuels, use in Europe, 4
Fracture, national costs,
 definition, 389
Fracture cost, relationship to total
 costs, 392
Freeze-thaw deterioration,
 concrete, 243

G

Galvalume
 corrosion film examined, 143
 SEM photomicrographs, 143
 skyward side, 144f
Galvanized steel
 automobile corrosion advantages, 211
 chromium presence on corrosion
 film, 142
 composition, 172
Galvanized steel--Continued
 control of deposition velocities
 during wet period, 188
 corrosion
 film, 197
 model formulation, 190
 products, 190
 rates, 152-161
 U.S. cities, 207
 developed model and linear corrosion
 model compared, 191-192
 dew composition in corrosion
 experiment, 181
 dry deposition corrosion model, 173
 electrochemical reactions on wet
 surface, 173
 exposure
 chamber in corrosion
 study, 175,176f
 conditions in corrosion
 study, 175-177
 system in corrosion study, 176f
 ISS analysis of the corrosion
 film, 142
 laboratory conditions in
 corrosion study, 175
 nitrogen oxide exposure
 results, 184t
 oxidant impact on corrosion, 172-192
 pollutant measurement in corrosion
 study, 177
 properties of potential corrosion
 stimulations, 189t
 rain rinse samples analysis in
 corrosion study, 177
 relative sensitivity factor to
 corrosion, 124
 SEM examination, 142
 skyward and groundward side, 144f
 sulfur dioxide-induced
 corrosion, 174
 surface roughness, 418
 time development of insoluble
 corrosion product layer, 190
 U.S. corrosion rates, 157
 visual comparison from test, 208f
 weight-loss data, 195,196t
Glass
 dissolution, 251-252
 hydrochloric acid attack, 252
 See also Silica glass
Granite, embossed
 due to natural weathering, 292f
 letter heights, Indiana
 study, 291t
Gravimetric analysis, marble and
 limestone, NAPAP study, 279-281
Gypsum
 carbonate-rock effect, 227
 effects on coquina, 305-307
 formation from cement, 242
 formation in lime mortar, 254

Gypsum--Continued
SEM photomicrograph within lime mortar, 256f

H

Heat transfer
boundary layer for structures, 415
circular cylinder in cross-flow, 423
coefficients, atmosphere, 420
enchancement on flat plate, 424
prisms, 423
spheres, 423
Hemicellulose, definition, 331
Humidity--See Relative humidity
Hydroxymethanesulfonate (HMS)
chromatogram for separation in fog water, 74
concentration versus measured sulfur(IV) and formaldehyde, 74f
kinetics of production, 72-75

I

Indiana
acid rain stripping of limestone, marble, and bronze, 285-298
average sulfate ion concentration, 287
linked to acid rain controversy, 285
pH average, 286
total water volume, 286
Indirect costs, description, 388
Insects, acid rain relationship, 336
Inventory accounting, building components in economic damage study, 402
Ion scattering spectroscopy (ISS)
analysis, corrosion films on metal surfaces, 132t
zinc corrosion film results, 136
Iron, catalysts for sulfate formation, 72

L

Lagrange function, benefits assessment for variable market size, 378
Laser-holographic moire profiling
design of sample assemblies, 271
marble results, 279
Latewood, definition 331
Law, materials damage, 360-368
Leachate, total acid, stone samples, 295t
Leaching, elements from foliage during acid rain, 334-334
Lead, catalysts for sulfate formation, 72
Lignin
definition, 311,331
paint adhesion, 328
sulfurous acid effect, 325
Lime, microtomed, acid treatment effects, 325
Limestone
acid rain data, 232t
damage assessment in NAPAP study, 268
dimension drawing of stone specimens in NAPAP study, 272f
dimensions
used in NAPAP study, 268
wedge-shaped briquettes in NAPAP study, 270
dissolution by acid rain, 226-237
elemental analysis of simulated acid rain stripping, 285-298
gravimetric analysis
after atmospheric exposure, 281t
NAPAP study, 279-281
IC analysis in NAPAP study, 274,275f
incident rainfall, rain pH, hydrogen ion loading, and surface recession, 234t
net sulfate removed and surface recession, 235t
rock exposure-site locations, 228f
roughening, recession, and chemical alteration, 266-283
sections of briquettes in lucite holders in NAPAP study, 272f
stone briquettes arrayed for exposure in NAPAP study, 270
stone specimen mounted in shaping machine for surface sampling in NAPAP study, 272f
sulfate net concentration in rock runoff, 237
sulfate profiles in NAPAP study, 274
surface chemistry methods used in NAPAP study, 269
surface recession and roughness after atmospheric exposure, 280t
surface roughening, recession measurements in NAPAP study, 269
weight-change measurement in NAPAP study, 273
See also Calcium carbonate
See also Salem limestone
Linden, acid rain effect on strength, 339
Liquid water content (LWC), ionic concentrations in fog, 84

Local free stream velocity, deposition
velocity controlled, 427
Lumen, definition, 331

M

Magnesium, leaching rates for micrite, marble, and Salem limestone, 293
Manganese, catalysts for sulfate formation, 72
Marble
- acid rain
  - data, 231t
  - stripping results, 291
- calcium leaching rates, 294f
- damage function, sulfur dioxide term, 282-283
- dimension drawing of stone specimens in NAPAP study, 272f
- dimensions
  - used in NAPAP study, 268
  - wedge-shaped briquettes in NAPAP study, 270
- dissolution by acid rain, 226-237
- elemental analysis of simulated acid rain stripping, 285-298
- gravimetric analysis
  - after atmospheric exposure, 281t
  - NAPAP study, 279-281
- IC analysis results in NAPAP study, 27,274
- net sulfate removed and surface recession, 235t
- rock exposure-site locations, 228f
- roughening, recession, and chemical alteration, 266-283
- sections of briquettes in lucite holders in NAPAP study, 272f
- stone briquettes arrayed for exposure in NAPAP study, 270
- stone specimen mounted in shaping machine for surface sampling in NAPAP study, 272f
- sulfate net concentration in rock runoff, 237
- surface chemistry methods used in NAPAP study, 269
- surface recession
  - after atmospheric exposure, 280t
  - correlation with rainfall amount, 237
- tombstones deterioration, 267f
- total acid leachate, 295t
- weight-change measurement in NAPAP study, 273

Mass median aerodynamic diameter (MMD), definition, 48
Mass transfer, boundary layer for structures, 415
Mass transfer coefficient, definition, 419
Materials damage
- CAA, 361
- consumer behavior influenced, 372-373
- economic assessment factors, 372
- economic consequences of perception thresholds, 376
- economic trade-offs, 361-362
- law, 360-368
- objective of economic assessment, 371
- scientific, technical, and economic underpinnings, 367
- treatment of uncertainity, 366
- See also Materials degradation

Materials degradation
- computing construction material damage, 401-406
- definition of economic effects, 384-395
- description, 398-399
- economic assessment dependence, 411
- economic effects, 384-395
- economic features, 369-381
- economic perspective, 370
- engineering perspective, 370
- forecasting problems, 411
- overall cost, 392
- quantification, 397-398
- reduced economic effect, 380

Materials inventorying, building patterns, 412
Metal oxides, acidic media effect, 202
Metals
- acid deposition influence on corrosion, 104
- corrosion
  - contributors, 153
  - cost due to acid rain, 393
  - data, 152-153
  - dominating factors, 155
  - studies, summary, 119-120
  - study methods, 221
- site distribution for corrosion rate study, 153

Micrite
- calcium leaching rates, 294f
- engraved letter depth changes as function of simulated rain depth, 289t
- mechanism of engraved letter deterioration by acid rain, 289-290
- physical observations in Indiana study, 289
- total acid leachate, 295t

Microtomed spruce, acid rain effect on tensile strength, 324-325
MIST test, corrosion of anodized aluminum, 214-215

Molecular weight
  function of breaking strength, 350
  light influence on nylon, 350,351f
Momentum, boundary layer for
    structures, 415
Mortar
  acid deposition effect relative to
    other building materials, 407
  composition and
    deterioration, 254-255
  cost of repointing, 406
  deterioration due to salt
    crystallization, 256f

N

National Acid Precipitation Assessment
    Program (NAPAP), description, 25,39[illegible]
New source performance standards,
    applications to welfare-only type
    pollutants, 364
Nickel
  catalysts for sulfate formation, 72
  sulfur pollutants effect, 110
Nitrate
  atmospheric ratio to ammonium
    ions, 13
  concentration and deposition in the
    United States, 30-34
  concentration during seasons, 7
  surface accumulation rates due to
    anthropogenic influences, 220t
Nitric acid, acid precipitation
    role, 4
Nitrogen(III), profiles of
    concentration versus time from
    urban fog model, 78f
Nitrogen(V), profiles of
    concentration versus time from
    urban fog model, 78f
Nitrogen, pollutants
  influence, 110-113
  metal corrosion, 104
Nitrogen dioxide
  dew deposition, corrosion of
    steel, 181
  galvanized steel exposure, 180,183t
  water reaction, 181
Nitrogen oxides
  emission in Europe, 7
  emission sources, 4-7
  oxidation, 65
  oxidation of sulfur dioxide, 111
North America, mean acidity of
    precipitation, 13
Norway, water-soluble compounds in
    precipitation and aerosols, 16f
Norway spruce needles, acidified water
    effect on nutrient leaching, 335
Nylon
  acid rain degradation, 343-356
  amino group influenced by
    acid, 350-352
  breaking strength, 347-350
  change in molecular weight as
    function of light, 350
  exposure to dark conditions, 355
  immersion in strong acid, 355
  materials and methods in acid
    degradation study, 346
  SEM, 354f
  simulated acid rain results, 355
  synergistic effect by light,
    humidity, heat, and aqueous
    acid, 345
  synergistic effect of light and
    acid, 352
  weathering textiles, 344

O

Optical interferometry, surface
    recession measurement, 271
Oxidants, formation, 7
Ozone
  corrosion effect on weathering
    steel, 171
  formation, 7

P

Paint
  acid deposition effect relative to
    other building materials, 407
  carbonate-extender use, 405
  correlation of particulate matter
    with repainting, 318
  failure of latewood of flat-grained
    siding, 319f
  films, effects of acid
    deposition, 321
  latex, frosting, 328
  ozone and sulfur dioxide
    effects, 321
  performance on stainless steel, 324
  silicate-extender use, 405
  sulfur dioxide effects, 318
  synergistic effects of sulfur
    dioxide, nitrogen dioxide, and
    ozone, 324
Painted wood surfaces, effects of acid
    rain, 310-328
Petri dish
  dry deposition collection
    devices, 53

Petri dish--Continued
  results as dry deposition collection
    devices, 56
pH
  example of variability in region, 37
  Illinois precipitation site, 38
  U.S. precipitation, 34,36f
Photochemical oxidants, forest damage
  role, 19
Photooxidation reactions, nylon, 344
Pine, acid rain effect on
  strength, 339
Pitting
  acid rain results, 291
  aluminum, 213
  nylon, 352-355
Pollen
  germination effect by acid rain, 333
  tube growth effect by acid rain, 333
Pollutant fluxes, calculation,
  weathering steel study, 166-167
Pollutant transmission, CAPITA, Monte
  Carlo regional model, 154
Pollutant transport, importance of
  knowledge, 42
Pollutants
  concentrations, retrospective
    reconstruction, 154-155
  corrosion, time dependence of
    gas-phase concentration during
    dew formation, 178
  interactions with atmospheric
    parameters, 426-427
Pollution
  basic science studies of impacts on
    materials, 371
  consumer surplus change, 400f
Pore sizes, influence in masonry
  deterioration, 256-257
Porosity, limestone, 233
Portland cement concrete
  acid rain effect, 254
  dry deposition studies needed, 246
  pure rain effect, 254
  weathering studies needed, 247
  wet deposition studies needed, 246
  See also Concrete
Poultice
  analysis techniques in corrosion
    study, 204
  concentration of road salt and acid
    deposition ions, 205f
  corrosion mechanism, 202
  deposits listed, 200
  mechanism of auto steel
    corrosion, 203f
  proportion ions on an equivalent
    basis, 205f
  soluble species, 204
  wet chemistry, 202-205
Precipitation
  acidification, 2-20
Precipitation--Continued
  collection for chemical
    analysis, 25-26
  corrosion action, 153
  interval between the collection of
    samples, 26
  sampler described, 26
Precipitation chemistry, 23-40
  background, 24-25
  character, 25
  quality assurance, 27
  selected interpretative
    analyses, 27-34
  U.S. sampling activities, 24-25
  written documentation, 27
Producer quasi-rent
  changes, 374f
  definition, 372
  effect of air pollution
    increase, 373
Producer surplus, 400f
Propylene-nitrogen oxides, exposed
  to galvanized steel, 182-185
Propylene-nitrogen oxides-sulfur
  dioxide, exposure
  experiment, 186-188

Q

Quartz--See Authigenic quartz
Quartz arenite, composition, 263

R

Rainwater, concentrations of ions, 95t
Red spruce, predisposing to drought
  stress, 338
Red tide, description, 20
Regional air pollution study,
  weathering steel study data, 165
Relative humidity
  definition, 331
  dew-point relationship, 165
  spatial gradients prediction, 412
Relative sensitivity factor,
  composition of corrosion, 124-125
Reynolds analogy, heat and mass
  transfer, 418-419
Reynolds number effect, smooth
  cylinders at low turbulence, 425
Rock-exposure rack, isometric
  projection, 229f
Rotating arm collector
  collection efficiency versus
    generalized stokes number, 82f
  performance characterization, 80

Runoff
corrosion product losses, 122
losses
due to hydrogen ion loading, 149
zinc basis, 138
zinc corrosion film, 136

S

Salem limestone
acid rain stripping results, 291
calcium leaching rates, 294f
leaching rates as function of pH, 295-296
total acid leachate, 295t
San Joaquin Valley, variation of gas-phase components, mixing heights, and stratus base during stagnation episode, 69
Sandstone
chemical alteration of the bonding matrix, 264
composition in Mesa Verde National Park, 263
porosity, 264
Scanning electron microscopy (SEM)
chemical changes in sandstone studied, 261-262
examination of steel corrosion films, 129
nylon, 354f
nylon exposed to sulfur dioxide, 345
nylon under atmospheric conditions, 352-355
procedure used in nylon study, 347
Silica glass, dissolution, 251
Softwood, shrinkage values of domestic, 312-313t
Soil
acid rain effect on acidification, 335
acidity characterized by metal ion content, 17
effect of acid rain, 335-336
effect on soil acidity, 335
Spalling
cement paste, 242
corrosion product losses, 122
zinc corrosion film, 136
Springwood--See Earlywood
Stainless steel
corrosion rates, 209
corrosion ratings, 207
Steel
atmospheric corrosion rates in automotive corrosion study, 206t
automotive, corrosion rates of galvanizing, 207t
corrosion in air containing sulfur dioxide and/or nitrogen oxide, 112f
nitrogen pollutants effect, 110
rain influence on corrosion, 113
weight loss as function of sulfur dioxide deposition, 107f,109f
Stratus clouds, pH, 84
Sulfate
concentration and deposition
Illinois site, 38
United States, 34f-35f
correlation with humidity, 65
dew deposition rates compared, 97
evolution of equivalent fraction in fog water, 71f
example of variability in region, 37
limiting factors in the autoxidation pathways, 72
oxidants, 70
profile on limestone surfaces in NAPAP study, 274
source in automobile poultice, 210
surface accumulation rates due to anthropogenic influences, 220t
Sulfate attack
damage function, 245
effects on portland cement concrete, 245
mechanism, 245
Sulfur
costs due to ambient concentrations, 397
deposition on limestone and marble compared, 233
pollutants
European countries, 3
influence, 105-110
metal corrosion, 104
redistribution during carbonate-rock weathering, 233
transport modeling of pollutants, 7-9
Sulfur(IV)
aqueous-phase oxidants, 76
aqueous-phase oxidation, 85
dew deposition rates compared, 97
equilibrium relationship, function of partial pressure of sulfur dioxide, 73f
metal-catalyzed autoxidation, 76
oxidation by hydrogen peroxide, 75
plot, concentration versus formaldehyde concentration in fog water, 73f
Sulfur(IV)
plot, concentration versus time in fog water, 77
profiles of concentration versus time from urban fog model, 78f
Sulfur dioxide
aldehydes influence on chemistry, 72

Sulfur dioxide--Continued
annual mean concentration in Europe, 9
corrosion acceleration, time-of-wetness, 195
corrosion film role in galvanized steel, 174
damage function term for marble, 282-283
deposition velocity
affected by pollutants, 426
estimate, 415
smooth and corrugated buildings, 424
zinc, 422
distribution of emissions
Europe, 9
North America, 9
dry deposition, 9
effect on concrete, 244-245
effect on paint, 321
emission
maximum, 13
North America, 4
sources impact on surface concentrations, 154
United States and Canada, 397
estimation in metal corrosion study, 155
local sources, building damages, 407
mean monthly concentration in air, 326f
nighttime concentrations, 154
oxidation, 65
sulfate attack on cement, 245
Sulfur oxides, atmospheric chemistry, 202
Sulfuric acid
acid precipitation role, 4
attack on nylon, 345
concrete deterioration, 244
contribution to acid precipitation, 241
Sulfurous acid, effect on Corsican pine and microtomed lime, 325
Summerwood--See Latewood
Surface, indoor accumulation rates of specific chemical species, 220
Surface recession
definition, 230
influence of rain amount and hydrogen ion loading, 233
marble, correlation with rainfall amount, 237
measurement, 230
Surface roughness, data, 418t
Surrogate surfaces
designs, 49-51t
dry deposition data, 54-55t
existing design, 48
geometry affecting dry deposition, 56
Surrogate surfaces--Continued
measurement method for deposition, 43
particle size influencing deposition rate, 56
Sweden, early precipitation studies, 24

T

Tangential shrinkage, affected by direction of annual rings, 314f
Teflon plate, dry deposition collection devices, 52f
Time-of-wetness
corrosion effect, 221
definition, 165
Titanium dioxide, photooxidation influenced, 344
Tombstones, depths of engraved letters, Indiana study, 290
Total economic surplus, definition, 399
Trona, production in portland cement concrete, 254-255
Troposphere, concentrations of species associated with acid rain, 218t
Turbulence intensity, atmospheric boundary layer, 414
Turbulent flow, transition from laminar, 421

U

Ultraviolet radiation, nylon, 344
Urban dew
average ambient concentrations, 98t
dry deposition rate influenced, 92-100
Use value, distinction from existence value, 376

V

Velocity distribution, atmospheric boundary layer, 414
Viscosity, measurement in nylon study, 347

W

Waldsterben syndrome, West Germany, 338

Water, corrosion effect, 153
Weathering
  causes, 339
  definition, 331
  natural versus
      pollution-induced, 411
  rate influences, 226
Weathering steel
  corrosion
    data fit to model, 170f
    factors, 163-171
    film properties, 164-165
    rate, 168
    regression coefficients for
        theoretical model, 169t
    time function, 197
  diffusivity through oxide film, 171
  environmental rain data in corrosion
      study, 198t
  U.S. corrosion rates, 156
  weight-loss data, 195,196t
Welfare effects, costly exchange, 379f
Western red cedar, acid treatment
    effect on erosion rate, 327
Wet deposition
  amount, calculation, 179
  calculation on surfaces, 178
  definition for atmospheric
      transport, 42
  first-order rate constant on metal
      panels, 178
  partitioning effects on
      metals, 194-198
Wet deposition chemistry--See
    Precipitation chemistry
Whatman 541 filter paper
  dry deposition collection
      devices, 52f,53
  results as dry deposition collection
      devices, 56
Willingness-to-pay, definition, 399
Wind speed
  sulfur dioxide correlation, 427
  variations in atmospheric boundary
      layers, 414
Wind velocity, rural and urban
    conditions, 417t
Wood
  acid deposition effects, 324
  acid rain influence on fungi, 339
  characteristics for painting and
      finishing, 315-317t
  degradation by ozone, 325
  efficacy of surface treatments, 327
  moisture content, 311
  normal weathering, 311
  painted
    view of the end grain after
        swelling with water, 320f
    view of the painted face and end
        grain, 320f
  properties that affect coating
      performance, 311-318
Wood--Continued
  UV degradation, 311
  See also Softwood
Wood finishes
  erosion data, 322-323t
  protection against acid
      deposition, 318
Woody plants
  acid rain effect on
      ectomycorrhizae, 337
  ambient acid rain injury, 334
  effects of acid rain, 332-340
  foliar lesions after acid rain, 334t
  germination affected by acid
      rain, 333
  leaf injury, 334
  symptoms of injury from acid
      rain, 333-334
World
  definition used in corrosion
      studies, 385-386
  nonexistent degradative process,
      model, 387f

X

X-ray photoelectron spectroscopic
    analysis, corrosion film on metal
    surfaces, 131t

Z

Z-contrast, phase transformations in
    sandstone detected, 262
Zinc
  ammonium-to-sulfate ratios, 222
  corrosion
    film composition, 134
    galvanized steel, 180
    near coastal effect of
        wetness, 156
    product accumulation, 140
    relationship to rain pH, 188
    relative variability, 156
  corrosion rates, 152-161
  corrosion rates in short-term
      exposure study, 121
  dry deposition effect on
      degradation, 221-222
  indoor, acid rain effects, 216-223
  influence of hydrogen ion loading on
      corrosion product, 137f,138
  mass of corrosion product
      retained, 135f
  nitrogen pollutants effect, 110
  oxides as buffers, 221
  rain influence on corrosion, 113

Zinc--*Continued*
relationship of corrosion film to corrosion loss, 134
relative sensitivity factor to corrosion, 124t
runoff loss relationship to hydrogen ion load, 136
SEM results on corrosion film, 138-139
skyward and groundward sides, 139f,141f
soluble corrosion products, 188
sulfur dioxide deposition, 422,425,426
sulfur pollutants effect, 106
U.S. corrosion rates, 157
Zinc carbonate, water reactions, 197
Zinc hydroxide, galvanized steel corrosion film, 174

Production by Joan C. Cook
Indexing by Keith B. Belton
Jacket design by Pamela Lewis

Elements typeset by Hot Type Ltd., Washington, DC
Printed and bound by Maple Press Co., York, PA